AF606550

195 A Universal Construction for Groups Acting Freely on Real Trees

A complete list of books in the series can be found at www.cambridge.org/mathematics.
Recent titles include the following:

166. The Lévy Laplacian. By M. N. Feller
167. Poincaré Duality Algebras, Macaulay's Dual Systems, and Steenrod Operations. By D. Meyer and L. Smith
168. The Cube-A Window to Convex and Discrete Geometry. By C. Zong
169. Quantum Stochastic Processes and Noncommutative Geometry. By K. B. Sinha and D. Goswami
170. Polynomials and Vanishing Cycles. By M. Tibăr
171. Orbifolds and Stringy Topology. By A. Adem, J. Leida, and Y. Ruan
172. Rigid Cohomology. By B. Le Stum
173. Enumeration of Finite Groups. By S. R. Blackburn, P. M. Neumann, and G. Venkataraman
174. Forcing Idealized. By J. Zapletal
175. The Large Sieve and its Applications. By E. Kowalski
176. The Monster Group and Majorana Involutions. By A. A. Ivanov
177. A Higher-Dimensional Sieve Method. By H. G. Diamond, H. Halberstam, and W. F. Galway
178. Analysis in Positive Characteristic. By A. N. Kochubei
179. Dynamics of Linear Operators. By F. Bayart and É. Matheron
180. Synthetic Geometry of Manifolds. By A. Kock
181. Totally Positive Matrices. By A. Pinkus
182. Nonlinear Markov Processes and Kinetic Equations. By V. N. Kolokoltsov
183. Period Domains over Finite and p-adic Fields. By J.-F. Dat, S. Orlik, and M. Rapoport
184. Algebraic Theories. By J. Adámek, J. Rosický, and E. M. Vitale
185. Rigidity in Higher Rank Abelian Group Actions I: Introduction and Cocycle Problem. By A. Katok and V. Niţică
186. Dimensions, Embeddings, and Attractors. By J. C. Robinson
187. Convexity: An Analytic Viewpoint. By B. Simon
188. Modern Approaches to the Invariant Subspace Problem. By I. Chalendar and J. R. Partington
189. Nonlinear Perron–Frobenius Theory. By B. Lemmens and R. Nussbaum
190. Jordan Structures in Geometry and Analysis. By C.-H. Chu
191. Malliavin Calculus for Lévy Processes and Infinite-Dimensional Brownian Motion. By H. Osswald
192. Normal Approximations with Malliavin Calculus. By I. Nourdin and G. Peccati
193. Distribution Modulo One and Diophantine Approximation. By Y. Bugeaud
194. Mathematics of Two-Dimensional Turbulence. By S. Kuksin and A. Shirikyan
195. A Universal Construction for R-free Groups. By I. Chiswell and T. Müller
196. The Theory of Hardy's Z-Function. By A. Ivić
197. Induced Representations of Locally Compact Groups. By E. Kaniuth and K. F. Taylor
198. Topics in Critical Point Theory. By K. Perera and M. Schechter
199. Combinatorics of Minuscule Representations. By R. M. Green
200. Singularities of the Minimal Model Program. By J. Kollár

A Universal Construction for Groups Acting Freely on Real Trees

IAN CHISWELL
Queen Mary, University of London

THOMAS MÜLLER
Queen Mary, University of London

CAMBRIDGE
UNIVERSITY PRESS

CAMBRIDGE UNIVERSITY PRESS
Cambridge, New York, Melbourne, Madrid, Cape Town,
Singapore, São Paulo, Delhi, Mexico City

Cambridge University Press
The Edinburgh Building, Cambridge CB2 8RU, UK

Published in the United States of America by Cambridge University Press, New York

www.cambridge.org
Information on this title: www.cambridge.org/9781107024816

First published 2012

Printed and bound in the United Kingdom by the MPG Books Group

A catalogue record for this publication is available from the British Library

ISBN 978-1-107-02481-6 Hardback

Dedicated
to the memory of
KARL W. GRUENBERG
1928–2007

Contents

Preface

In summer 2004, V. N. Remeslennikov, during a visit to Queen Mary and Westfield College, gave a series of three talks in which he outlined the construction of a class of groups $\mathscr{R}\mathscr{F}(G)$, starting from the collection of (set-theoretic) functions $f : [0, \alpha] \to G$, where α is any non-negative real number and G is a given (discrete) group. Apparently, his main motivation was to imitate the construction of free groups in a continuous setting. He indicated that these new groups would have natural $\mathbb{R}$-tree actions associated with them, and he pointed out that it might be possible to study the centraliser of a hyperbolic element f in terms of (suitably defined) periods of f. However, no proofs were given.

Nevertheless, the picture emerging was felt to be interesting; the authors set out to try to fill in missing proofs, at first with the modest aim of establishing that the construction really produced groups. This task alone turned out to be rather difficult, leading to the development of a substantial body of cancellation theory (as given in Chapter 2 of the present book) before the actual proof that 'reduced multiplication' was associative could be given. By the time this task was accomplished (more than half a year later), the authors were already absorbed in what turned out to be a difficult but ultimately rewarding theory.

Now, several years further on, we present the fruits of our labour. To mention just a few highlights: the bounded subgroups of $\mathscr{R}\mathscr{F}(G)$ are determined; it is shown that $\mathscr{R}\mathscr{F}(G)$ (if non-trivial) is not generated by its elliptic elements and that the quotient of $\mathscr{R}\mathscr{F}(G)$ by the span $E(G)$ of the elliptic elements has an isomorphism type depending at most on two cardinal numbers, the number of involutions in G as well as the cardinality of its complement. Moreover, the conjugacy relation for hyperbolic elements is characterised, thereby yielding a continuous analogue of the classical conjugacy theorem for free groups; cf. Theorem 1.3 in Magnus, Karrass, and Solitar [31].

Also, Remeslennikov's prediction concerning the centralisers of hyperbolic elements ultimately turns out to be substantially true, with some modification, but to prove this involves a considerable amount of work. Further, the last section of Chapter 10 contains the beginnings of a structure theory for $\mathscr{R}\mathscr{F}(G)$ and its quotient $\mathscr{R}\mathscr{F}(G)/E(G)$, while Chapter 4 explains our recent finding that $\mathscr{R}\mathscr{F}$-groups and their associated $\mathbb{R}$-trees are universal (with respect to inclusion) for free $\mathbb{R}$-tree actions.

Something has been accomplished, yet much remains to be done. Nevertheless, as far as the case $\Lambda = \mathbb{R}$ is concerned the theory is beginning to shape nicely, despite the fact that there are still a large number of open problems (see Appendix B for a sample); thus it seemed a good idea, and the right time, to present our findings obtained so far in the hope of stimulating further research in what the authors feel is an exciting new area.

1
Introduction

In the first four sections we show how, starting with the usual description of free groups by means of reduced words, it is possible to arrive at a definition of the groups $\mathscr{RF}(G)$ and their associated $\mathbb{R}$-trees $\mathbf{X}_G$, which are the objects of study in this book. The final section summarises the contents of the following chapters.

1.1 Finite words and free groups

In constructing free groups, one may start from the collection of all finite *words*

$$w = x_{i_1}^{e_1} x_{i_2}^{e_2} \cdots x_{i_n}^{e_n}$$

over an alphabet $X \cup X^{-1}$, where X is some given set, $e_1, \ldots, e_n \in \{1, -1\}$, and

$$X^{-1} = \{x^{-1} : x \in X\}$$

is a set in one-to-one correspondence with X via the map $x \mapsto x^{-1}$ such that $X \cap X^{-1} = \varnothing$. We extend this map to an involution of $X \cup X^{-1}$ by setting $(x^{-1})^{-1} = x$. A word w can be thought of as a function

$$\{1, 2, \ldots, n\} \to X \cup X^{-1},$$

for some integer $n \geq 0$, the unique word of length 0 being the *empty word* ε. A word $w = x_{i_1}^{e_1} x_{i_2}^{e_2} \cdots x_{i_n}^{e_n}$ is called *reduced* if we have $x_{i_j}^{e_j} \neq x_{i_{j+1}}^{-e_{j+1}}$ for all indices j with $1 \leq j \leq n-1$, that is, if w does not contain a subword of the form $x_i^e x_i^{-e}$. Clearly, the empty word ε itself is reduced.

1.2 Words over a discretely ordered abelian group Λ

One can generalise the above set-up by taking an arbitrary discretely ordered abelian group[1] Λ, and considering 'infinite words' $w : [1,\alpha] \to X \cup X^{-1}$ for $\alpha \geq 0$, where

$$[1,\alpha] = \{\beta \in \Lambda : 1 \leq \beta \leq \alpha\}$$

and where 1 denotes the least positive element of Λ, the case $\alpha = 0$ corresponding to the empty word ε. This has indeed been done; see Myasnikov, Remeslennikov and Serbin [40]. In this setting the concept of reducedness still makes sense: a word w as above is *reduced*, if there does not exist $\beta \in [1, \alpha - 1]$ such that $w(\beta+1) = w(\beta)^{-1}$. Clearly, the empty word ε is reduced. Let $R(\Lambda, X)$ be the set of all reduced words. We define the *inverse* of a word w on $[1,\alpha]$ as the function w^{-1} given on the same domain $[1,\alpha]$ by

$$w^{-1}(\beta) = w(\alpha - \beta + 1)^{-1}, \quad 1 \leq \beta \leq \alpha.$$

One can check immediately that if w is reduced then so is w^{-1}.

The concatenation of two words u, v on domains $[1,\alpha]$ and $[1,\beta]$, respectively, is defined in a natural way as the word $u \circ v$ with domain $[1, \alpha+\beta]$ given by

$$(u \circ v)(\xi) = \left.\begin{cases} u(\xi), & 1 \leq \xi \leq \alpha \\ v(\xi - \alpha), & \alpha + 1 \leq \xi \leq \alpha + \beta \end{cases}\right\} \quad (\xi \in [1, \alpha+\beta]).$$

In this situation one can define a partial multiplication (reduced concatenation) on $R(\Lambda, X)$ in a way that is analogous to multiplication in a free group. We first define, for $u, v \in R(\Lambda, X)$, $\mathrm{com}(u,v)$ to be the largest common initial segment of u and v, more precisely, $\mathrm{com}(u,v) = u|_{[1,\gamma]}$ with $\gamma \in \Lambda$ and $\gamma \geq 0$ such that

$$u(\xi) = v(\xi), \quad \xi \in [1,\gamma],$$

and either $\gamma = \min\{\alpha, \beta\}$ or $u(\gamma+1) \neq v(\gamma+1)$. The problem with this definition is, of course, that $\mathrm{com}(u,v)$ does not always exist, for which reason we shall only be able to define a partial multiplication on $R(\Lambda, X)$. Suppose that $w := \mathrm{com}(u^{-1}, v)$ is defined. Then we can write $u^{-1} = w \circ u_1$, $v = w \circ v_1$, so that $u = u_1^{-1} \circ w^{-1}$, and we define the reduced product uv of the reduced words u and v by setting

$$uv = u_1^{-1} \circ v_1.$$

[1] By an ordered abelian group, we shall always mean a *totally ordered* abelian group.

Since u and v are reduced, so is uv. In this way, we obtain a partial multiplication on $R(\Lambda,X)$, which one can show is associative if it is defined; that is, if uv and vw are defined, then $(uv)w$ is defined if and only if $u(vw)$ is defined, in which case $(uv)w = u(vw)$. (Unfortunately, none of the elegant constructions of a free group that circumvent the need for establishing associativity work directly in this situation.)

Note that the *empty word* ε (corresponding to $\alpha = 0$) is a two-sided identity element, that is,

$$\varepsilon u = u = u\varepsilon, \quad u \in R(\Lambda,X).$$

Also, we have

$$uu^{-1} = \varepsilon = u^{-1}u, \quad u \in R(\Lambda,X).$$

Apart from the fact that reduced multiplication is only a partial operation, another marked difference from the free group case is that there can be words w with $w \neq \varepsilon$ but $w^2 = \varepsilon$.

Example 1.1 Let $\Lambda = \mathbb{Z}^2$ with right lexicographic ordering, so that the least positive element is $(1,0)$. Let $\alpha = (0,1)$ and fix $x \in X$. Define a word w on $[(1,0),\alpha]$ via

$$w(\beta) = \begin{cases} x, & \beta = (s,0),\ s \geq 1, \\ x^{-1}, & \beta = (s,1),\ s \leq 0. \end{cases}$$

Then w is reduced and non-trivial, and $w^2 = \varepsilon$.

There is also a notion of a *cyclically reduced* word: a word $w \in R(\Lambda,X)$ is cyclically reduced if $w(1) \neq w(\alpha)^{-1}$. Let

$$\begin{aligned} &CDR(\Lambda,X) \\ &\quad = \{w \in R(\Lambda,X) : w = u \circ v \circ u^{-1} \text{ for some cyclically reduced word } v\}. \end{aligned}$$

One can show that

$$CDR(\Lambda,X) = \{w \in R(\Lambda,X) : w^2 \text{ is defined and } w^2 \neq \varepsilon\} \cup \{\varepsilon\};$$

see Lemma 3.6 in [40].

We say that $G \subseteq CDR(\Lambda,X)$ is a subgroup of $CDR(\Lambda,X)$, if $u,v \in G$ implies that uv is defined and that $uv \in G$, if $u \in G$ implies that $u^{-1} \in G$, and if $\varepsilon \in G$. If G is a subgroup of $CDR(\Lambda,X)$, one can show that the function $L : G \to \Lambda$ given by $L(w) = \alpha$, where the domain of w is $[1,\alpha]$, and $L(\varepsilon) = 0$, is a Lyndon length function on G and gives rise to an action of G on a Λ-tree that is free and without inversions. (These terms are explained in Appendix A.) This generalises the fact that a free group, and so any subgroup, acts freely on its Cayley graph with respect to a basis; this graph is a tree. In fact, one can prove the following.

Theorem 1.2 *Let Λ be a discretely ordered abelian group. A group G acts freely and without inversions on a Λ-tree if and only if G is a subgroup of $CDR(\Lambda, X)$ for some set X.*

This is shown in [11]; the backward implication also appears in [40].

1.3 The case where Λ is densely ordered

At this stage, the question arises: can something analogous be done if instead we start from a *densely ordered* abelian group Λ? The first problem is that there is no longer a least positive element, so we replace a domain $[1,\alpha]$ with an interval $[0,\alpha]$ where $\alpha \geq 0$. A more serious problem, however, is that concatenation can no longer be defined as above. Our solution is to replace the set $X \cup X^{-1}$ by a (discrete) group G. Let

$$\mathscr{F}(\Lambda, G) := \bigcup_{\substack{\alpha\in\Lambda \\ \alpha\geq 0}} G^{[0,\alpha]} = \{f : [0,\alpha] \to G : \alpha \in \Lambda, \alpha \geq 0\}$$

be the set of all functions with values in G defined on an interval of Λ of the form $[0,\alpha]$ for some $\alpha \geq 0$. Concatenation is then replaced by an operation denoted $*$, the *star product*, defined as follows: if $f, g \in \mathscr{F}(\Lambda, G)$ are functions with domains $[0,\alpha]$ and $[0,\beta]$, respectively, then $f * g$ is the function given on the interval $[0,\alpha+\beta]$ of Λ via

$$(f*g)(\xi) = \left\{\begin{array}{ll} f(\xi), & 0 \leq \xi < \alpha \\ f(\alpha)g(0), & \xi = \alpha \\ g(\xi-\alpha), & \alpha < \xi \leq \alpha+\beta \end{array}\right\} \quad (\xi \in [0,\alpha+\beta]).$$

The function $\mathbf{1}_G$ defined on the interval $[0,0] = \{0\}$ by $\mathbf{1}_G(0) = 1_G$ (where 1_G is the identity element of G) is a two-sided identity element with respect to the star operation; that is, we have

$$f * \mathbf{1}_G = f = \mathbf{1}_G * f, \quad f \in \mathscr{F}(\Lambda, G).$$

We also have a notion of the *formal inverse* f^{-1} of a function $f \in \mathscr{F}(\Lambda, G)$: if f is defined on the domain $[0, \alpha]$ then f^{-1} is the function given on the same interval $[0, \alpha]$ by

$$f^{-1}(\xi) = \big(f(\alpha - \xi)\big)^{-1}, \quad 0 \le \xi \le \alpha.$$

In this setting there is also a notion of a reduced function, which necessarily needs to be somewhat more elaborate. A function $f \in \mathscr{F}(\Lambda, G)$ defined on the interval $[0, \alpha]$ of Λ is called *reduced* if, for each point $\xi_0 \in (0, \alpha)$ with $f(\xi_0) = 1_G$ and every element $\varepsilon \in \Lambda$ with $0 < \varepsilon \le \min\{\alpha - \xi_0, \xi_0\}$, there exists some $\delta \in \Lambda$ such that $0 < \delta \le \varepsilon$ and such that $f(\xi_0 + \delta) \neq \big(f(\xi_0 - \delta)\big)^{-1}$. The set of all reduced functions in $\mathscr{F}(\Lambda, G)$ is denoted by $\mathscr{RF}(\Lambda, G)$. Given a function $f : [0, \alpha] \to G$ in $\mathscr{F}(\Lambda, G)$, let us call an ε-neighbourhood

$$[\xi_0 - \varepsilon, \xi_0 + \varepsilon] \subseteq [0, \alpha]$$

of a point $\xi_0 \in (0, \alpha)$, with $f(\xi_0) = 1_G$, a *cancelling neighbourhood around* ξ_0 if $f(\xi_0 - \delta) = \big(f(\xi_0 + \delta)\big)^{-1}$ for all $0 < \delta \le \varepsilon$. Then we can say that a function $f \in \mathscr{F}(\Lambda, G)$ as above is reduced if and only if there does not exist a cancelling neighbourhood around any interior point of the domain $[0, \alpha]$ of f satisfying $f(\xi_0) = 1_G$.

For $u, v \in \mathscr{F}(\Lambda, G)$, an analogue of $\mathrm{com}(u, v)$ can be defined and, if the element $\mathrm{com}(u^{-1}, v) =: w$ exists, so that $u^{-1} = w * u_1$ and $v = w * v_1$, we may define the reduced product uv of u and v by $uv = u_1^{-1} * v_1$. This gives a partial multiplication on $\mathscr{F}(\Lambda, G)$ that is associative when defined. It can also be shown that the product of two reduced functions, when it exists, is again reduced.

1.4 The case where $\Lambda = \mathbb{R}$

In this book, we shall confine our attention to the case where $\Lambda = \mathbb{R}$, taking the view that this is already quite difficult to deal with (in particular, the proof of the associativity of reduced multiplication is non-trivial). In this case (reduced) multiplication is always defined, and we obtain a group denoted by $\mathscr{RF}(G)$, with the formal inverse of a reduced function f acting as a two-sided inverse of f and with $\mathbf{1}_G$ as the neutral element.

There is a construction of an $\mathbb{R}$-tree $\mathbf{X}_G$ on which $\mathscr{RF}(G)$ acts with point stabilisers isomorphic to G. More precisely, by definition, each element f of $\mathscr{RF}(G)$ has a real number $L(f)$ assigned to it, namely the length α of its domain $[0,\alpha]$; it is not hard to see that the function $L : \mathscr{RF}(G) \to \mathbb{R}$ defined in this way is a Lyndon length function. It follows that $\mathscr{RF}(G)$ has a canonical action by isometries on an $\mathbb{R}$-tree

$$\mathbf{X}_G = (X_G, d)$$

with a distinguished base-point x_0 such that $L_{x_0} = L$, where L_{x_0} is the displacement function

$$L_{x_0}(f) = d(x_0, fx_0), \quad f \in \mathscr{RF}(G)$$

associated with this action and such that

$$G_0 := \mathrm{stab}_{\mathscr{RF}(G)}(x_0) = \{f \in \mathscr{RF}(G) : L(f) = 0\} \cong G.$$

It turns out that $\mathbf{X}_G$ is always metrically complete and that the action of $\mathscr{RF}(G)$ on $\mathbf{X}_G$ is transitive.

As always in such situations, the action of $\mathscr{RF}(G)$ on $\mathbf{X}_G$ leads to a classification of the elements of $\mathscr{RF}(G)$ according to whether they are elliptic (that is, have a fixed point) or hyperbolic (that is, act as a fixed-point free isometry). Hyperbolic elements have some local geometry associated with them, leading, in particular, to another type of length function on $\mathscr{RF}(G)$: if $f \in \mathscr{RF}(G)$ is hyperbolic then there exists an isometric copy $A_f \subseteq \mathbf{X}_G$ of the real line (the so-called *axis* of f) such that f acts on A_f as a non-trivial translation; in particular, hyperbolic elements have infinite order. The translation length of a hyperbolic element f along its axis A_f is called the *hyperbolic length* of f, denoted $\ell(f)$, and ℓ is extended to the whole of $\mathscr{RF}(G)$ by setting $\ell(f) = 0$ for an elliptic function f.

With a view to investigating further the action of $\mathscr{RF}(G)$, we shall introduce and study an analogue of cyclic reduction in free groups; this allows us, among other things, to characterize hyperbolic elements in a purely algebraic way and to compute hyperbolic length in terms of the length function L.

It follows from the transitivity of the action that the set of elliptic elements in $\mathscr{RF}(G)$ coincides with the union of all conjugates of G_0; in particular, $\mathscr{RF}(G)$ has torsion if and only if the group G has. We will establish a stronger result to the effect that a subgroup of $\mathscr{RF}(G)$ is bounded (with respect to the length function L) if and only if it is conjugate to a subgroup of G_0. This result shows in particular that every finite subgroup of $\mathscr{RF}(G)$ is conjugate to a subgroup of G_0, a result reminiscent of the bounded subgroup theorem for free products with amalgamation; see, for instance, Theorem 8 of Chapter I in Serre [45]. It also follows that the trivial group $\{\mathbf{1}_G\}$ is the only bounded subnormal subgroup of $\mathscr{RF}(G)$.

1.5 Contents of the book

We now turn to a discussion of individual chapters.

Chapters 2 *and* 3. Here we give the basic definitions, introduce the groups $\mathscr{RF}(G)$, and develop some cancellation theory needed for (among other things) a proof of the associativity of reduced multiplication. We then study the geometry associated with $\mathscr{RF}(G)$ via its action on the $\mathbb{R}$-tree $\mathbf{X}_G$ and the classification of the group elements effected by this action, covering the ground indicated in Section 1.4 above and more.

Chapter 4. This chapter reflects a rather exciting new development, reporting on the authors' recent discovery of two basic *embedding theorems*. We show that a group G acting freely and without inversions on a Λ-tree $\mathbf{X}$ (for an arbitrary ordered abelian group Λ) can be embedded into a group $\hat{G}$, acting freely, without inversions, and *transitively* on the Λ-tree $\hat{\mathbf{X}}$, which isometrically and G-equivariantly embeds $\mathbf{X}$. This result is of considerable independent interest, in particular shedding new light on the class of infinitely generated $\mathbb{R}$-free groups.

We then proceed to discuss a second, more specialised, embedding theorem concerning free and transitive $\mathbb{R}$-tree actions: we show that a group G acting freely and transitively on an $\mathbb{R}$-tree $\mathbf{X}$ can be embedded into $\mathscr{RF}(H)$ for some suitable group H such that $\mathbf{X}$ embeds isometrically and G-equivariantly into $\mathbf{X}_H$, the $\mathbb{R}$-tree canonically associated with $\mathscr{RF}(H)$.

Combining these two results we infer that $\mathscr{RF}$-groups and their associated $\mathbb{R}$-trees are in fact *universal* (with respect to inclusion) for free $\mathbb{R}$-tree actions.

Chapter 5. Very little is known at present concerning homomorphisms involving $\mathscr{RF}$-groups. In this chapter a certain homomorphism

$$e_g : \mathscr{RF}(G) \to \mathbb{R}$$

is defined for each element $g \in G$ by means of Lebesgue measure theory. The construction of these maps e_g is analogous to and inspired by the exponent sum maps of a free group relative to a basis element. By construction, the elliptic elements of $\mathscr{RF}(G)$ are contained in the kernel of e_g for every g, and if $g \in G$ is not an involution then the corresponding map e_g is surjective; this shows in particular that if G is not an elementary abelian 2-group then $\mathscr{RF}(G)$ is not generated by its elliptic elements. For G an elementary abelian 2-group, the question remains open at this stage since all exponent sums of G are trivial.

The problem is taken up and resolved in Chapter 9 as part of the theory of test functions (see below).

Chapter 6. In this chapter we explore various aspects of functoriality of the $\mathscr{RF}$-construction. The most striking result obtained here is that if two groups G and H have the same (cardinal) number of involutions and the same number of non-involutions then we have

$$\mathscr{RF}(G)/E(G) \cong \mathscr{RF}(H)/E(H),$$

where $E(G)$ is the subgroup of $\mathscr{RF}(G)$ generated by the elliptic elements, that is, the normal closure of G_0. With slight imprecision the last result may be rephrased as follows.

The isomorphism type of the group $\mathscr{RF}(G)/E(G)$ depends only on the two cardinal numbers $|\mathrm{Inv}(G)|$ and $|G - \mathrm{Inv}(G)|$.

The proof of this surprising and rather deep lying *rigidity result* is long and somewhat technical. However, the techniques developed in this chapter also allow us to obtain at least a partial result concerning the automorphism group of the quotient group $\mathscr{RF}(G)/E(G)$; see Proposition 6.7 and Corollary 6.8.

Chapter 7. A guiding principle when investigating $\mathscr{RF}$-groups appears to be the following.

Hyperbolic elements of a non-trivial $\mathscr{RF}$-group behave analogously to the non-trivial elements of a (large) free group.

This principle manifests itself for instance in the *conjugacy theorem for hyperbolic elements* established in Chapter 7, which (except for its proof) is an exact continuous analogue of the corresponding result for free groups. We also show there that the centraliser of a hyperbolic element $f \in \mathscr{RF}(G)$ has index at most 2 in the normaliser of the infinite cyclic group $\langle f \rangle$ in $\mathscr{RF}(G)$.

Chapter 8. It is easy to see that, for a non-trivial element $g \in G_0$, we have

$$C_{\mathscr{RF}(G)}(g) = C_{G_0}(g).$$

Consequently the centralisers of elliptic elements in $\mathscr{RF}(G)$ are determined, up to isomorphism, by the isomorphism types of the centralisers in the group G itself; hence, in general (that is, without restricting the structure of G), nothing more can be said here.

The situation is very different, and much more interesting, for hyperbolic elements, and the present chapter provides a penetrating study of their centralisers. We establish a criterion characterising those hyperbolic elements whose centraliser is cyclic and obtain considerable insight into the centraliser structure in the general case; in particular, we show that centralisers of hyperbolic elements are abelian and relatively 'small', in that they always embed into the additive reals. As suggested by Remeslennikov, the centraliser $C_{\mathscr{RF}(G)}(f)$ for hyperbolic f is controlled by (a subset of) the periods of the function f; the reader is referred to Chapter 8 for details.

As an application of the main result of that chapter (Theorem 8.16), we show that $\mathscr{RF}$-groups enjoy an analogue of the *centraliser partition property* of free groups: the binary relation $\leftrightarrow$ given by

$$f \leftrightarrow g \;:\Longleftrightarrow\; f \text{ and } g \text{ commute}$$

is an equivalence relation on the set

$$\mathscr{RF}(G) - \bigcup_{t \in \mathscr{RF}(G)} tG_0t^{-1}$$

of hyperbolic elements of $\mathscr{RF}(G)$; see Proposition 8.23 and Corollary 8.24. This result provides a further illustration of the philosophy concerning hyperbolic elements expressed above. As another application of Theorem 8.16, we show that $\mathscr{RF}$-groups do not contain non-trivial soluble normal subgroups. A completely different approach to this last result, using the theory of test functions, is given in Chapter 10.

Chapters 9 *and* 10. These two chapters provide an introduction to the theory of test functions and its applications, as developed originally in Müller [36] and Müller and Schlage-Puchta [38]. Roughly speaking, a test function is a mapping $f : [0, \alpha] \to G$ of positive length $L(f) = \alpha$, such that f does not look locally like its own inverse. More precisely, we require that there do not exist $\varepsilon > 0$ and points $\xi_1, \xi_2 \in (0, \alpha)$ such that

$$f(\xi_1 + \eta) = f^{-1}(\xi_2 + \eta), \quad |\eta| < \varepsilon.$$

Test functions do in fact always exist; for instance, the function f_0 of length 1 given by

$$f_0(\xi) = \begin{Bmatrix} x, & \xi^2 \in \mathbb{Q} \\ 1_G, & \xi^2 \notin \mathbb{Q} \end{Bmatrix} \quad (0 \le \xi \le 1),$$

where x is any non-trivial element of G, can be shown to be a test function; see Section 9.3. Test functions are automatically (cyclically) reduced and give rise to a further class of homomorphisms $\mathscr{RF}(G) \to \mathbb{R}$. Roughly speaking, given a test function $f \in \mathscr{RF}(G)$ of length $\alpha > 0$, the idea is to compare ('test') functions $g \in \mathscr{F}(G)$ locally against f and f^{-1}, in this way obtaining two sets $\mathscr{M}_f^+(G), \mathscr{M}_f^-(G) \subseteq (0, L(g))$. To be more explicit, we set

$$\mathscr{M}_f^{\pm}(g) := \Big\{\xi \in (0, L(g)) : \exists \varepsilon > 0, \exists \xi' \in (0, \alpha) \text{ such that} \\ g(\xi + \eta) = f^{\pm}(\xi' + \eta) \text{ for all } |\eta| < \varepsilon\Big\},$$

observing that $\mathscr{M}_f^+(g)$ and $\mathscr{M}_f^-(g)$ are open sets and thus Lebesgue measurable, and define a function $\lambda_f : \mathscr{RF}(G) \to \mathbb{R}$ by

$$\lambda_f(g) = \mu\big(\mathscr{M}_f^+(g)\big) - \mu\big(\mathscr{M}_f^-(g)\big),$$

where μ denotes Lebesgue measure. We show that λ_f is a surjective homomorphism whose kernel contains $E(G)$, in this way demonstrating in particular that $\mathscr{RF}(G)$ is never generated by its elliptic elements; see Theorem 9.8 and Corollary 9.9.

A second important idea introduced in Chapter 9 is that of *local compatibility* and *incompatibility*. Roughly speaking, given functions $f : [0, \alpha] \to G$ and $g : [0, \beta] \to G$, we say that f and g are locally compatible if f looks locally like g or g^{-1}. To be more precise, f and g as above are termed locally compatible if there exist $\varepsilon > 0$ and points $\xi \in (0, \alpha)$, $\zeta \in (0, \beta)$ such that either

$$f(\xi + \eta) = g(\zeta + \eta), \quad |\eta| < \varepsilon,$$

or

$$f(\xi + \eta) = g^{-1}(\zeta + \eta), \quad |\eta| < \varepsilon.$$

If f and g both have positive length but are not locally compatible then they are called locally incompatible. Locally incompatible functions have no cancellation against each other, and if f, g are locally incompatible then so are f^{-1} and g as well as f^{-1} and g^{-1}.

We call a subgroup $\mathscr{H} \leq \mathscr{RF}(G)$ *hyperbolic* if the set $\mathscr{H} - \{\mathbf{1}_G\}$ consists entirely of hyperbolic elements. As a further application of test function theory (as developed so far), in Section 9.6 we show among other things that the family of centralisers $\{C_{\mathscr{RF}(G)}(f_\sigma)\}_{\sigma \in S}$ corresponding to a family $\{f_\sigma\}_{\sigma \in S}$ of pairwise locally incompatible test functions generates a hyperbolic subgroup of $\mathscr{RF}(G)$ isomorphic to the free product $\ast_{\sigma \in S} C_{\mathscr{RF}(G)}(f_\sigma)$; see Corollary 9.24.

The most striking applications of test function theory to date, however, stem from a rather deep result (Theorem 10.1) asserting the existence of large families of pairwise locally incompatible test functions with prescribed centraliser: *given a non-trivial group G and a proper subgroup $\Lambda \leq (\mathbb{R}, +)$, there exists a family $\mathfrak{F}$ of pairwise locally incompatible test functions in $\mathscr{RF}(G)$ such* *that* $|\mathfrak{F}| = |G|^{(\mathbb{R}:\Lambda)}$ *and such that the length function L induces an isomorphism* $C_{\mathscr{R}\mathscr{F}(G)}(f) \to \Lambda$ *for each function* $f \in \mathfrak{F}$. The long and fairly technical proof of Theorem 10.1 is sketched in Section 10.3. Among the direct consequences of this remarkable result we mention (i) a calculation of the cardinality of $\mathscr{R}\mathscr{F}(G)$,

$$|\mathscr{R}\mathscr{F}(G)| = |G|^{2^{\aleph_0}},$$

where $\aleph_0$ is the first infinite cardinal, and (ii) a proof of the fact that all proper real groups [2] are realised as centralisers of hyperbolic elements. Further, combining Theorem 10.1 with other results of test function theory, we obtain a number of important structural conclusions concerning the group $\mathscr{R}\mathscr{F}(G)$ and its quotient group $\mathscr{R}\mathscr{F}_0(G) := \mathscr{R}\mathscr{F}(G)/E(G)$. Among other things, we show in Section 10.7 that:

- both $\mathscr{R}\mathscr{F}(G)$ and $\mathscr{R}\mathscr{F}_0(G)$ contain free subgroups of rank $|G|^{2^{\aleph_0}}$ but are not free;
- the abelianised groups

  $$\overline{\mathscr{R}\mathscr{F}(G)} = \mathscr{R}\mathscr{F}(G)/[\mathscr{R}\mathscr{F}(G), \mathscr{R}\mathscr{F}(G)]$$

  and

  $$\overline{\mathscr{R}\mathscr{F}_0(G)} = \mathscr{R}\mathscr{F}_0(G)/[\mathscr{R}\mathscr{F}_0(G), \mathscr{R}\mathscr{F}_0(G)]$$

  both contain (the additive group of) a $\mathbb{Q}$-vector space of dimension $|G|^{2^{\aleph_0}}$; in particular $\overline{\mathscr{R}\mathscr{F}(G)}$ and $\overline{\mathscr{R}\mathscr{F}_0(G)}$ are not free abelian, while still containing large free abelian subgroups;
- every non-trivial normal subgroup $\mathscr{N} \trianglelefteq \mathscr{R}\mathscr{F}(G)$ contains a free subgroup of rank $|G|^{2^{\aleph_0}}$; in particular $|\mathscr{N}| = |\mathscr{R}\mathscr{F}(G)|$ and $\mathscr{N}$ is not soluble.

At first sight, these results may appear rather weak, while nevertheless lying fairly deep; their proofs involve the full power of the theory explained in Chapters 9 and 10.

We conclude our survey of results in Chapters 9 and 10 by mentioning the following interesting embedding theorem.

Fix a non-trivial group G, and let $\{\Lambda_i\}_{i\in I}$ *be a family of non-trivial real groups. Then the free product* $\Lambda = \ast_{i\in I}\Lambda_i$ *embeds as a hyperbolic subgroup in* $\mathscr{R}\mathscr{F}(G)$ *whenever* Λ *can be embedded as a mere subset, that is, whenever* $|I| \le |G|^{2^{\aleph_0}}$.

[2] In this book, 'real group' means a subgroup of $(\mathbb{R}, +)$.

This is Theorem 10.5, which first appeared as Theorem 1.1 in Müller [36]. As mentioned earlier, $\mathscr{RF}$-groups and their associated $\mathbb{R}$-trees are universal with respect to inclusion for free $\mathbb{R}$-tree actions, which implies that every free product Λ as above embeds as a hyperbolic subgroup in $\mathscr{RF}(G)$ for some suitable group G. The result just stated resolves the modified embedding problem, where some non-trivial group G is given in advance and the following question is asked. *Which free products Λ of non-trivial real groups embed as hyperbolic subgroups in $\mathscr{RF}(G)$ for this given group G?*

Chapter 11. In a seminal paper of $\mathbb{R}$-tree theory, Alperin and Moss [1] gave a construction of a family of groups acting freely on an $\mathbb{R}$-tree and used this to disprove a conjecture of Roger Lyndon to the effect that $\mathbb{R}$-free groups should be isomorphic to subgroups of a free power of the additive reals. Our final chapter explores a common generalisation of their construction and also of the construction of the groups $\mathscr{RF}(G)$, where, starting from a Brandt groupoid (that is, a small category (S,G) with object set S and morphism set G such that each $f \in G$ is an isomorphism), we construct a new groupoid

$$\mathfrak{G} = \big(S, \mathscr{ARF}(S,G)\big)$$

on the same set of objects. We show that $\mathfrak{G}$ has a natural Lyndon length function (the concept suitably generalised) associated with it, which in turn gives rise to an $\mathbb{R}$-tree action of $\mathfrak{G}$, that is, a covariant functor from $\mathfrak{G}$ to the category $\mathscr{T}_{\mathbb{R}}$ whose objects are all $\mathbb{R}$-trees and whose morphisms are the isometries between them, written on the right. In particular, the vertex group of $\mathfrak{G}$ attached to the vertex $s \in S$ acts (on the right) as isometries on its associated $\mathbb{R}$-tree $\mathbf{X}_s$. This $\mathbb{R}$-tree action of $\mathfrak{G}$ turns out to be transitive, in the sense that each vertex group acts transitively on its associated $\mathbb{R}$-tree. Moreover, we show that two key ideas of earlier chapters, namely that of cyclic reduction and that of an exponent sum, generalise to this groupoid context, and we discuss the functoriality of the $\mathscr{ARF}$-construction.

The book closes with two appendices: Appendix A provides an introduction to the basic ideas of Λ-tree theory, thus making the book more or less self contained, while in Appendix B we collect and comment on a number of open problems for future research in this area, which we believe is exciting and an important contribution to the theory of $\mathbb{R}$-trees.

2

The group $\mathscr{RF}(G)$

This chapter introduces our main object of study, the group $\mathscr{RF}(G)$ associated with an arbitrary (discrete) group G. Among other things, we shall develop some cancellation theory, which, while being crucial for many of our later arguments, in particular enables us to show that the group operation introduced for $\mathscr{RF}(G)$ is indeed associative, a fact that turns out to be surprisingly hard to establish.

2.1 The monoid $(\mathscr{F}(G), *)$

Let G be any group, and consider (set-theoretic) functions $f : [0, \alpha] \to G$ defined on some closed real interval $[0, \alpha]$ with $\alpha \geq 0$. Let $\mathscr{F}(G)$ be the collection of all these functions (for arbitrary α). The real number α will be called the *length* of the function f and denoted $L(f)$. For two functions $f, g \in \mathscr{F}(G)$ of lengths α and β, respectively, let $f * g$ be the function of length $\alpha + \beta$ defined by

$$(f * g)(\xi) := \left\{ \begin{array}{ll} f(\xi), & 0 \leq \xi < \alpha \\ f(\alpha)g(0), & \xi = \alpha \\ g(\xi - \alpha), & \alpha < \xi \leq \alpha + \beta \end{array} \right\} \quad (\xi \in [0, \alpha + \beta]).$$

Proposition 2.1 *The set $\mathscr{F}(G)$ equipped with the multiplication $*$ is a cancellative monoid.*

Proof One immediately checks that the function $\mathbf{1}_G$ defined on the one-point segment $[0,0]$ with $\mathbf{1}_G(0) = 1_G$, the right-hand side denoting the identity element of G, is a two-sided neutral element of $(\mathscr{F}(G), *)$.

Next, if

$$f = f_1 * f_2 = f_1 * f_2',$$

where $L(f_i) = \alpha_i$ and $L(f_2') = \alpha_2'$, then, by the definition of $*$, we have $\alpha_2 = \alpha_2'$ and, for $0 \le \xi \le L(f)$,

$$f(\xi) = \begin{cases} f_1(\xi), & 0 \le \xi < \alpha_1, \\ f_1(\alpha_1) f_2(0), & \xi = \alpha_1, \\ f_2(\xi - \alpha_1), & \alpha_1 < \xi \le \alpha_1 + \alpha_2, \end{cases}$$

$$= \begin{cases} f_1(\xi), & 0 \le \xi < \alpha_1, \\ f_1(\alpha_1) f_2'(0), & \xi = \alpha_1, \\ f_2'(\xi - \alpha_1), & \alpha_1 < \xi \le \alpha_1 + \alpha_2. \end{cases}$$

Comparing values gives

$$f_1(\alpha_1) f_2(0) = f(\alpha_1) = f_1(\alpha_1) f_2'(0),$$

that is, $f_2(0) = f_2'(0)$, as well as

$$f_2(\xi) = f(\xi + \alpha_1) = f_2'(\xi), \quad 0 < \xi \le \alpha_2,$$

whence $f_2 = f_2'$. Since f_1 was arbitrary this shows that $(\mathscr{F}(G), *)$ is left cancellative, and a similar argument yields that every element of $\mathscr{F}(G)$ is right cancellative; that is, $(\mathscr{F}(G), *)$ is also right cancellative and thus cancellative.

Finally, it remains to check the associativity of the star product. However, if f, g, h are elements of $\mathscr{F}(G)$ of lengths α, β, and γ, respectively, then a straightforward calculation shows that, for $0 \le \xi \le \alpha+\beta+\gamma$,

$$((f*g)*h)(\xi) = (f*(g*h))(\xi) = \begin{cases} f(\xi), & 0 \le \xi < \alpha, \\ f(\alpha)g(0), & \xi = \alpha, \\ g(\xi-\alpha), & \alpha < \xi < \alpha+\beta, \\ g(\beta)h(0), & \xi = \alpha+\beta, \\ h(\xi-\alpha-\beta), & \alpha+\beta < \xi \le \alpha+\beta+\gamma \end{cases} \tag{2.1}$$

if $\beta > 0$; whereas, for $\beta = 0$ we have

$$((f*g)*h)(\xi) = (f*(g*h))(\xi) = \begin{cases} f(\xi), & 0 \le \xi < \alpha, \\ f(\alpha)g(0)h(0), & \xi = \alpha, \\ h(\xi-\alpha), & \alpha < \xi \le \alpha+\gamma. \end{cases} \tag{2.2}$$

□

The following result, which computes the values of long star products and generalises equation (2.1), will be used in various places throughout our work.

Lemma 2.2 *For $k \ge 1$, let $f_1, f_2, \dots, f_{k+1} \in \mathscr{F}(G)$ be functions such that $L(f_j) > 0$ for $2 \le j \le k$, and, for $1 \le j \le k+1$, set $\xi_j := \sum_{1\le i\le j} L(f_i)$. Then, for $0 \le \xi \le \xi_{k+1}$, we have*

$$(f_1 * f_2 * \cdots * f_{k+1})(\xi) = \begin{cases} f_1(\xi), & 0 \le \xi < \xi_1, \\ f_j(\xi - \xi_{j-1}), & \xi_{j-1} < \xi < \xi_j,\ 2 \le j \le k, \\ f_{k+1}(\xi - \xi_k), & \xi_k < \xi \le \xi_{k+1}, \\ f_j(L(f_j))f_{j+1}(0), & \xi = \xi_j,\ 1 \le j \le k. \end{cases} \tag{2.3}$$

Proof The proof is by induction on k. For $k = 1$, we have two functions with no further condition, and formula (2.3) boils down to the definition of the multiplication $*$. Suppose that (2.3) holds for $k = K-1$ with some $K \ge 2$, let $f_1, f_2, \dots, f_{K+1}$ be functions such that $L(f_j) > 0$ for $2 \le j \le K$, and set

$g := f_1 * \cdots * f_K$. Then, for $0 \leq \xi \leq \xi_{K+1}$, we have

$$
\begin{aligned}
(f_1 * \cdots * f_K * f_{K+1})(\xi) &= (g * f_{K+1})(\xi) \\
&= \begin{cases} g(\xi), & 0 \leq \xi < \xi_K, \\ g(\xi_K) f_{K+1}(0), & \xi = \xi_K, \\ f_{K+1}(\xi - \xi_K), & \xi_K < \xi \leq \xi_{K+1}, \end{cases} \\
&= \begin{cases} f_1(\xi), & 0 \leq \xi < \xi_1, \\ f_j(\xi - \xi_{j-1}), & \xi_{j-1} < \xi < \xi_j,\ 2 \leq j \leq K-1, \\ f_K(\xi - \xi_{K-1}), & \xi_{K-1} < \xi < \xi_K, \\ f_j(L(f_j)) f_{j+1}(0), & \xi = \xi_j,\ 1 \leq j \leq K-1, \\ f_K(L(f_K)) f_{K+1}(0), & \xi = \xi_K, \\ f_{K+1}(\xi - \xi_K), & \xi_K < \xi \leq \xi_{K+1}, \end{cases} \\
&= \begin{cases} f_1(\xi), & 0 \leq \xi < \xi_1, \\ f_j(\xi - \xi_{j-1}), & \xi_{j-1} < \xi < \xi_j,\ 2 \leq j \leq K, \\ f_{K+1}(\xi - \xi_K), & \xi_K < \xi \leq \xi_{K+1}, \\ f_j(L(f_j)) f_{j+1}(0), & \xi = \xi_j,\ 1 \leq j \leq K. \end{cases}
\end{aligned}
$$

Here, the fact that $L(f_j) > 0$ for $2 \leq j \leq K-1$ allows us to evaluate the function g inductively, while the fact that $L(f_K) > 0$ ensures that $\xi_{K-1} < \xi_K$, which was used in entry 4 of step 3 (the third right-hand side of the equation) above. □

Definition 2.3 The *formal inverse* f^{-1} of an element $f \in \mathscr{F}(G)$ is the function defined on the same interval $[0, \alpha]$ as f via

$$f^{-1}(\xi) = \left(f(\alpha - \xi)\right)^{-1}, \quad 0 \leq \xi \leq \alpha.$$

By definition,

$$L((f^{-1})^{-1}) = L(f^{-1}) = L(f)$$

and, for $0 \leq \xi \leq L(f)$,

$$\begin{aligned} ((f^{-1})^{-1})(\xi) &= \left((f^{-1})(L(f^{-1}) - \xi)\right)^{-1} \\ &= \left((f(\xi))^{-1}\right)^{-1} \\ &= f(\xi); \end{aligned}$$

thus

$$(f^{-1})^{-1} = f, \quad f \in \mathscr{F}(G).$$

2.2 Reduced functions and reduced multiplication

Definition 2.4

(i) A function $f \in \mathscr{F}(G)$ is termed *reduced* if, for every interior point ξ_0 in the domain of f with $f(\xi_0) = 1_G$ and every real number ε satisfying $0 < \varepsilon \leq \min\{L(f) - \xi_0, \xi_0\}$, there exists δ such that $0 < \delta \leq \varepsilon$ and such that $f(\xi_0 + \delta) \neq \left(f(\xi_0 - \delta)\right)^{-1}$.

(ii) The set of all reduced functions in $\mathscr{F}(G)$ will be denoted $\mathscr{R}\mathscr{F}(G)$.

Remarks 2.5

(i) Given $f \in \mathscr{F}(G)$ we shall sometimes refer to an ε-neighbourhood $[\xi_0 - \varepsilon, \xi_0 + \varepsilon] \subseteq [0, L(f)]$ of a point $\xi_0 \in (0, L(f))$, with $f(\xi_0) = 1_G$ and $f(\xi_0 - \delta) f(\xi_0 + \delta) = 1_G$ for all $0 < \delta \leq \varepsilon$, as a *cancelling neighbourhood* around ξ_0. Definition 2.4 can thus be rephrased as follows: *the function f is reduced if and only if there exists no cancelling neighbourhood around any interior point ξ_0 of the domain $[0, L(f)]$ satisfying $f(\xi_0) = 1_G$.*

(ii) Every element in $\mathscr{F}(G)$ of length 0 is reduced.

(iii) If f is reduced then f is not identically equal to 1_G on any non-degenerate subinterval of its domain; in particular, we have

$$\mathscr{R}\mathscr{F}(\{1_G\}) = \{\mathbf{1}_G\}.$$

(iv) If f is reduced then so is its formal inverse f^{-1}.

(v) If $f \in \mathscr{F}(G)$ has positive length then $f * f^{-1}$ is not reduced; in particular, the star product of two reduced functions need not be reduced.

We now proceed to define another multiplication on $\mathscr{F}(G)$ with the property that the product of two reduced functions is also reduced.

Given $f,g \in \mathscr{F}(G)$ of respective lengths α and β, let

$$\varepsilon_0 = \varepsilon_0(f,g) := \begin{cases} \sup \mathscr{E}(f,g), & f(\alpha) = g(0)^{-1}, \\ 0 & \text{otherwise}, \end{cases}$$

where

$$\mathscr{E}(f,g) := \Big\{ \varepsilon \in [0, \min\{\alpha,\beta\}] : f(\alpha-\delta) = g(\delta)^{-1} \text{ for all } \delta \in [0,\varepsilon] \Big\},$$

and define fg on the interval $[0, \alpha+\beta-2\varepsilon_0]$ via

$$(fg)(\xi) := \begin{cases} f(\xi), & 0 \le \xi < \alpha - \varepsilon_0, \\ f(\alpha-\varepsilon_0)g(\varepsilon_0), & \xi = \alpha - \varepsilon_0, \\ g(\xi - \alpha + 2\varepsilon_0), & \alpha - \varepsilon_0 < \xi \le \alpha + \beta - 2\varepsilon_0. \end{cases} \tag{2.4}$$

Definition 2.6 The function fg defined in (2.4) is called the *(reduced) product* of the functions f and g in $\mathscr{F}(G)$.

Lemma 2.7 *The reduced product fg of two reduced functions $f,g \in \mathscr{RF}(G)$ is also reduced.*

Proof Let $L(f) = \alpha$ and $L(g) = \beta$. If $f(\alpha-\varepsilon_0)g(\varepsilon_0) \neq 1_G$, our claim clearly holds. So, suppose that

$$f(\alpha-\varepsilon_0)g(\varepsilon_0) = 1_G$$

and that there exists ε' such that

$$0 < \varepsilon' \le \min\{\alpha-\varepsilon_0, \beta-\varepsilon_0\}$$

and

$$(fg)(\alpha-\varepsilon_0-\delta)(fg)(\alpha-\varepsilon_0+\delta) = 1_G, \quad 0 < \delta \le \varepsilon'; \tag{2.5}$$

in particular,

$$\varepsilon_0 = \sup \mathscr{E}(f,g) \in \mathscr{E}(f,g).$$

By the definition of fg, equation (2.5) implies that

$$f(\alpha-\eta)g(\eta) = 1_G, \quad \varepsilon_0 < \eta \le \varepsilon_0 + \varepsilon', \tag{2.6}$$

while the fact that $\varepsilon_0 \in \mathscr{E}(f,g)$ gives

$$f(\alpha-\eta)g(\eta) = 1_G, \quad 0 \le \eta \le \varepsilon_0. \tag{2.7}$$

Combining assertions (2.6) and (2.7) we conclude that $\varepsilon_0 + \varepsilon' \in \mathscr{E}(f,g)$, implying $\varepsilon' \leq 0$, a contradiction. Hence, fg is reduced as claimed. □

As a consequence of the product definition plus the fact that the product of two reduced functions is also reduced, we have the following.

Lemma 2.8 *For $f,g \in \mathscr{RF}(G)$, the following assertions are equivalent:*

(i) $\varepsilon_0(f,g) = 0$;

(ii) $fg = f * g$;

(iii) $f * g$ *is reduced.*

Proof (i) ⇒ (ii). If $\varepsilon_0(f,g) = 0$ then

$$L(fg) = L(f) + L(g) = L(f * g),$$

and, by the definitions of fg and $f * g$, we have, for $0 \leq \xi \leq L(f) + L(g)$, that

$$(fg)(\xi) = \left\{ \begin{array}{ll} f(\xi), & 0 \leq \xi < L(f) \\ f(L(f))g(0), & \xi = L(f) \\ g(\xi - L(f)), & L(f) < \xi \leq L(f) + L(g) \end{array} \right\} = (f * g)(\xi);$$

hence, $fg = f * g$ as claimed.

(ii) ⇒ (iii). If $fg = f * g$ then $f * g$ is reduced, by Lemma 2.7.

(iii) ⇒ (i). Suppose that $\varepsilon_0(f,g) > 0$. Then $\alpha := L(f)$ and $\beta := L(g)$ are strictly positive, $f(\alpha) = g(0)^{-1}$, and $\varepsilon_0(f,g) = \sup \mathscr{E}(f,g)$; in particular, α is an interior point of the interval $[0, \alpha + \beta]$ and $(f * g)(\alpha) = 1_G$. Moreover, there exists $\varepsilon \in \mathscr{E}(f,g)$ with $\varepsilon > 0$ and, for this ε, we have

$$f(\alpha - \eta)g(\eta) = 1_G, \qquad 0 \leq \eta \leq \varepsilon. \tag{2.8}$$

However, for $\eta > 0$, equation (2.8) can be rewritten as

$$(f * g)(\alpha - \eta)(f * g)(\alpha + \eta) = 1_G, \qquad 0 < \eta \leq \varepsilon. \tag{2.9}$$

Equation (2.9) says that $[\alpha - \varepsilon, \alpha + \varepsilon]$ is a cancelling neighbourhood for $f * g$ around the interior point α, so that $f * g$ is not reduced, contradicting assertion (iii). Hence $\varepsilon_0(f,g) = 0$ as required. □

Remark 2.9 We note that the implication (i) ⇒ (ii) of Lemma 2.8 holds, in fact, for $f,g \in \mathscr{F}(G)$.

Definition 2.10 For $f, g \in \mathscr{F}(G)$, we shall write $f \circ g$ to mean $f * g$ together with the information that $\varepsilon_0(f, g) = 0$, so that we have

$$f \circ g = f * g = fg$$

by Remark 2.9.

Remark 2.11 One can think of the circle operation of Definition 2.10 as a partial multiplication on $\mathscr{F}(G)$, $f \circ g$ being defined if and only if $\varepsilon_0(f, g) = 0$, in which case it equals $f * g$.

It follows from Remark 2.9 that $\mathbf{1}_G$ is a (two-sided) neutral element for $\mathscr{F}(G)$ with respect to (reduced) multiplication. Also, it is not hard to see that

$$ff^{-1} = \mathbf{1}_G = f^{-1}f, \quad f \in \mathscr{F}(G).$$

Furthermore, we observe the following relations between the star and circle operations, respectively, and inversion.

Lemma 2.12 (Inversion of star products)

(i) *Let* $f_1, f_2 \in \mathscr{F}(G)$; *then* $\varepsilon_0(f_1, f_2) = \varepsilon_0(f_2^{-1}, f_1^{-1})$.

(ii) *Let* $f_1, f_2 \in \mathscr{F}(G)$, *and let* $f = f_1 * f_2$. *Then* $f^{-1} = f_2^{-1} * f_1^{-1}$; *in particular,* $f = f_1 \circ f_2$ *implies* $f^{-1} = f_2^{-1} \circ f_1^{-1}$.

Proof (i) We note that

$$f_1(L(f_1))f_2(0) = \left(f_2^{-1}(L(f_2^{-1}))f_1^{-1}(0)\right)^{-1};$$

in particular,

$$f_1(L(f_1))f_2(0) = 1_G \iff f_2^{-1}(L(f_2^{-1}))f_1^{-1}(0) = 1_G.$$

Thus, if $f_1(L(f_1))f_2(0) \neq 1_G$ then $\varepsilon_0(f_1, f_2) = \varepsilon_0(f_2^{-1}, f_1^{-1}) = 0$; hence, we may assume that $f_1(L(f_1))f_2(0) = 1_G$, implying

$$\varepsilon_0(f_1, f_2) = \sup \mathscr{E}(f_1, f_2)$$

as well as

$$\varepsilon_0(f_2^{-1}, f_1^{-1}) = \sup \mathscr{E}(f_2^{-1}, f_1^{-1}).$$

But, for $\varepsilon \in [0, \min\{L(f_1), L(f_2)\}]$,

$$\begin{aligned} \varepsilon \in \mathscr{E}(f_1, f_2) &\iff f_1(L(f_1) - \delta)f_2(\delta) = 1_G \text{ for } 0 \leq \delta \leq \varepsilon \\ &\iff f_2^{-1}(L(f_2^{-1}) - \delta)f_1^{-1}(\delta) = 1_G \text{ for } 0 \leq \delta \leq \varepsilon \\ &\iff \varepsilon \in \mathscr{E}(f_2^{-1}, f_1^{-1}); \end{aligned}$$

that is, $\mathscr{E}(f_1,f_2)=\mathscr{E}(f_2^{-1},f_1^{-1})$ and thus $\varepsilon_0(f_1,f_2)=\varepsilon_0(f_2^{-1},f_1^{-1})$, as claimed.

(ii) Clearly,

$$L(f^{-1})=L(f)=L(f_1)+L(f_2)=L(f_1^{-1})+L(f_2^{-1})=L(f_2^{-1}*f_1^{-1}).$$

Moreover, for $0\leq\xi\leq L(f)$,

$$\begin{aligned}
f^{-1}(\xi) &= \big(f(L(f)-\xi)\big)^{-1}\\
&= \begin{cases} \big(f_2(L(f_2)-\xi)\big)^{-1}, & 0\leq\xi<L(f_2),\\ \big(f_1(L(f_1))f_2(0)\big)^{-1}, & \xi=L(f_2),\\ \big(f_1(L(f)-\xi)\big)^{-1}, & L(f_2)<\xi\leq L(f), \end{cases}\\
&= \begin{cases} f_2^{-1}(\xi), & 0\leq\xi<L(f_2^{-1}),\\ f_2^{-1}(L(f_2^{-1}))f_1^{-1}(0), & \xi=L(f_2^{-1}),\\ f_1^{-1}(\xi-L(f_2^{-1})), & L(f_2^{-1})<\xi\leq L(f^{-1}), \end{cases}\\
&= (f_2^{-1}*f_1^{-1})(\xi);
\end{aligned}$$

hence $f^{-1}=f_2^{-1}*f_1^{-1}$, as claimed. The particular statement follows from this together with part (i). □

Example and Remark It is easy to see that reduced multiplication is not associative on the whole of $\mathscr{F}(G)$. For instance, let f,g be the functions, of lengths 1 and $\frac{1}{2}$, respectively, given by

$$f(\xi):=\left\{\begin{matrix} x, & 0\leq\xi<\frac{1}{2}\\ 1_G, & \xi=\frac{1}{2}\\ x^{-1}, & \frac{1}{2}<\xi\leq 1 \end{matrix}\right\} \quad\text{and}\quad g(\xi):=\left\{\begin{matrix} x, & 0\leq\xi<\frac{1}{2}\\ 1_G, & \xi=\frac{1}{2} \end{matrix}\right\},$$

where x is some fixed element of G. Certainly $f\neq\mathbf{1}_G$. Moreover, for every ε in the range $0\leq\varepsilon\leq\frac{1}{2}$, we have

$$f(1-\delta)g(\delta)=1_G,\quad 0\leq\delta\leq\varepsilon;$$

hence

$$\mathscr{E}(f,g)=\big[0,\tfrac{1}{2}\big]$$

and

$$\varepsilon_0(f,g) = \sup \mathscr{E}(f,g) = \tfrac{1}{2}.$$

It follows that $L(fg) = \frac{1}{2}$ and that, for $0 \le \xi \le \frac{1}{2}$,

$$(fg)(\xi) = \begin{cases} f(\xi), & 0 \le \xi < \frac{1}{2}, \\ f(\frac{1}{2})g(\frac{1}{2}), & \xi = \frac{1}{2}, \end{cases}$$

$$= \begin{cases} x, & 0 \le \xi < \frac{1}{2}, \\ 1_G, & \xi = \frac{1}{2}, \end{cases}$$

$$= g(\xi),$$

that is, $fg = g$. We conclude that

$$(fg)g^{-1} = gg^{-1} = \mathbf{1}_G \ne f = f(gg^{-1}),$$

which shows that reduced multiplication is indeed not associative on $\mathscr{F}(G)$.

2.3 Cancellation theory for $\mathscr{RF}(G)$

Our principal aim for the remainder of this chapter is to show that the restriction of reduced multiplication to the subset $\mathscr{RF}(G)$ is associative, so that we have the following.

Theorem 2.13 *For every group G, the set $\mathscr{RF}(G)$ forms a group under reduced multiplication.*

The following results, which (from a later point of view) mainly develop the cancellation theory of the groups $\mathscr{RF}(G)$, will be crucial for the proof of Theorem 2.13; they will also be needed many times in the rest of this book. The proof of Theorem 2.13 itself will be given in Section 2.4.

Lemma 2.14 **(Dissection of reduced functions)** *Let $f : [0,\alpha] \to G$ be a reduced function, and let β be a real number such that $0 \le \beta \le \alpha$. Then there exist reduced functions $f_1 : [0,\beta] \to G$ and $f_2 : [0,\alpha-\beta] \to G$ such that $f = f_1 \circ f_2$. Moreover, f_1 and f_2 with these properties are uniquely determined once one of the values $f_1(\beta)$, $f_2(0)$ has been specified; one of these values may be chosen arbitrarily in G.*

Proof The equation $f = f_1 * f_2$ with $L(f_1) = \beta$ and $L(f_2) = \alpha - \beta$ is equivalent to the conjunction of

$$f_1(\xi) = f(\xi), \qquad 0 \le \xi < \beta,$$

$$f_2(\xi) = f(\xi + \beta), \quad 0 < \xi \le \alpha - \beta,$$

and

$$f_1(\beta) f_2(0) = f(\beta).$$

Hence f_1 and f_2 living on the domains specified above and solving the equation $f = f_1 * f_2$ do exist; they are uniquely determined once one of the values $f_1(\beta), f_2(0)$ has been specified, and one of these values can be chosen arbitrarily. Moreover, f_1 and f_2 are reduced (independently of our choice of $f_1(\beta), f_2(0)$) since f is reduced, and $\varepsilon_0(f_1, f_2) = 0$ holds for the same reason in view of Lemma 2.8. □

Lemma 2.15 **(Visibility of cancellation)** *Let $f, g \in \mathscr{RF}(G)$ be reduced functions. Then there exist $f_1, g_1, u \in \mathscr{RF}(G)$ such that $f = f_1 \circ u$, $g = u^{-1} \circ g_1$, and $fg = f_1 \circ g_1$.*

Proof If $\varepsilon_0 := \varepsilon_0(f, g) = 0$ then our conclusion is satisfied for $f_1 := f$, $g_1 := g$, and $u := \mathbf{1}_G$; hence, we may assume that $\varepsilon_0 > 0$. By Lemma 2.14 we can find reduced functions $f_1 : [0, L(f) - \varepsilon_0] \to G$, $g_1 : [0, L(g) - \varepsilon_0] \to G$, $u : [0, \varepsilon_0] \to G$, and $v : [0, \varepsilon_0] \to G$ such that $f = f_1 \circ u$ and $g = v \circ g_1$. By the definition of ε_0 plus the fact that $\varepsilon_0 > 0$, we have

$$f(L(f) - \delta) = (g(\delta))^{-1}, \quad 0 \le \delta < \varepsilon_0. \tag{2.10}$$

However, in this range,

$$f(L(f) - \delta) = u(\varepsilon_0 - \delta)$$

and

$$g(\delta) = v(\delta),$$

so that (2.10) gives

$$u^{-1}(\delta) = \big(u(\varepsilon_0 - \delta)\big)^{-1} = v(\delta), \quad 0 \le \delta < \varepsilon_0.$$

Moreover, again by Lemma 2.14, we can choose $u(0)$ and $v(\varepsilon_0)$ arbitrarily (and independently of each other); in particular, we can arrange that

$$u^{-1}(\varepsilon_0) = \big(u(0)\big)^{-1} = v(\varepsilon_0),$$

so that $v = u^{-1}$. Finally, we have

$$\begin{aligned} L(fg) &= L(f) + L(g) - 2\varepsilon_0 \\ &= L(f_1) + L(g_1) \\ &= L(f_1 * g_1) \end{aligned}$$

and, for $0 \le \xi \le L(f) + L(g) - 2\varepsilon_0$,

$$(fg)(\xi) = \begin{cases} f(\xi), & 0 \le \xi < L(f) - \varepsilon_0, \\ f(L(f) - \varepsilon_0) g(\varepsilon_0), & \xi = L(f) - \varepsilon_0, \\ g(\xi - L(f) + 2\varepsilon_0), & L(f) - \varepsilon_0 < \xi \le L(f) + L(g) - 2\varepsilon_0, \end{cases}$$

$$= \begin{cases} f_1(\xi), & 0 \le \xi < L(f_1), \\ f_1(L(f_1)) g_1(0), & \xi = L(f_1), \\ g_1(\xi - L(f_1)), & L(f_1) < \xi \le L(f_1) + L(g_1), \end{cases}$$

$$= (f_1 * g_1)(\xi),$$

which shows that $fg = f_1 \circ g_1$ by Lemma 2.8, since fg is reduced. □

Lemma 2.16 (Visible cancellation)

Suppose that $f = f_1 \circ u$ and $g = u^{-1} \circ g_1$, with $f_1, g_1, u \in \mathscr{RF}(G)$. Then $fg = f_1 g_1$. Moreover, if $L(u) = \varepsilon_0(f, g)$ then $fg = f_1 \circ g_1$.

Proof The crux of the proof is the equation

$$\varepsilon_0(f, g) = L(u) + \varepsilon_0(f_1, g_1), \tag{2.11}$$

which we shall discuss first.

Case 1: $L(u) = 0$. Then $L(f) = L(f_1)$, $L(g) = L(g_1)$, and

$$\begin{aligned} f(L(f))g(0) &= f_1(L(f_1))u(0)u^{-1}(0)g_1(0) \\ &= f_1(L(f_1))g_1(0); \end{aligned}$$

in particular, (2.11) holds if $f(L(f))g(0) \ne 1_G$. Thus, we may assume that $f(L(f))g(0) = 1_G$ and therefore that

$$\varepsilon_0(f, g) = \sup \mathscr{E}(f, g)$$

as well as

$$\varepsilon_0(f_1,g_1) = \sup\mathscr{E}(f_1,g_1).$$

Moreover, for $\varepsilon \in [0, \min\{L(f), L(g)\}]$,

$$\begin{aligned} \varepsilon \in \mathscr{E}(f,g) &\iff f(L(f)-\delta)g(\delta) = 1_G \text{ for } 0 < \delta \leq \varepsilon \\ &\iff f_1(L(f_1)-\delta)g_1(\delta) = 1_G \text{ for } 0 < \delta \leq \varepsilon \\ &\iff \varepsilon \in \mathscr{E}(f_1,g_1), \end{aligned}$$

that is, $\mathscr{E}(f,g) = \mathscr{E}(f_1,g_1)$. It follows that $\varepsilon_0(f,g) = \varepsilon_0(f_1,g_1)$, so that (2.11) holds again.

Case 2: $L(u) > 0$. Then $L(f_1) < L(f)$ and

$$f(L(f))g(0) = u(L(u))u^{-1}(0) = 1_G,$$

so that $\varepsilon_0(f,g) = \sup\mathscr{E}(f,g)$. Now consider the equation

$$f(L(f)-\delta)g(\delta) = 1_G. \tag{2.12}$$

For $0 \leq \delta < L(u)$,

$$f(L(f)-\delta)g(\delta) = u(L(u)-\delta)u^{-1}(\delta) = 1_G$$

while, for $\delta = L(u)$,

$$\begin{aligned} f(L(f)-\delta)g(\delta) &= f(L(f_1))g(L(u)) \\ &= f_1(L(f_1))u(0)u^{-1}(L(u))g_1(0) \\ &= f_1(L(f_1))g_1(0). \end{aligned}$$

Thus, if $f(L(f_1))g(L(u)) \neq 1_G$ then $\varepsilon_0(f_1,g_1) = 0$ and $\varepsilon_0(f,g) = L(u)$, so that (2.11) holds. It remains to discuss the case where $L(u) > 0$ and $f(L(f_1))g(L(u)) = 1_G$. But then equation (2.12) holds for $\delta = L(u)$, $\varepsilon_0(f_1,g_1) = \sup\mathscr{E}(f_1,g_1)$, and, by the definition of $\varepsilon_0(f_1,g_1)$,

$$f_1(L(f_1)-\eta)g_1(\eta) = 1_G, \quad 0 < \eta < \varepsilon_0(f_1,g_1),$$

implying equation (2.12) in the range $L(u) < \delta < L(u) + \varepsilon_0(f_1,g_1)$; hence

$$\varepsilon_0(f,g) \geq L(u) + \varepsilon_0(f_1,g_1).$$

However, by the definition of $\varepsilon_0(f,g)$, equation (2.12) holds in the range $L(u) < \delta < \varepsilon_0(f,g)$, which leads to

$$f_1(L(f_1)-\eta)g_1(\eta) = 1_G, \quad 0 \leq \eta < \varepsilon_0(f,g) - L(u);$$

and so

$$\varepsilon_0(f_1,g_1) \geq \varepsilon_0(f,g) - L(u),$$

establishing equation (2.11) in case 2 as well.

Having (2.11) at our disposal, the fact that $fg = f_1g_1$ follows by straightforward calculation. First, we see that

$$\begin{aligned} L(fg) &= L(f) + L(g) - 2\varepsilon_0(f,g) \\ &= L(f_1) + L(g_1) + 2L(u) - 2\big(L(u) + \varepsilon_0(f_1,g_1)\big) \\ &= L(f_1) + L(g_1) - 2\varepsilon_0(f_1,g_1) \\ &= L(f_1g_1). \end{aligned}$$

Then, using again the facts that $f = f_1 \circ u$ and $g = u^{-1} \circ g_1$, we obtain, for $0 \leq \xi \leq L(fg)$,

$$\begin{aligned} (fg)(\xi) &= \begin{cases} f(\xi), & 0 \leq \xi < L(f) - \varepsilon_0(f,g), \\ f(L(f) - \varepsilon_0(f,g))g(\varepsilon_0(f,g)), & \xi = L(f) - \varepsilon_0(f,g), \\ g(\xi - L(f) + 2\varepsilon_0(f,g)), & L(f) - \varepsilon_0(f,g) < \xi \leq L(fg), \end{cases} \\ &= \begin{cases} f_1(\xi), & 0 \leq \xi < L(f_1) - \varepsilon_0(f_1,g_1), \\ f_1(L(f_1) - \varepsilon_0(f_1,g_1))g_1(\varepsilon_0(f_1,g_1)), & \xi = L(f_1) - \varepsilon_0(f_1,g_1), \\ g_1(\xi - L(f_1) + 2\varepsilon_0(f_1,g_1)), & L(f_1) - \varepsilon_0(f_1,g_1) < \xi \leq L(f_1g_1), \end{cases} \\ &= (f_1g_1)(\xi), \end{aligned}$$

hence $fg = f_1g_1$, as claimed. The only entry requiring some justification is the second entry in step two, for which we note that

$$\begin{aligned} f(L(f) - \varepsilon_0(f,g))g(\varepsilon_0(f,g)) &= f(L(f_1) - \varepsilon_0(f_1,g_1))g(L(u) + \varepsilon_0(f_1,g_1)) \\ &= \begin{cases} f_1(L(f_1) - \varepsilon_0(f_1,g_1))g_1(\varepsilon_0(f_1,g_1)), & \varepsilon_0(f_1,g_1) > 0 \\ f_1(L(f_1))u(0)u^{-1}(L(u))g_1(0), & \varepsilon_0(f_1,g_1) = 0 \end{cases} \\ &= f_1(L(f_1) - \varepsilon_0(f_1,g_1))g_1(\varepsilon_0(f_1,g_1)), \end{aligned}$$

independently of whether or not $\varepsilon_0(f_1,g_1) = 0$.

Finally, if $L(u) = \varepsilon_0(f,g)$ then $\varepsilon_0(f_1,g_1) = 0$ by equation (2.11); hence $fg = f_1 \circ g_1$ by Lemma 2.8. □

Lemma 2.17 *Let $f,g,h \in \mathscr{F}(G)$, and suppose that $L(g) > 0$.*

(i) *If $\varepsilon_0(f,g) = 0$ then*

$$\varepsilon_0(fg,h) = 0 \iff \varepsilon_0(g,h) = 0.$$

(ii) *If $\varepsilon_0(g,h) = 0$ then*

$$\varepsilon_0(f,gh) = 0 \iff \varepsilon_0(f,g) = 0.$$

Proof (i) Suppose that $\varepsilon_0(g,h) > 0$. Then

$$g(L(g)-\delta) = (h(\delta))^{-1}, \quad 0 \le \delta < \varepsilon_0(g,h). \tag{2.13}$$

Also, by the definition of the star product,

$$(f * g)(L(f)+L(g)-\delta) = g(L(g)-\delta), \quad 0 \le \delta < L(g). \tag{2.14}$$

Combining (2.13) and (2.14) and using the fact that $\varepsilon_0(f,g) = 0$ together with Remark 2.9, we get

$$(fg)(L(fg)-\delta) = (h(\delta))^{-1}, \quad 0 \le \delta < \min\{\varepsilon_0(g,h), L(g)\};$$

hence

$$\varepsilon_0(fg,h) = \sup \mathscr{E}(fg,h) \ge \min\{\varepsilon_0(g,h), L(g)\} > 0$$

since $L(g) > 0$. Conversely, suppose that $\varepsilon_0(fg,h) > 0$. Then, since $\varepsilon_0(f,g) = 0$,

$$(fg)(L(f)+L(g)-\delta) = (h(\delta))^{-1}, \quad 0 \le \delta < \varepsilon_0(fg,h).$$

Using (2.14) again plus the fact that $\varepsilon_0(f,g) = 0$, we get

$$g(L(g)-\delta) = (h(\delta))^{-1}, \quad 0 \le \delta < \min\{\varepsilon_0(fg,h), L(g)\},$$

whence

$$\varepsilon_0(g,h) = \sup \mathscr{E}(g,h) \ge \min\{\varepsilon_0(fg,h), L(g)\} > 0.$$

(ii) This is similar to the proof of (i), and is omitted (see Exercise 2.3). □

Corollary 2.18 **(Associativity of the circle product)** *Let $f,g,h \in \mathscr{F}(G)$. Then*

$$(f \circ g) \circ h = f \circ (g \circ h).^1$$

[1] This equation is to be interpreted in the manner standard for partial operations: if one of the products $(f \circ g) \circ h$ and $f \circ (g \circ h)$ is defined then so is the other, and the two products are equal.

Proof We first consider the case where $L(g) > 0$. Suppose that $(f \circ g) \circ h$ is defined, that is,

$$\varepsilon_0(f,g) = \varepsilon_0(f \circ g, h) = 0.$$

Then, by part (i) of Lemma 2.17 and Remark 2.9, $\varepsilon_0(g,h) = 0$. Applying Lemma 2.17(ii) and the remark then gives

$$\varepsilon_0(f, g \circ h) = \varepsilon_0(f, gh) = 0.$$

Hence, $f \circ (g \circ h)$ is defined. In a similar way one finds that $(f \circ g) \circ h$ is defined once $f \circ (g \circ h)$ is defined. Finally, if both products are defined then they are equal by the associativity of the star operation.

It remains to consider the case where $L(g) = 0$. Again suppose that $(f \circ g) \circ h$ is defined, implying $\varepsilon_0(f \circ g, h) = 0$. Note that $\varepsilon_0(f,g) = 0$ holds automatically in this situation. Indeed, either $f(L(f))g(0) \neq 1_G$, so that $\varepsilon_0(f,g) = 0$ straight from the definition, or $f(L(f))g(0) = 1_G$, so that $\varepsilon_0(f,g) = \sup \mathscr{E}(f,g)$; then, however, $\mathscr{E}(f,g) = \{0\}$ and thus $\varepsilon_0(f,g) = 0$ again. A similar argument shows that $\varepsilon_0(g,h) = 0$. We may suppose that $f(L(f))(g \circ h)(0) = 1_G$, so that $\varepsilon_0(f, g \circ h) = \sup \mathscr{E}(f, g \circ h)$. Now,

$$f(L(f) - \eta)(g \circ h)(\eta) = (f \circ g)(L(f) - \eta)h(\eta), \quad 0 \leq \eta \leq \min\{L(f), L(h)\}; \tag{2.15}$$

in particular

$$(f \circ g)(L(f))h(0) = 1_G,$$

so that

$$0 = \varepsilon_0(f \circ g, h) = \sup \mathscr{E}(f \circ g, h).$$

It follows that, for every $\varepsilon > 0$, there exists $\eta \in (0, \varepsilon]$ such that the right-hand side of equation (2.15) does not equal 1_G. Hence $\mathscr{E}(f, g \circ h) = \{0\}$; thus $\varepsilon_0(f, g \circ h) = 0$ and $f \circ (g \circ h)$ is defined. An analogous argument shows that, for $L(g) = 0$, $(f \circ g) \circ h$ is defined if $f \circ (g \circ h)$ is defined, and the equality of these two expressions (if both are defined) follows as before. □

2.4 Proof of Theorem 2.13

It remains only to establish the associativity of the group law. Let $f, g, h \in \mathscr{RF}(G)$ be reduced functions. By Lemma 2.15 (visibility of cancellation), there exist elements $f_1, g_1, g_2, h_1, u, v \in \mathscr{RF}(G)$ such that

$$f = f_1 \circ u, \; g = u^{-1} \circ g_1, \; g = g_2 \circ v, \; h = v^{-1} \circ h_1,$$

$$fg = f_1 \circ g_1, \text{ and } gh = g_2 \circ h_1.$$

Case 1: $L(u) < L(g_2)$. Then, using Lemma 2.14 (dissection of reduced functions) plus the case assumption, we can decompose g_2 as $g_2 = u^{-1} \circ g_3$ for some $g_3 \in \mathscr{RF}(G)$. Indeed, taking $f = g_2$ and $\beta = L(u)$ in Lemma 2.14, we obtain $g_2 = w \circ g_3$ with reduced functions w, g_3 and $L(w) = L(u)$. Moreover we are at liberty to define $w(L(u))$ as we like. Making use of the equations $g_2 = w \circ g_3$, $g = g_2 \circ v$, and $g = u^{-1} \circ g_1$, we find that

$$w(\xi) = g_2(\xi) = g(\xi) = u^{-1}(\xi), \quad 0 \le \xi < L(u),$$

and, defining

$$w(L(u)) := u^{-1}(L(u)),$$

we obtain $w = u^{-1}$, as required. Note that

$$L(g_3) = L(g_2) - L(u) > 0$$

by the case assumption. Next, we have on the one hand

$$g = g_2 \circ v = (u^{-1} \circ g_3) \circ v = u^{-1} \circ (g_3 \circ v),$$

where we have used Corollary 2.18 (associativity of the circle product) in the last step. On the other hand we have $g = u^{-1} \circ g_1$, hence $g_1 = g_3 \circ v$ by Proposition 2.1. The last equation in conjunction with Corollary 2.18 now yields

$$fg = f_1 \circ g_1 = f_1 \circ (g_3 \circ v) = (f_1 \circ g_3) \circ v;$$

in particular, $\varepsilon_0(f_1, g_3) = 0$. Application of Lemma 2.16 (visible cancellation) gives

$$(fg)h = ((f_1 \circ g_3) \circ v)(v^{-1} \circ h_1) = (f_1 \circ g_3)h_1.$$

We claim that $\varepsilon_0(f_1 \circ g_3, h_1) = 0$, so that

$$(fg)h = (f_1 \circ g_3) \circ h_1.$$

Indeed, applying Lemma 2.17(i) to the functions u^{-1}, g_3, h_1 and using the fact that

$$\varepsilon_0(u^{-1}, g_3) = \varepsilon_0(g_2, h_1) = 0,$$

we get $\varepsilon_0(g_3, h_1) = 0$. Now we apply Lemma 2.17(i) to the functions f_1, g_3, h_1 to obtain

$$\varepsilon_0(f_1 \circ g_3, h_1) = \varepsilon_0(f_1 g_3, h_1) = 0,$$

as required. Analysing $f(gh)$ in a similar way, we find that

$$\begin{aligned} f(gh) &= f(g_2 \circ h_1) \\ &= (f_1 \circ u)((u^{-1} \circ g_3) \circ h_1) \\ &= (f_1 \circ u)(u^{-1} \circ (g_3 \circ h_1)) \\ &= f_1(g_3 \circ h_1) \\ &= f_1 \circ (g_3 \circ h_1). \end{aligned}$$

More specifically, we have used Corollary 2.18 in step 3, Lemma 2.16 in step 4, and Lemma 2.17(ii) plus

$$\varepsilon_0(f_1, g_3) = \varepsilon_0(g_3, h_1) = 0$$

in the last step. That $(fg)h = f(gh)$ in the present case now follows from the associativity of the star product (or from Corollary 2.18).

Case 2: $L(u) \geq L(g_2)$. Proceeding in a similar way to that for case 1, we now use Lemma 2.14 together with the equations $g = u^{-1} \circ g_1$ and $g = g_2 \circ v$ to decompose u^{-1} as $u^{-1} = g_2 \circ u_1$, for some $u_1 \in \mathscr{RF}(G)$.

Case 2(a): $L(u_1) > 0$. Making use of Lemma 2.12 (inversion of star products) and Corollary 2.18 (associativity of the circle product), we write

$$f = f_1 \circ u = f_1 \circ (u_1^{-1} \circ g_2^{-1}) = (f_1 \circ u_1^{-1}) \circ g_2^{-1};$$

in particular, $\varepsilon_0(f_1, u_1^{-1}) = 0$. Using Corollary 2.18 again,

$$g = u^{-1} \circ g_1 = (g_2 \circ u_1) \circ g_1 = g_2 \circ (u_1 \circ g_1).$$

However, we also have that $g = g_2 \circ v$, hence $v = u_1 \circ g_1$ by Proposition 2.1. Now

$$\begin{aligned} (fg)h &= (f_1 \circ g_1)(v^{-1} \circ h_1) \\ &= (f_1 \circ g_1)((g_1^{-1} \circ u_1^{-1}) \circ h_1) \\ &= (f_1 \circ g_1)(g_1^{-1} \circ (u_1^{-1} \circ h_1)) \\ &= f_1(u_1^{-1} \circ h_1) \\ &= f_1 \circ (u_1^{-1} \circ h_1). \end{aligned}$$

Here we have used Lemma 2.12 in step 2, Corollary 2.18 in step 3, Lemma 2.16 in step 4, and Lemma 2.17(ii), together with

$$\varepsilon_0(f_1, u_1^{-1}) = \varepsilon_0(u_1^{-1}, h_1) = 0$$

plus the subcase assumption in the final step, the fact that $\varepsilon_0(u_1^{-1}, h_1) = 0$ coming from step 3. However, similar reasoning employing Lemma 2.12, Corollary 2.18, Lemma 2.16, and Lemma 2.17(i) gives

$$\begin{aligned} f(gh) &= (f_1 \circ u)(g_2 \circ h_1) \\ &= (f_1 \circ (u_1^{-1} \circ g_2^{-1}))(g_2 \circ h_1) \\ &= ((f_1 \circ u_1^{-1}) \circ g_2^{-1})(g_2 \circ h_1) \\ &= (f_1 \circ u_1^{-1})h_1 \\ &= (f_1 \circ u_1^{-1}) \circ h_1, \end{aligned}$$

and so $(fg)h = f(gh)$ follows as before.

Case 2(b): $L(u_1) = 0$. Changing the values $g_2(L(u))$, $v(0)$, and $h_1(0)$ if necessary, we can ensure that $g_2 = u^{-1}$ while keeping the equations $g = g_2 \circ v$ and $h = v^{-1} \circ h_1$ intact, the former now reading $g = u^{-1} \circ v$. Since also $g = u^{-1} \circ g_1$, we conclude by Proposition 2.1 that $v = g_1$ in this case. Consequently, by Lemma 2.16,

$$\begin{aligned} (fg)h &= (f_1 \circ g_1)(v^{-1} \circ h_1) \\ &= (f_1 \circ v)(v^{-1} \circ h_1) \\ &= f_1 h_1 \end{aligned}$$

and

$$\begin{aligned} f(gh) &= (f_1 \circ u)(g_2 \circ h_1) \\ &= (f_1 \circ u)(u^{-1} \circ h_1) \\ &= f_1 h_1, \end{aligned}$$

completing the proof of the theorem.

2.5 The subgroup G_0

Definition 2.19 Given a group G, we set

$$G_0 := \left\{ f \in \mathscr{RF}(G) : L(f) = 0 \right\},$$

a subgroup of $\mathscr{RF}(G)$ canonically isomorphic to G.

Proposition 2.20

(i) *If $f \in G_0$ is a non-trivial element then we have*

$$C_{\mathscr{RF}(G)}(f) = C_{G_0}(f).$$

(ii) *For every non-trivial subgroup U of G_0, we have*

$$N_{\mathscr{RF}(G)}(U) = N_{G_0}(U). \tag{2.16}$$

In particular, the subgroup G_0 is self-normalizing in $\mathscr{RF}(G)$.

Proof (i) Suppose that an element $g \in \mathscr{RF}(G)$ has positive length and commutes with f. Then

$$(fg)(0) = (f \circ g)(0) = f(0)g(0),$$

while

$$(gf)(0) = (g \circ f)(0) = g(0).$$

Since $fg = gf$, we conclude that $f = \mathbf{1}_G$, a contradiction. Hence

$$C_{\mathscr{RF}(G)}(f) \subseteq C_{G_0}(f),$$

and the reverse inclusion holds trivially.

(ii) Let $f \in \mathscr{RF}(G)$ and $g \in U$ be elements such that $fgf^{-1} \in G_0$ and $g \neq \mathbf{1}_G$, and let

$$h := f * g * f^{-1}.$$

By equation (2.2), we have

$$\begin{aligned} h(L(f)) &= f(L(f))g(0)f^{-1}(0) \\ &= f(L(f))g(0)\big(f(L(f))\big)^{-1}, \end{aligned}$$

which is non-trivial as $g(0) \neq 1_G$. Since the inverse of a reduced function is also reduced, we obtain that h itself is reduced; hence, by Lemma 2.8,

$$h = f * g * f^{-1} = (f \circ g)f^{-1} = fgf^{-1} \in G_0,$$

so that

$$0 = L(h) = 2L(f),$$

that is, $f \in G_0$.

Now let $f \in N_{\mathscr{RF}(G)}(U)$. Then, in particular, $fgf^{-1} \in G_0$, where $g \in U$ is some non-trivial element; so, by what we have already shown, $L(f) = 0$; thus $f \in N_{G_0}(U)$. This gives

$$N_{\mathscr{RF}(G)}(U) \subseteq N_{G_0}(U),$$

and the reverse inclusion holds trivially. Finally, the particular statement at the end of (ii) is an immediate consequence of (2.16). □

2.6 Appendix to Chapter 2

The following examples are intended to demonstrate that, while we have a well-defined concept of reducedness, there is in general no finite or transfinite analogue of the *process of reduction* in free groups.

Example 2.21 Suppose for the moment that there is a semigroup homomorphism φ from $\mathscr{F}(G)$ to $\mathscr{RF}(G)$ that is constant on reduced functions and satisfies $L(f) \geq L(\varphi(f))$ for all $f \in \mathscr{F}(G)$, together with other reasonable properties. Then the map φ defines a notion of an infinite product in G in the following way. Given a sequence of elements $\{g_\nu\}_{\nu \geq 0}$ in G and a fixed element $g \in G$, define a function $f \in \mathscr{F}(G)$ of length 1 via

$$f(\xi) = \begin{cases} g_\nu, & \xi = 1 - 2^{-\nu},\ \nu \geq 0, \\ g, & 1 - 2^{-\nu} < \xi < 1 - 3 \times 2^{-\nu-2},\ \nu \geq 0, \\ 1_G, & \xi = 1 - 3 \times 2^{-\nu-2},\ \nu \geq 0, \\ g^{-1}, & 1 - 3 \times 2^{-\nu-2} < \xi < 1 - 2^{-\nu-1},\ \nu \geq 0, \\ 1_G, & \xi = 1. \end{cases}$$

Then $L(\varphi(f)) = 0$ and we can set

$$\prod_{\nu=0}^{\infty} g_\nu := \varphi(f)(0).$$

However, in general G will not admit a natural and useful concept of infinite product (compare, for instance, the next example); hence, in general a reasonable reduction process φ does not exist.

Example 2.22 Let $G = \{1, \zeta\}$ be a cyclic group of order 2. Define an equivalence relation on subsets of the set of natural numbers $\mathbb{N}$ via

$$A \sim B :\Longleftrightarrow |A \Delta B| < \infty,$$

and choose a system of representatives for the equivalence classes such that finite sets are represented by the empty set. Then we define the product $\prod_{\nu=0}^{\infty} g_\nu$ of a sequence $(g_\nu)_{\nu\geq 0}$ in G as follows: let A be the set of indices ν with $g_\nu = \zeta$, let B be the representative of A, and set $\prod_{\nu=0}^{\infty} g_\nu = \zeta^{|A\Delta B|}$. This infinite product concept satisfies the usual requirements, for instance the empty product equals 1 and changing finitely many terms changes the product in the expected way, but it is clearly highly unnatural since it depends heavily on the choice of representatives.

2.7 Exercises

2.1. Show that if $f \in \mathscr{F}(G)$ is reduced then so is its formal inverse f^{-1}.

2.2. Give the detailed working to show that, for $f \in \mathscr{F}(G)$, we have

$$\varepsilon_0(f, f^{-1}) = L(f) = \varepsilon_0(f^{-1}, f)$$

and that, consequently,

$$ff^{-1} = \mathbf{1}_G = f^{-1}f.$$

2.3. Write out the proof of Lemma 2.17(ii) in full.

2.4. Let G be a group, and suppose that $G \cong G_1 * G_2$, that is, G is the free product of two subgroups G_1 and G_2. Show that the subgroup

$$\mathscr{H} := \langle \mathscr{RF}(G_1), \mathscr{RF}(G_2) \rangle$$

of $\mathscr{RF}(G)$ satisfies $\mathscr{H} \cong \mathscr{H}_1 * \mathscr{H}_2$ where, for $i = 1, 2$, $\mathscr{H}_i := \mathscr{RF}(G_i)$.

3

The $\mathbb{R}$-tree $\mathbf{X}_G$ associated with $\mathscr{RF}(G)$

3.1 Introduction

In this chapter we explore the geometry associated with $\mathscr{RF}(G)$ via its length function L. We show that L is a (real) Lyndon length function, thus giving rise in a canonical way to a pointed $\mathbb{R}$-tree $(\mathbf{X}_G, x_0)$ and an isometric action of $\mathscr{RF}(G)$ on $\mathbf{X}_G$.[1] We prove that $\mathbf{X}_G$ is complete as a metric space and that the action of $\mathscr{RF}(G)$ is transitive on the points of $\mathbf{X}_G$.

As always in such situations, the action of $\mathscr{RF}(G)$ on $\mathbf{X}_G$ leads to a classification of the elements of $\mathscr{RF}(G)$ according to whether they are elliptic (that is, have a fixed point) or hyperbolic (that is, act as a fixed-point free isometry). A third class of elements, the inversions (that is, elements that are fixed-point free but whose square has a fixed point), does not arise over $\mathbb{R}$. Hyperbolic elements have some local geometry associated with them, leading to another type of length function on $\mathscr{RF}(G)$: if $f \in \mathscr{RF}(G)$ is hyperbolic then there exists an isometric copy $A_f \subseteq \mathbf{X}_G$ of the real line (the so-called axis of f), such that f acts on A_f as a non-trivial translation; in particular, hyperbolic elements have infinite order. The translation length of a hyperbolic element f on its axis A_f is called the hyperbolic length of f and is denoted $\ell(f)$; one can extend ℓ to the whole of $\mathscr{RF}(G)$ by setting $\ell(f) = 0$ for elliptic f.

With a view to investigating further the action of $\mathscr{RF}(G)$, we shall introduce and study an analogue of cyclic reduction in free groups; this allows us among other things to characterise hyperbolic elements in a purely algebraic way and to compute hyperbolic length in terms of the length function L.

[1] The reader who is unfamiliar with these ideas can find an account of them in Appendix A. We shall refer to relevant parts of this appendix in what follows.

It follows from the transitivity of the action that the set of elliptic elements in $\mathscr{RF}(G)$ coincides with the union of all conjugates of G_0; we shall show that a subgroup of $\mathscr{RF}(G)$ is bounded (with respect to L) if and only if it is conjugate to a subgroup of G_0. It also follows that the trivial group $\{\mathbf{1}_G\}$ is the only bounded subnormal subgroup of $\mathscr{RF}(G)$.

3.2 Construction of $\mathbf{X}_G$

The concept of a *Lyndon length function.* is defined in Appendix A, Section A.3. From now on, the term length function will always mean Lyndon length function. We are concerned with real-valued length functions $L:\Gamma\to\mathbb{R}$, where Γ is a group, and as usual we define

$$c(\gamma_1,\gamma_2)=\tfrac{1}{2}\big(L(\gamma_1)+L(\gamma_2)-L(\gamma_1^{-1}\gamma_2)\big)$$

for $\gamma_1,\gamma_2\in\Gamma$.

Proposition 3.1 *The map $L:\mathscr{RF}(G)\to\mathbb{R}$ which assigns to each function $f\in\mathscr{RF}(G)$ the length $L(f)$ of its domain, that is, the interval $[0,L(f)]$, is a (real) length function.*

Proof Axioms (i) and (ii) for a length function at the start of Section A.3 obviously hold by the definition of $\mathscr{RF}(G)$; hence we may focus on axiom (iii). Since, by definition of the group operation in $\mathscr{RF}(G)$,

$$L(f^{-1}g)=L(f)+L(g)-2\varepsilon_0(f^{-1},g)$$

it follows that

$$c(f,g)=\varepsilon_0(f^{-1},g),\quad f,g\in\mathscr{RF}(G).$$

Let $f,g,h\in\mathscr{RF}(G)$, and suppose that $c(f,g),c(g,h)>0$ (otherwise there is nothing to prove). Then

$$c(f,g)=\varepsilon_0(f^{-1},g)=\sup\mathscr{E}(f^{-1},g)$$

while

$$c(g,h)=\varepsilon_0(g^{-1},h)=\sup\mathscr{E}(g^{-1},h),$$

and we have

$$f^{-1}(L(f)-\eta)g(\eta)=1_G,\quad 0\le\eta<\varepsilon_0(f^{-1},g)$$

as well as

$$g^{-1}(L(g)-\eta)h(\eta)=1_G, \quad 0\le\eta<\varepsilon_0(g^{-1},h).$$

From these two equations it follows that, for non-negative η such that $\eta<\min\{\varepsilon_0(f^{-1},g),\varepsilon_0(g^{-1},h)\}$, we have

$$f^{-1}(L(f)-\eta)h(\eta)=f^{-1}(L(f)-\eta)g(\eta)g^{-1}(L(g)-\eta)h(\eta)=1_G.$$

Since $\min\{\varepsilon_0(f^{-1},g),\varepsilon_0(g^{-1},h)\}>0$, we conclude that

$$\begin{aligned}c(f,h)&=\varepsilon_0(f^{-1},h)\\&=\sup\mathscr{E}(f^{-1},h)\\&\ge\min\{\varepsilon_0(f^{-1},g),\varepsilon_0(g^{-1},h)\}\\&=\min\{c(f,g),c(g,h)\},\end{aligned}$$

as required. □

Combining Proposition 3.1 with Theorem A.29 yields the existence of an $\mathbb{R}$-tree $\mathbf{X}_G=(X_G,d_G)$ on which $\mathscr{RF}(G)$ acts, with a canonical base-point x_0 and such that $L=L_{x_0}$, where

$$L_{x_0}(f):=d_G(x_0,fx_0), \quad f\in\mathscr{RF}(G)$$

is the displacement function associated with the action of $\mathscr{RF}(G)$ on $(\mathbf{X}_G,x_0)$. Moreover, $\mathbf{X}_G$ is spanned by the orbit of x_0, that is,[2]

$$X_G=\bigcup_{f\in\mathscr{RF}(G)}[x_0,fx_0].$$

It follows in particular from these remarks that the stabiliser $\mathrm{stab}_{\mathscr{RF}(G)}(x_0)$ of the point x_0 under the action of $\mathscr{RF}(G)$ is given by

$$\mathrm{stab}_{\mathscr{RF}(G)}(x_0)=G_0,$$

the subgroup of $\mathscr{RF}(G)$ introduced in Definition 2.19.

We shall briefly describe the construction of $\mathbf{X}_G$ as a pointed metric space, since this will be helpful later.[3] Introduce an equivalence relation $\approx$ on $\mathscr{RF}(G)$ by

$$f\approx g \;:\Longleftrightarrow\; L(f^{-1}g)=0, \tag{3.1}$$

[2] See Exercise A.3 in Section A.7.

[3] See Theorems A.23 and A.29 for more details.

and denote by $\langle f\rangle$ the equivalence class of $f \in \mathscr{RF}(G)$. One easily sees that

$$f \approx g \iff \big(L(f) = L(g) \text{ and } f|_{[0,L(f))} = g|_{[0,L(g))}\big)$$

$$\iff fG_0 = gG_0,$$

so that $\mathscr{RF}(G)/\approx$ is nothing other than the coset space $\mathscr{RF}(G)/G_0$. Next, we form the set

$$Y_G := \Big\{(\langle f\rangle, \alpha) : f \in \mathscr{RF}(G),\ \alpha \in \mathbb{R},\ 0 \le \alpha \le L(f)\Big\},$$

and introduce an equivalence relation $\sim$ on Y_G via

$$(\langle f\rangle, \alpha) \sim (\langle g\rangle, \beta) \ :\iff\ \varepsilon_0(f^{-1}, g) \ge \alpha = \beta.$$

We denote the equivalence class of $(\langle f\rangle, \alpha)$ by $\langle f, \alpha\rangle$, observing that we always have $\langle f, \alpha\rangle = \langle f|_{[0,\alpha]}, \alpha\rangle$. Then

$$X_G = Y_G/\sim$$

and

$$d_G(\langle f, \alpha\rangle, \langle g, \beta\rangle) = \alpha + \beta - 2\min\Big\{\alpha, \beta, \varepsilon_0(f^{-1}, g)\Big\},$$

while

$$x_0 = \langle \mathbf{1}_G, 0\rangle.$$

It is not hard to check that the segment $[x_0, \langle f, \alpha\rangle]$ spanned by the base-point x_0 and an arbitrary point $\langle f, \alpha\rangle$ of $\mathbf{X}_G$ is given by

$$[x_0, \langle f, \alpha\rangle] = \Big\{\langle f|_{[0,\beta]}, \beta\rangle : 0 \le \beta \le \alpha\Big\}. \tag{3.2}$$

3.3 Completeness and transitivity

Here we establish two important properties of $\mathbf{X}_G$ and the associated action of $\mathscr{RF}(G)$. First we show that $\mathbf{X}_G$ is a *complete* metric space. Then we show that the action of $\mathscr{RF}(G)$ on $\mathbf{X}_G$ is transitive. To prove completeness, we shall make use of a *completeness criterion* for $\mathbb{R}$-trees, stated in Lemma 3.2 below, which appears to be well known although we were unable to locate it in the literature. In order to state this result we need the concept of a *monotone increasing* sequence in an $\mathbb{R}$-tree, introduced in Definition A.42. If $\{y_i\}$ is a sequence in X, where (X, d) is an $\mathbb{R}$-tree, and $x_0 \in X$ then $\{y_i\}$ is called *monotone increasing with respect to* x_0 if $i < j$ implies that $y_i \in [x_0, y_j]$.

The completeness criterion we are going to use here is the following.

Lemma 3.2 *Let (X,d) be an $\mathbb{R}$-tree. Then (X,d) is metrically complete if and only if every monotone increasing Cauchy sequence with respect to some base-point converges.*

A more elaborate version of this completeness criterion for $\mathbb{R}$-trees, also involving a geometric characterisation, is given in Proposition A.43.

With the help of Lemma 3.2 we can now prove the following.

Proposition 3.3 *The $\mathbb{R}$-tree $\mathbf{X}_G$ associated with $\mathscr{RF}(G)$ via the length function L is metrically complete.*

Proof If $\{y_\mu\}_{\mu\in\mathbb{N}}$ with $y_\mu = \langle f_\mu, L(f_\mu)\rangle$ is a Cauchy sequence in $\mathbf{X}_G$ that is monotone increasing with respect to x_0 then, by the construction of $\mathbf{X}_G$ described in the previous section, we have $\alpha_\mu := L(f_\mu) \nearrow \alpha$ and (after changing the values of the functions f_μ at the right-hand endpoints if necessary) $\mu < \nu$ implies that $f_\mu = f_\nu|_{[0,L(f_\mu)]}$. Define a function $f : [0,\alpha] \to G$ by

$$f(\xi) = \begin{cases} f_\mu(\xi), & L(f_{\mu-1}) < \xi \leq L(f_\mu), \\ f_0(0), & \xi = 0, \\ 1_G, & \xi = \alpha. \end{cases}$$

Then f is reduced and, setting $y := \langle f, \alpha\rangle$, we have

$$d_G(y, y_\mu) = \alpha - \alpha_\mu \to 0 \text{ as } \mu \to \infty;$$

that is, the sequence $\{y_\mu\}_{\mu\in\mathbb{N}}$ converges. Invoking Lemma 3.2 we find that $\mathbf{X}_G$ is complete, as claimed. □

Finally, we shall prove that the action of $\mathscr{RF}(G)$ on $\mathbf{X}_G$ is transitive. Most of the work for this is carried out in Appendix A.

Proposition 3.4 (Transitivity) *The group $\mathscr{RF}(G)$ acts transitively on the points of $\mathbf{X}_G$.*

Proof By Lemma 2.14 (dissection of reduced functions), the length function L on $\mathscr{RF}(G)$ is *strongly regular* (see Definition A.31). Hence, by Proposition A.37, $\mathscr{RF}(G)$ acts transitively on the span of $\mathscr{RF}(G)x_0$, which we have observed is the whole of $\mathbf{X}_G$. □

3.4 Cyclic reduction

In order to investigate further the action of $\mathscr{RF}(G)$ on $\mathbf{X}_G$, we need to introduce the concept of a *cyclically reduced* function.

Definition 3.5 A function $f \in \mathscr{RF}(G)$ is called *cyclically reduced*, if it satisfies $\varepsilon_0(f,f) = 0$; or, equivalently, if $L(f^2) = 2L(f)$.

Lemma 3.6

(i) *Every function $f \in \mathscr{RF}(G)$ of length 0 (that is, each element of the subgroup G_0) is cyclically reduced.*

(ii) *If $f \in \mathscr{RF}(G)$ is cyclically reduced then so is f^k for every $k \in \mathbb{Z}$, and we have*

$$L(f^k) = |k|L(f), \quad k \in \mathbb{Z}. \tag{3.3}$$

Proof (i) This is clear by the definition of $\varepsilon_0(f,f)$.

(ii) Observe first that (3.3) implies that f^k is cyclically reduced for all $k \in \mathbb{Z}$, since

$$L(f^{2k}) = 2|k|L(f) = 2L(f^k).$$

Hence, it suffices to establish equation (3.3). Let $f \in \mathscr{RF}(G)$ be cyclically reduced. Then we claim that

$$f^k = \underbrace{f \circ f \circ \cdots \circ f}_{k\,\text{times}}, \quad k \in \mathbb{N}_0. \tag{3.4}$$

This is trivial if $L(f) = 0$; thus, we may suppose that $L(f) > 0$. Also, (3.4) holds trivially for $k = 0, 1, 2$. So, let us suppose, by way of induction, that equation (3.4) holds with k replaced by $k-1$ and that $k \geq 3$. If $\varepsilon_0(f^{k-1}, f)$ were strictly positive then there would exist ε such that $0 < \varepsilon < L(f)$ and

$$f^{k-1}(L(f^{k-1}) - \eta)f(\eta) = 1_G, \quad 0 \leq \eta \leq \varepsilon. \tag{3.5}$$

Applying Lemma 2.2 together with the inductive hypothesis we see that

$$f^{k-1}(L(f^{k-1}) - \eta) = f(L(f) - \eta), \quad 0 \leq \eta < L(f),$$

so that (3.5) becomes

$$f(L(f) - \eta)f(\eta) = 1_G, \quad 0 \leq \eta \leq \varepsilon,$$

implying that

$$\varepsilon_0(f,f) = \sup \mathscr{E}(f,f) \geq \varepsilon > 0,$$

which contradicts our assumption that f is cyclically reduced. Hence we have $\varepsilon_0(f^{k-1}, f) = 0$ and (3.4) holds. With (3.4) at our disposal, it now follows that, for $k \in \mathbb{Z}$,

$$L(f^k) = L(f^{|k|}) = |k|L(f),$$

as required. □

Our next result is crucial: it establishes, in analogy with the case of free groups, the existence and uniqueness of cyclically reduced cores for the elements of $\mathcal{RF}(G)$. This version of cyclic reduction in $\mathcal{RF}(G)$ will turn out to be important in many respects; see, for instance, Proposition 3.13 and Corollary 3.21.

Lemma 3.7

(i) *Let $f \in \mathcal{RF}(G)$. Then there exist $t, f_1 \in \mathcal{RF}(G)$, with f_1 cyclically reduced, such that*

$$f = t \circ f_1 \circ t^{-1}.$$

(ii) *If*

$$f = t \circ f_1 \circ t^{-1} = s \circ f_2 \circ s^{-1}, \tag{3.6}$$

where $t, s, f_1, f_2 \in \mathcal{RF}(G)$ and f_1, f_2 are cyclically reduced, then $s = tg$ and $f_2 = g^{-1} f_1 g$ for some $g \in G_0$; in particular, $L(f_1) = L(f_2)$.

Proof (i) For $f \in \mathcal{RF}(G)$, set

$$\varepsilon_0(f) := \min\left\{\varepsilon_0(f,f), \tfrac{1}{2}L(f)\right\}.$$

By Lemma 2.14 (dissection of reduced functions), we can find $t, f_0 \in \mathcal{RF}(G)$ such that $f = t \circ f_0$ and $L(t) = \varepsilon_0(f)$. Then

$$L(f_0) = L(f) - L(t) = L(f) - \varepsilon_0(f) \geq \tfrac{1}{2}L(f) \geq \varepsilon_0(f);$$

hence, we can apply Lemma 2.14 again to obtain $f_0 = f_1 \circ u$ where $L(u) = \varepsilon_0(f)$. By Corollary 2.18 (associativity of the circle product) we have $\varepsilon_0(t, f_1) = 0$, so that $t \circ f_1$ is defined and

$$L(t \circ f_1) = L(t) + L(f_1) = L(f) - \varepsilon_0(f).$$

For $0 \leq \eta < \varepsilon_0(f) \leq \varepsilon_0(f,f)$, we have

$$f(L(f) - \eta) f(\eta) = 1_G;$$

also, $f(\eta) = t(\eta)$ for such an η. Moreover, for η in this range,

$$L(f) - \eta > L(f) - \varepsilon_0(f) = L(t \circ f_1),$$

so that $f(L(f) - \eta) = u(\varepsilon_0(f) - \eta)$; thus, $t \approx u^{-1}$ (see (3.1) for the definition of $\approx$). If $L(f_1) > 0$ then we can use the freedom in choosing $t(\varepsilon_0(f))$ and $u(0)$ incorporated in Lemma 2.14, adapting the initial and terminal values of f_1 to arrange that $u(0) = t(\varepsilon_0(f))^{-1}$, so that indeed $t = u^{-1}$. If $L(f_1) = 0$, that is, $\varepsilon_0(f) = \frac{1}{2}L(f)$, then the values of t at $\xi = \varepsilon_0(f)$ and of u at $\xi = 0$ are governed by the single equation

$$f(\tfrac{1}{2}L(f)) = t(\varepsilon_0(f))f_1(0)u(0),$$

and again $t(\varepsilon_0(f))$ and $u(0)$ can be chosen independently of each other, their values now determining $f_1(0)$. In both cases, we have found that $f = t \circ f_1 \circ t^{-1}$.

It remains to show that f_1 is cyclically reduced and, in doing so, we may assume that $L(f_1) > 0$. Then, since $L(f_1) = L(f) - 2\varepsilon_0(f)$, we have $\varepsilon_0(f) < \frac{1}{2}L(f)$, so that

$$L(t) = \varepsilon_0(f) = \varepsilon_0(f, f).$$

Now, by Lemma 2.16 (visible cancellation) and Corollary 2.18,

$$\begin{aligned} f^2 &= ((t \circ f_1) \circ t^{-1})(t \circ (f_1 \circ t^{-1})) \\ &= (t \circ f_1) \circ (f_1 \circ t^{-1}) \\ &= t \circ f_1^2 \circ t^{-1} \end{aligned}$$

and hence

$$L(f^2) = 2L(f_1) + 2\varepsilon_0(f) = L(f_1^2) + 2\varepsilon_0(f);$$

that is, $L(f_1^2) = 2L(f_1)$, as required.

(ii) Assume without loss of generality that $L(t) \geq L(s)$. Then, using Lemma 2.14 together with equations (3.6), we can write $t = s \circ u$ for some $u \in \mathscr{RF}(G)$. By Lemma 2.12 (inversion of star products), we have $t^{-1} = u^{-1} \circ s^{-1}$ and so

$$s \circ f_2 \circ s^{-1} = f = s \circ u \circ f_1 \circ u^{-1} \circ s^{-1}.$$

Applying Proposition 2.1 twice, this yields $f_2 = u \circ f_1 \circ u^{-1}$. The last equation, together with the fact that f_1 and f_2 are cyclically reduced, gives

$$L(f_2^2) = 2L(f_2) = 2\big(2L(u) + L(f_1)\big)$$

and also

$$L(f_2^2) = L(uf_1^2u^{-1}) \leq 2L(u) + L(f_1^2) = 2\big(L(u) + L(f_1)\big),$$

implying that $L(u) \leq 0$; thus $L(u) = 0$ and $t \approx s$. Setting $s = t \circ g$ with $g \in G_0$, Corollary 2.18 gives

$$\begin{aligned} f &= s \circ f_2 \circ s^{-1} \\ &= (t \circ g) \circ f_2 \circ (g^{-1} \circ t^{-1}) \\ &= t \circ (gf_2g^{-1}) \circ t^{-1} \end{aligned}$$

and hence $f_1 = gf_2g^{-1}$ by Proposition 2.1. □

We pause to record a simple but useful consequence of Lemma 3.7.

Corollary 3.8 *Let $x \in \bigcup_{t \in \mathscr{RF}(G)} tG_0t^{-1}$, and suppose that $x \neq \mathbf{1}_G$. Then x lies in precisely one conjugate of G_0.*

Proof Suppose that $x \in tG_0t^{-1} \cap sG_0s^{-1}$. Then

$$x = t \circ g \circ t^{-1} = s \circ h \circ s^{-1}$$

for some $g, h \in G_0$, since x is non-trivial. By part (ii) of Lemma 3.7 we have $s = tk$ for some $k \in G_0$, so $sG_0s^{-1} = tG_0t^{-1}$. □

Remark 3.9 Corollary 3.8 may be rephrased by saying that the set

$$\{tG_0t^{-1} : t \in \mathscr{RF}(G)\}$$

of conjugates of G_0 forms an *amalgam with trivial intersection*.

Definition 3.10

(i) The cyclically reduced function f_1 found in Lemma 3.7 for given $f \in \mathscr{RF}(G)$, which is unique up to conjugation by a G_0-element, is called the (cyclically reduced) *core* of f, denoted $c(f)$. The passage from f to f_1 itself is termed *cyclic reduction* of f.

(ii) The extra assumption that $c(f)(0) = 1_G$ singles out a uniquely defined core $c_0(f)$ of $f \in \mathscr{RF}(G)$, termed the *normalized core* of f. The function $c_0(f)$ exists, and is uniquely determined by f through the conditions that $f = t \circ c_0(f) \circ t^{-1}$ for some $t \in \mathscr{RF}(G)$, that $c_0(f)$ is cyclically reduced, and that $c_0(f)(0) = 1_G$.

3.5 Classification of elements

Next we briefly recall the classification of elements of a group Γ acting by isometries on a Λ-tree $(\mathbf{X}, d)$; see Section A.5 for more details.

An element $\gamma \in \Gamma$ is called

- *elliptic*, if γ has a fixed point,
- an *inversion*, if γ has no fixed point but γ^2 has a fixed point,
- *hyperbolic*, if γ is neither elliptic nor an inversion.

Moreover, we remark that in the case where $\Lambda = \mathbb{R}$ the group Γ does not contain any inversion. See the observation after Definition A.48.

Definition 3.11 We denote by $E(G)$ the subgroup of $\mathscr{RF}(G)$ generated by the elliptic elements with respect to the canonical action on $\mathbf{X}_G$.

Remark 3.12 We note that $E(G)$ is normal in $\mathscr{RF}(G)$, since a conjugate of an elliptic element is again elliptic. More precisely we have, in view of Proposition 3.4, that

$$E(G) = \left\langle \bigcup_{t \in \mathscr{RF}(G)} tG_0t^{-1} \right\rangle = \langle\langle G_0 \rangle\rangle,$$

that is, $E(G)$ is the normal closure of the subgroup G_0.

Our next result characterizes the elliptic and hyperbolic elements of $\mathscr{RF}(G)$ in terms of their cyclically reduced cores, as introduced in Definition 3.10.

Proposition 3.13

(i) *An element of $\mathscr{RF}(G)$ is elliptic if and only if it is conjugate to an element of G_0.*

(ii) *An element $f \in \mathscr{RF}(G)$ is hyperbolic if and only if $L(c(f)) > 0$.*

Proof (i) If $f \in \mathscr{RF}(G)$ is elliptic then, by Proposition 3.4 (transitivity), there exists $g \in \mathscr{RF}(G)$ such that $(g^{-1}fg)x_0 = x_0$, so f is conjugate to an element of G_0. Conversely, if $f = hgh^{-1}$ with $g \in G_0$ then f fixes the point hx_0, since

$$fhx_0 = (hg)x_0 = h(gx_0) = hx_0;$$

hence f is elliptic.

(ii) Let $f \in \mathscr{RF}(G)$ be a hyperbolic element. By Part (i) of Lemma 3.7, we can write $f = t \circ f_1 \circ t^{-1}$ with elements $t, f_1 \in \mathscr{RF}(G)$ and f_1 cyclically reduced; that is, $f_1 = c(f)$. If we had $L(f_1) = 0$, that is, $f_1 \in G_0$, then $f = tf_1t^{-1}$ would be elliptic by part (i), contradicting the fact that f is hyperbolic; hence, we must have $L(f_1) > 0$. Conversely, suppose that $L(c(f)) > 0$, that is, that $f = t \circ f_1 \circ t^{-1}$ with $L(f_1) > 0$ and f_1 cyclically reduced. If f were elliptic then f_1 would also be elliptic; thus, by part (i) above, $f_1^k \in sG_0s^{-1}$ for some $s \in \mathscr{RF}(G)$ and all $k \geq 1$. However, the elements of sG_0s^{-1} have lengths bounded above by $2L(s)$ while, according to part (ii) of Lemma 3.6, the length of f_1^k becomes unbounded as $k \to \infty$. Hence, f must be hyperbolic. □

Remark 3.14 Corollary 3.8 can now be rephrased as follows: every non-trivial elliptic element of $\mathscr{RF}(G)$ lies in precisely one conjugate of G_0. Equivalently, every non-trivial elliptic element stabilises exactly one point of $\mathbf{X}_G$. We also note that the centralisers of elliptic elements are described, up to conjugation, by the first part of Proposition 2.20.

If $[x,y]$ is a segment in a Λ-tree $\mathbf{X} = (X,d)$ with $x \neq y$ and G is a group acting on $\mathbf{X}$ by isometries, then the pointwise stabiliser of $[x,y]$ is $\operatorname{stab}(x) \cap \operatorname{stab}(y)$; such a subgroup of G will be called a *pointwise arc stabiliser*. By Remark 3.14, every (pointwise) arc stabiliser is trivial.

Lemma 3.15

(i) *If $f = t \circ f_1 \circ t^{-1}$ with f_1 cyclically reduced and $L(f_1) > 0$ (that is, $f \in \mathscr{RF}(G)$ is hyperbolic) then*

$$f^k = t \circ f_1^k \circ t^{-1}, \quad k \in \mathbb{Z} \backslash \{0\}, \tag{3.7}$$

is the cyclically reduced decomposition of f^k; in particular, we have

$$L(f^k) = 2L(t) + |k| L(f_1). \tag{3.8}$$

(ii) *A hyperbolic element $f \in \mathscr{RF}(G)$ has at most one kth root, for every $k \geq 2$.*

(iii) *Let $f, g \in \mathscr{RF}(G)$ be hyperbolic elements such that $[f,g] = \mathbf{1}_G$, and let r, s be roots of f respectively g. Then $[r,s] = \mathbf{1}_G$.*

Proof (i) In view of Lemma 3.6(ii), it suffices to establish equation (3.7), and by Lemma 2.12 (inversion of star products), it is enough to prove (3.7) for $k \geq 1$. It was shown in the proof of Lemma 3.6(ii) that

$$f_1^k = \underbrace{f_1 \circ \cdots \circ f_1}_{k\,\text{times}},$$

implying that $\varepsilon_0(f_1, f_1^{k-1}) = 0$ by Corollary 2.18 (associativity of the circle product). Moreover, by the assumption plus Corollary 2.18, we also have $\varepsilon_0(t, f_1) = 0$. Applying part (ii) of Lemma 2.17 to the functions t, f_1, f_1^{k-1} now yields $\varepsilon_0(t, f_1^k) = 0$; thus

$$t f_1^k = t \circ \underbrace{f_1 \circ \cdots \circ f_1}_{k \text{ times}}.$$

Combining Corollary 2.18 with the last decomposition, we deduce that

$$\varepsilon_0(t f_1^{k-1}, f_1) = 0,$$

while $\varepsilon_0(f_1, t^{-1}) = 0$ holds by the assumption plus Corollary 2.18. Applying part (i) of Lemma 2.17 to the functions $t f_1^{k-1}, f_1, t^{-1}$ now gives $\varepsilon_0(t f_1^k, t^{-1}) = 0$, whence equation (3.7).

(ii) A root of a hyperbolic element is also hyperbolic by part (i) of Proposition 3.13, so let g and h be two hyperbolic elements such that $g^k = h^k$ for some $k \geq 2$. Setting $g = t \circ g_1 \circ t^{-1}$ and $h = s \circ h_1 \circ s^{-1}$, with g_1 and h_1 cyclically reduced and $L(g_1), L(h_1) > 0$ (according to part (ii) of Proposition 3.13), and applying part (i) of the present lemma we find that

$$t \circ g_1^k \circ t^{-1} = s \circ h_1^k \circ s^{-1}.$$

By Lemma 3.6(ii), g_1^k and h_1^k are also cyclically reduced, so Lemma 3.7(ii) gives $s = tx$ and $g_1^k = x h_1^k x^{-1}$ for some $x \in G_0$; in particular $L(g_1) = L(h_1)$. Noting that both h_1 and $x h_1 x^{-1}$ are cyclically reduced, and using part (ii) of Lemma 3.6 to decompose g_1^k as

$$g_1^k = \underbrace{g_1 \circ \cdots \circ g_1}_{k \text{ times}}$$

and $x h_1^k x^{-1}$ as

$$x h_1^k x^{-1} = \underbrace{x h_1 x^{-1} \circ \cdots \circ x h_1 x^{-1}}_{k \text{ times}},$$

application of Lemma 2.2 yields, for $0 \leq \xi \leq L(g_1)$, that

$$g_1(\xi) = \left\{ \begin{array}{ll} x(0) h_1(0), & \xi = 0 \\ h_1(\xi), & 0 < \xi < L(g_1) \\ h_1(L(g_1)) x^{-1}(0), & \xi = L(g_1) \end{array} \right\} = (x h_1 x^{-1})(\xi);$$

that is, $g_1 = x h_1 x^{-1}$. It follows now that

$$g = t g_1 t^{-1} = (s x^{-1})(x h_1 x^{-1})(x s^{-1}) = s h_1 s^{-1} = h,$$

as claimed.

(iii) Let $f = r^k$ and $g = s^\ell$ for positive integers k and ℓ. By our hypothesis,

$$r^k s^\ell = s^\ell r^k.$$

Thus

$$s^\ell = r^{-k} s^\ell r^k = (r^{-k} s r^k)^\ell,$$

so s and $r^{-k}sr^k$ are both ℓth roots of the hyperbolic element $g = r^{-k}s^\ell r^k$; hence, by part (ii),

$$[s, r^k] = \mathbf{1}_G.$$

It follows that

$$(srs^{-1})^k = sr^k s^{-1} = r^k,$$

so r and srs^{-1} are two kth roots of the hyperbolic element $f = sr^k s^{-1}$; therefore $[r,s] = \mathbf{1}_G$ as claimed. □

Remark 3.16 In the next section, having introduced and discussed the characteristic set A_f associated with an element $f \in \mathscr{RF}(G)$, we shall be able to give short geometric proofs of parts (ii) and (iii) of Lemma 3.15; see Remark 3.19 below.

3.6 Bounded subgroups

We begin by recalling the following definition.

Definition 3.17 Let Γ be a group acting by isometries on a Λ-tree $\mathbf{X} = (X, d)$, and let $\gamma \in \Gamma$. The set

$$A_\gamma := \left\{x \in X : x \in [\gamma^{-1}x, \gamma x]\right\}$$

is called the *characteristic set* of γ, where, for points $x,y \in X$, $[x,y]$ denotes the segment joining x and y.

Our next result, which describes the structure of characteristic sets, is well known. It combines parts of several results given in Section A.5, whose proofs can be found in Chapter 3 of Chiswell [10].

Lemma 3.18

(i) *If γ is elliptic then $A_\gamma = \mathbf{X}^\gamma$ is the set of fixed points of γ.*

(ii) *If γ is an inversion then $A_\gamma = \varnothing$.*

(iii) *If γ is hyperbolic then A_γ is a non-empty closed $\langle\gamma\rangle$-invariant subtree of $\mathbf{X}$. Furthermore, A_γ is a linear tree (that is, it is metrically isomorphic to a subtree of Λ), and, in this identification, the image of γ restricted to A_γ is a translation*

$$a \mapsto a + \ell(\gamma)$$

for some $\ell(\gamma) \in \Lambda$ with $\ell(\gamma) > 0$.

We recall some definitions given in Section A.5 after Theorem A.51. In case (iii) of Lemma 3.18 the set A_γ is called the *axis* of γ (which explains the notation). When γ is elliptic or an inversion, $\ell(\gamma)$ is defined to be 0. For $\gamma \in \Gamma$, $\ell(\gamma)$ is called the *hyperbolic length* of γ. As is well known, ℓ is invariant under conjugation; see Remark A.53.

Remark 3.19 Before proceeding, we note that there are simpler proofs of parts (ii) and (iii) of Lemma 3.15, using the geometry of $\mathbb{R}$-trees, as follows.

(ii) Suppose that $f^k = g^k$ is hyperbolic. Then, by Lemma A.54(i), we have

$$\ell(f) = \ell(g) > 0;$$

thus f, g are hyperbolic and f, g have the same axis. Hence $f^{-1}g$ fixes every point of this axis, so $f^{-1}g = \mathbf{1}_G$ by Remark 3.14.

(iii) By Lemma A.54(ii), f and g have the same axis and this is also the axis of r and s, again by Lemma A.54(i). Consequently, $[r, s]$ fixes every point of this common axis; hence, as before, $[r,s] = \mathbf{1}_G$, as claimed.

Lemma 3.20 *If $f \in \mathscr{RF}(G)$ then we have $x_0 \in A_f$ if and only if f is cyclically reduced.*

Proof In the notation of Definition A.6, we have $x_0 \in A_f$ if and only if

$$d(x_0, Y(x_0, f^{-1}x_0, fx_0)) = c_{x_0}(f^{-1}, f) = 0$$

(see Example A.28). Moreover, as we have observed in the proof of Proposition 3.1, we have $c_{x_0}(f,g) = \varepsilon_0(f^{-1}, g)$. Hence we conclude that $x_0 \in A_f$ if and only if $\varepsilon_0(f,f) = 0$; that is, if and only if f is cyclically reduced. □

Corollary 3.21 *If $f \in \mathcal{RF}(G)$ is decomposed as $f = t \circ f_1 \circ t^{-1}$ with f_1 cyclically reduced (according to Lemma 3.7) then $\ell(f) = L(f_1)$; that is, we have*

$$\ell(f) = L(c(f)), \quad f \in \mathcal{RF}(G). \tag{3.9}$$

Proof If, on the one hand, f is elliptic then $L(f_1) = \ell(f) = 0$ by definition of the hyperbolic length and Proposition 3.13. If, on the other hand, f is hyperbolic then, since $x_0 \in A_{f_1}$ by Lemma 3.20 and ℓ is invariant under conjugation, we have

$$\ell(f) = \ell(f_1) = L_{x_0}(f_1) = L(f_1)$$

by the last statement in part (iii) of Lemma 3.18. □

Lemma 3.22 *Let a and b be two elliptic elements of $\mathcal{RF}(G)$ which do not lie in the same conjugate of G_0. Then their product ab is hyperbolic.*

Proof Let $a = t \circ g \circ t^{-1}$ and $b = s \circ h \circ s^{-1}$ with $g, h \in G_0 - \{\mathbf{1}_G\}$. Then, according to part (i) of Lemma 3.18, the characteristic set of a is $X^a = \{tx_0\}$ while that of b is $X^b = \{sx_0\}$ and our hypothesis is equivalent to saying that $t \not\approx s$, that is, $t^{-1}sx_0 \neq x_0$. If follows that $X^a \cap X^b = \varnothing$; hence, a and b satisfy the hypotheses of Lemma 2.2 in Chapter 3 of Chiswell [10] and, consequently,

$$\ell(ab) = 2d_G(tx_0, sx_0) > 0,$$

so that ab is hyperbolic as claimed. □

Remark 3.23 The proof of part (ii) of Proposition 3.29 in Section 3.7 will provide another approach, this time by an algebraic argument, to the assertion of Lemma 3.22.

Definition 3.24 We call a subgroup $\mathcal{H} \leq \mathcal{RF}(G)$ *bounded* if there exists $c \in \mathbb{R}$ such that $L(f) \leq c$ for all $f \in \mathcal{H}$.

Proposition 3.25 *Let $\mathcal{H}$ be a subgroup of $\mathcal{RF}(G)$. Then the following assertions are equivalent:*

(i) *$\mathcal{H}$ is bounded;*

(ii) *$\mathcal{H}$ consists entirely of elliptic elements;*

(iii) *$\mathcal{H}$ is conjugate to a subgroup of G_0.*

Proof Obviously (iii) implies (i). Next we show that (i) implies (ii). Indeed, if $f \in \mathscr{H}$ is hyperbolic then, by part (ii) of Proposition 3.13, we have $f = t \circ f_1 \circ t^{-1}$ with $t, f_1 \in \mathscr{RF}(G)$, f_1 cyclically reduced, and $L(f_1) > 0$; thus, by equation (3.8) in part (i) of Lemma 3.15,

$$L(f^n) = 2L(t) + nL(f_1) \to \infty \quad \text{as } n \to \infty,$$

contradicting assertion (i). Finally, suppose that $\mathscr{H}$ does not contain a hyperbolic element. In showing that (ii) implies (iii), we may suppose that $\mathscr{H} \neq \{\mathbf{1}_G\}$. Fix a non-trivial element $a = t \circ g \circ t^{-1}$ in $\mathscr{H}$ with $L(g) = 0$, and let $b \in \mathscr{H}$ be an arbitrary element. According to Corollary 3.8, a lies in tG_0t^{-1} and in no other conjugate of G_0. Hence, our assumption that $\mathscr{H}$ consists entirely of elliptic elements together with Lemma 3.22 implies that $b \in tG_0t^{-1}$; so $\mathscr{H} \subseteq tG_0t^{-1}$, since b is arbitrary. □

Corollary 3.26 *Every finite subgroup of $\mathscr{RF}(G)$ is conjugate to a subgroup of G_0; in particular, $\mathscr{RF}(G)$ is torsion-free if and only if G is torsion-free.*

Proof A finite subgroup of $\mathscr{RF}(G)$ is automatically bounded and hence contained in some conjugate of G_0 by Proposition 3.25. If $f \in \mathscr{RF}(G)$ is a non-trivial torsion element then $\langle f \rangle$, the cyclic subgroup generated by f, is finite and thus conjugate to a subgroup of G_0. Hence G itself contains a non-trivial torsion element. Conversely, if G contains a non-trivial torsion element then so does G_0, and hence $\mathscr{RF}(G)$. □

Corollary 3.27 *The only bounded subnormal subgroup of $\mathscr{RF}(G)$ is the trivial group $\{\mathbf{1}_G\}$.*

Proof Let $\mathscr{N} \leq \mathscr{RF}(G)$ be a non-trivial, bounded, and subnormal subgroup of $\mathscr{RF}(G)$. Then $G \neq \{1_G\}$, so $G_0 < \mathscr{RF}(G)$ (for instance, $\mathscr{RF}(G)$ contains functions of positive length with constant value equal to some element $x \in G - \{1_G\}$). Further, by Proposition 3.25 we have $\mathscr{N} \leq tG_0t^{-1}$ for some $t \in \mathscr{RF}(G)$. Let

$$\mathscr{N} = N_0 \lhd N_1 \lhd \cdots \lhd N_r \lhd \mathscr{RF}(G)$$

be a strict subnormal series connecting $\mathscr{N}$ with $\mathscr{RF}(G)$. Then $r \geq 0$, since $\mathscr{RF}(G)$ itself is not bounded.[4] Conjugating by t, we find that $\mathscr{N}^t \leq G_0$ and

[4] Consider, for instance, a function f of length 1 given by

$$f(\xi) = \begin{Bmatrix} x, & 0 \leq \xi < 1 \\ 1_G, & \xi = 1 \end{Bmatrix} \quad (0 \leq \xi \leq 1),$$

where x is some non-trivial element of G. Then f is cyclically reduced and of positive length, hence hyperbolic by part (ii) of Proposition 3.13. Now use part (i) of Lemma 3.15 to conclude that $\mathscr{RF}(G)$ contains elements of arbitrarily large length.

that

$$\{\mathbf{1}_G\} \neq \mathscr{N}^t = N_0^t \triangleleft N_1^t \triangleleft \cdots \triangleleft N_r^t = N_r \triangleleft \mathscr{R}\mathscr{F}(G).$$

Suppose that $N_i^t \leq G_0$ for some $0 \leq i < r$. Then, making use of part (ii) of Proposition 2.20 plus the fact that $N_i^t \neq \{\mathbf{1}_G\}$, we obtain

$$N_{i+1}^t \leq N_{\mathscr{R}\mathscr{F}(G)}(N_i^t) = N_{G_0}(N_i^t) \leq G_0.$$

Since $N_0^t = \mathscr{N}^t \leq G_0$, this shows that $N_r \leq G_0$. Applying part (ii) of Proposition 2.20 again, it follows that

$$\mathscr{R}\mathscr{F}(G) = N_{\mathscr{R}\mathscr{F}(G)}(N_r) = N_{G_0}(N_r) \leq G_0 < \mathscr{R}\mathscr{F}(G),$$

a contradiction. Hence, such an $\mathscr{N}$ does not exist. □

Remark 3.28 We can give an alternative proof for the implication (i) ⇒ (iii) in Proposition 3.25 as follows. Let $\mathscr{H} \leq \mathscr{R}\mathscr{F}(G)$ be a bounded subgroup. By Proposition 3.3, the $\mathbb{R}$-tree $\mathbf{X}_G$ is complete; thus we can apply a result of Wilkens[5] to obtain that $\mathscr{H}$ has a global fixed point $x \in \mathbf{X}_G$. Since the action of G on $\mathbf{X}_G$ is transitive by Proposition 3.4, we can find an element $f \in \mathscr{R}\mathscr{F}(G)$ such that $\mathscr{H}^f$ stabilizes the base-point x_0, that is, $\mathscr{H}^f \subseteq G_0$ as required.

3.7 Presenting $E(G)$

As before, let

$$E(G) = \Big\langle \bigcup_{t \in \mathscr{R}\mathscr{F}(G)} tG_0t^{-1} \Big\rangle = \langle\langle G_0 \rangle\rangle$$

be the subgroup of $\mathscr{R}\mathscr{F}(G)$ generated by the elliptic elements, that is, the normal closure of G_0 in $\mathscr{R}\mathscr{F}(G)$, and let

$$\pi : \mathop{\ast}_{t} tG_0t^{-1} \to E(G) \tag{3.10}$$

be the natural projection induced by the inclusions $tG_0t^{-1} \hookrightarrow E(G)$, where t runs through a system of representatives for the left cosets of $\mathscr{R}\mathscr{F}(G)$ modulo

[5] See, for instance, Lemma 2.5 in Chapter 4 of Chiswell [10].

G_0. We note that

$$tG_0t^{-1} = sG_0s^{-1} \iff t^{-1}sG_0s^{-1}t = G_0$$
$$\iff t^{-1}s \text{ normalizes } G_0$$
$$\iff t^{-1}s \in G_0;$$

the last equivalence comes from part (ii) of Proposition 2.20 or, more precisely, from the fact that the subgroup G_0 is self-normalizing in $\mathscr{RF}(G)$. Hence the free product in (3.10) contains each conjugate of G_0 in $\mathscr{RF}(G)$ exactly once as a free factor.

It appears natural to try to study the structure of $E(G)$ by presenting it over the free product $*_t\, tG_0t^{-1}$ via the map π; the relations for $E(G)$ are then given by the elements of the kernel K_0 of π. More precisely, a relation of $E(G)$ corresponds to an equation

$$t_1g_1t_1^{-1}t_2g_2t_2^{-1}\cdots t_mg_mt_m^{-1} = \mathbf{1}_G \tag{3.11}$$

in $\mathscr{RF}(G)$, where $t_i^{-1}t_{i+1} \notin G_0$ for $i = 1,2,\ldots,m-1$ and $g_1,g_2,\ldots,g_m \in G_0 - \{\mathbf{1}_G\}$. The left-hand side of (3.11) is called a *(normal form) relation of* $E(G)$ *of length* m. Such a relation is called *non-trival* if $m > 0$, that is, if the left-hand side is a non-empty product. Our next result summarizes what little is known at present concerning the normal form relations of $E(G)$.

Proposition 3.29 *Let G be a non-trivial group.*

- (i) *Every non-trivial relation of $E(G)$ involves at least three distinct conjugates of G_0; that is, any two distinct conjugates generate their free product.*
- (ii) *The group $E(G)$ does not have relations of length* 3.
- (iii) *Given any two distinct conjugates of G_0, there exists a third conjugate such that $E(G)$ admits a normal form relation of length* 4 *involving these three conjugates; in particular,* π (see (3.10)) *is not an isomorphism.*

Clearly, part (ii) of Proposition 3.29 is equivalent to the assertion of Lemma 3.22; however, we are going to give a different proof via a cancellation argument based on the following simple observation.

Lemma 3.30 *Given any three elements $f,g,h \in \mathscr{RF}(G)$ such that $fgh = \mathbf{1}_G$, there exist elements $u,v,w \in \mathscr{RF}(G)$ such that $f = u \circ v$, $g = v^{-1} \circ w$, and $h = w^{-1} \circ u^{-1}$.*

Proof By Lemma 2.15 (visibility of cancellation), we can find elements

$u, v, w \in \mathscr{RF}(G)$ such that $f = u \circ v$, $g = v^{-1} \circ w$, and $fg = u \circ w$. By Lemma 2.12 (inversion of star products), we then have

$$h = (fg)^{-1} = (u \circ w)^{-1} = w^{-1} \circ u^{-1},$$

as required. □

Proof of Proposition 3.29 (i) Clearly, there cannot be a non-trivial normal form relation involving only one conjugate of G_0. Suppose we have a normal form relation

$$t_1 g_1 t_1^{-1} t_2 g_2 t_2^{-1} t_1 g_3 t_1^{-1} \cdots t_1 g_{2k-1} t_1^{-1} t_2 g_{2k} t_2^{-1} = \mathbf{1}_G$$

with $t_1^{-1} t_2 \notin G_0$ and $g_i \in G_0 - \{\mathbf{1}_G\}$ for $i = 1, 2, \ldots, 2k$. Conjugating by t_1^{-1}, we obtain a corresponding relation

$$g_1 s g_2 s^{-1} g_3 s g_4 s^{-1} \cdots g_{2k-1} s g_{2k} s^{-1} = \mathbf{1}_G \tag{3.12}$$

with $s := t_1^{-1} t_2$ of positive length and $g_1, g_2, \ldots, g_{2k}$ as before. The left-hand side of (3.12) can be rewritten as

$$R = g_1 \circ s \circ g_2 \circ s^{-1} \circ g_3 \circ s \circ g_4 \circ s^{-1} \circ \cdots \circ g_{2k-1} \circ s \circ g_{2k} \circ s^{-1};$$

hence $L(R) = 2kL(s)$, implying $k = 0$ since $L(s) > 0$. A normal form relation

$$t_1 g_1 t_1^{-1} t_2 g_2 t_2^{-1} t_1 g_3 t_1^{-1} \cdots t_1 g_{2k-1} t_1^{-1} = \mathbf{1}_G$$

is handled in the same way. Hence, a normal form relation between two conjugates has to be trivial (in other words, any two distinct G_0-conjugates generate their free product).

(ii) Suppose we have a normal form relation

$$t_1 g_1 t_1^{-1} t_2 g_2 t_2^{-1} t_3 g_3 t_3^{-1} = \mathbf{1}_G$$

of length 3. Conjugating by t_1^{-1}, we obtain a corresponding relation

$$g_1 s g_2 s^{-1} t g_3 t^{-1} = \mathbf{1}_G$$

with the same elements g_1, g_2, g_3, $s := t_1^{-1} t_2$, and $t := t_1^{-1} t_3$; in particular, $g_i \neq \mathbf{1}_G$ and $L(s) > 0$. Since, by part (i), the three conjugates $t_i G_0 t_i^{-1}$ $(i = 1, 2, 3)$ are distinct, we also have $L(t) > 0$. Set

$$f := g_1 s = u \circ v,$$

$$g := g_2 s^{-1} = v^{-1} \circ w,$$

$$h := t g_3 t^{-1} = w^{-1} \circ u^{-1},$$

according to Lemma 3.30, observing that

$$L(f) = L(g) = L(s)$$

and that

$$L(h) = L(t \circ g_3 \circ t^{-1}) = 2L(t)$$

since g_3 is non-trivial. Suppose first that $L(u) = 0$. Then $L(v) = L(s)$. Hence $L(w) = 0$ and thus

$$L(h) = L(u) + L(w) = 0 = 2L(t);$$

that is, $L(t) = 0$, a contradiction. Hence, we must have $L(u) > 0$. But then $L(v) < L(s)$, so $L(w) > 0$ as well. Now w ends like s^{-1}, while u begins like $g_1 s$; in particular, we have

$$w(L(w)) = s^{-1}(L(s)) = (s(0))^{-1}$$

and

$$u(0) = g_1(0)s(0) \neq s(0),$$

since g_1 is non-trivial. Invoking Lemma 2.12 (inversion of star products), it follows that, on the one hand,

$$\varepsilon_0(u^{-1}, w^{-1}) = \varepsilon_0(w, u) = 0$$

and consequently that $\varepsilon_0(h, h) = 0$. On the other hand, we have $h = t \circ g_3 \circ t^{-1}$ and so

$$\varepsilon_0(h, h) \geq L(t) > 0,$$

a final contradiction.

(iii) Let sG_0s^{-1} and tG_0t^{-1} be two distinct conjugates of G_0, and choose non-trivial elements $sg_1s^{-1} \in sG_0s^{-1}$ and $tg_2t^{-1} \in tG_0t^{-1}$. Then the element

$$sg_1s^{-1}tg_2t^{-1}sg_1^{-1}s^{-1}$$

of $\mathscr{RF}(G)$ is elliptic and non-trivial (being the conjugate of a non-trivial elliptic element) and is not contained in sG_0s^{-1}, for otherwise we would have

$$tg_2t^{-1} \in sG_0s^{-1} \cap tG_0t^{-1} = \{\mathbf{1}_G\}$$

by Corollary 3.8, implying $g_2 = \mathbf{1}_G$, and so contradicting our choice of tg_2t^{-1}. Hence, there exists a conjugate uG_0u^{-1} of G_0 such that $uG_0u^{-1} \neq sG_0s^{-1}$ and

$$sg_1s^{-1}tg_2t^{-1}sg_1^{-1}s^{-1} = ug_3u^{-1} \in uG_0u^{-1} - \{\mathbf{1}_G\}.$$

It follows that

$$(sg_1s^{-1})(tg_2t^{-1})(sg_1^{-1}s^{-1})(ug_3^{-1}u^{-1}) = \mathbf{1}_G$$

is a normal form relation of length 4 involving the conjugates sG_0s^{-1}, tG_0t^{-1}, and uG_0u^{-1}. Finally, the fact that $uG_0u^{-1} \neq tG_0t^{-1}$ is now immediate from part (i).

3.8 A remark concerning universality

In view of the fact that the construction of the groups $\mathscr{RF}(G)$ as explained in Chapter 2 is a continuous analogue of a construction leading to free groups, the question arises whether this construction of $\mathscr{RF}(G)$ and its action on $\mathbf{X}_G$ is in any sense universal, either for inclusion or for quotients. Recently the authors have shown that if G is a group acting freely on an $\mathbb{R}$-tree $\mathbf{X} = (X,d)$ then there are a group H, an embedding $G \to \mathscr{RF}(H)$, and an equivariant isometry $\mathbf{X} \to \mathbf{X}_H$. The proof will be summarised in the next chapter and is in two stages. In the first stage we show that one can assume that the action of G is transitive, and this part works also for arbitrary Λ-trees if one assumes in addition that the action is without inversions.

For non-free actions one might ask, for instance, whether the construction of $\mathscr{RF}(G)$ is universal for transitive actions on $\mathbb{R}$-trees. One way of making this question precise is to ask whether, if $\Gamma \to \mathrm{Iso}(\mathbf{X})$ is a transitive action on an $\mathbb{R}$-tree $\mathbf{X}$ with stabilisers isomorphic to a group G, either Γ embeds into $\mathscr{RF}(G)$ and $\mathbf{X}$ embeds into $\mathbf{X}_G$ or Γ and $\mathbf{X}$ are quotients of $\mathscr{RF}(G)$ and $\mathbf{X}_G$, respectively. There are, however, examples of non-trivial groups acting freely and transitively on $\mathbb{R}$-trees, which suggests a negative answer to the above question: if $G = \{1_G\}$ then $\mathscr{RF}(G)$ is trivial and thus cannot embed or project onto a non-trivial group; similarly, $\mathbf{X}_G$ in this case consists of a single point and so again cannot embed or project onto a non-trivial $\mathbb{R}$-tree.

One such instance of a free transitive action is the additive group of $\mathbb{R}$ acting on itself by translations, but this example, being abelian, might not strike the reader as particularly interesting. It follows, however, that the free product of any number of copies of $\mathbb{R}$ has a free, transitive, action on an $\mathbb{R}$-tree, a fact which certainly appears to shatter our hopes concerning the universality of $\mathscr{RF}(G)$ and its associated $\mathbb{R}$-tree $\mathbf{X}_G$ in the sense outlined in the previous paragraph.

To prove this statement about free products, we shall establish a more general result concerning length functions. Let Λ be an ordered abelian group, let $\{G_i : i \in I\}$ be a family of groups, and let $L_i : G_i \to \Lambda$ be a length function, for each $i \in I$. Let $G = \ast_{i\in I} G_i$ be the free product of the G_i. Then there is a length function $L : G \to \Lambda$ given by $L(x_1 \cdots x_n) = \sum_{i=1}^{n} L_{j_i}(x_i)$, where $x_i \in G_{j_i}$ and $x_1 \cdots x_n$ is a reduced word relative to the free product G; see Lemma A.57. For the idea of strong regularity in the next lemma, see Definition A.31.

Lemma 3.31 *If the length functions L_i are all strongly regular then so is L.*

Proof Viewing G_i as a subgroup of G, we have $L_i = L|_{G_i}$; thus we can omit the subscript i on L_i. Let $x \in G$ be written in reduced form as $x = x_1 \cdots x_n$, and let $c \in \Lambda$ be such that $0 \le c \le L(x)$. We need to find $y, z \in G$ such that $x = yz$, $L(y) = c$, and $L(y) + L(z) = L(x)$. If $c = 0$, we can take $y = 1$, $z = x$ and if $c = L(x)$ then we may set $y = x$, $z = 1$. Thus we may assume that $0 < c < L(x)$.

Let k be maximal, subject to the property that $\sum_{i=1}^{k} L(x_i) < c$; hence $0 \le k < n$ and

$$c \le \sum_{i=1}^{k+1} L(x_i) = \sum_{i=1}^{k} L(x_i) + L(x_{k+1});$$

consequently,

$$0 < c - \sum_{i=1}^{k} L(x_i) \le L(x_{k+1}).$$

Suppose that $x_{k+1} \in G_j$. Then, since $L|_{G_j}$ is strongly regular by assumption, we can find $y', z' \in G_j$ such that $x_{k+1} = y'z'$, where $L(y') = c - \sum_{i=1}^{k} L(x_i)$ and $L(y') + L(z') = L(x_{k+1})$. Set $y := x_1 \cdots x_k y'$ and $z := z' x_{k+2} \cdots x_n$. Then

$$\begin{aligned} yz &= (x_1 \cdots x_k y')(z' x_{k+2} \cdots x_n) \\ &= x_1 \cdots x_k x_{k+1} x_{k+2} \cdots x_n \\ &= x \end{aligned}$$

and, since $x_1 \cdots x_k y'$ is a reduced word,

$$\begin{aligned} L(y) &= L(x_1) + \cdots + L(x_k) + L(y') \\ &= L(x_1) + \cdots + L(x_k) + \left(c - \sum_{i=1}^{k} L(x_i)\right) \\ &= c. \end{aligned}$$

Furthermore, we have

$$\begin{aligned} L(x) &= \sum_{i=1}^{n} L(x_i) \\ &= \sum_{i=1}^{k} L(x_i) + L(x_{k+1}) + \sum_{i=k+2}^{n} L(x_i) \\ &= \left(\sum_{i=1}^{k} L(x_i) + L(y')\right) + \left(L(z') + \sum_{i=k+2}^{n} L(x_i)\right) \\ &= L(y) + L(z) \end{aligned}$$

(this is valid even if $z' = 1$ or $n = k+1$). Since y and z satisfy all requirements and since $x \in G$ and c with $0 \leq c \leq L(x)$ were chosen arbitrarily, it follows that the length function L is strongly regular, as claimed. □

Before stating the next result the reader is reminded that in an action on an $\mathbb{R}$-tree there are no inversions, as observed in the remark after Definition A.48.

Corollary 3.32 *If $\{G_i : i \in I\}$ is a family of groups, each G_i having a free and transitive action on an $\mathbb{R}$-tree, then the free product $\ast_{i\in I} G_i$ also has a free and transitive action on an $\mathbb{R}$-tree.*

Proof Suppose that G_i has a free and transitive action on an $\mathbb{R}$-tree $X_i, d_i)$ and choose a base-point $x_i \in X_i$, for each $i \in I$. The length function $L_i := L_{x_i}$ is free by Lemma A.56 (see Definition A.55 for the definition of a free length function), and it is strongly regular by Proposition A.37. Therefore the length function L in Lemma 3.31 is strongly regular and is free by Lemma A.57.

By Lemma A.56 there are an action of $\ast_{i\in I} G_i$ on an $\mathbb{R}$-tree (X,d) and a point $x \in X$ such that $L = L_x$. The action is free by Lemma A.56, and restricting to the subtree spanned by the orbit of x gives the required action by Proposition A.37. □

In particular, it follows that a free product of copies of $\mathbb{R}$ has a free and transitive action on an $\mathbb{R}$-tree, as claimed above.

Remarks 3.33

(i) Alperin and Moss [1] gave a construction which is claimed, without a complete proof, to be universal for free actions on $\mathbb{R}$-trees with respect to inclusion. The present authors had some difficulty in trying to give a detailed proof along the lines suggested by Alperin and Moss. However, we shall adapt their construction in the next chapter to show that the class of groups $\mathscr{RF}(G)$, where G is a group, is universal (with respect to inclusion) for free actions on $\mathbb{R}$-trees. We shall return to their construction, and generalise it, in Chapter 11. In Chiswell [11] it was shown that a construction given in Myasnikov, Remeslennikov, and Serbin [40] is universal for free actions on Λ-trees with respect to inclusion, for any Λ.

(ii) The fact that $(\mathbb{R},+)$ acts freely by translation on the real line implies that every torsion-free abelian group of rank $r \leq 2^{\aleph_0}$ has a free $\mathbb{R}$-tree action, since these are precisely the groups that embed into the additive reals.

3.9 The degree of vertices of $\mathbf{X}_G$

We want to establish that the number of directions at the base-point x_0 of $\mathbf{X}_G$ equals $|\mathscr{RF}(G)|$.[6] Since such a direction is defined by a segment $[x_0, fx_0]$ (as the action of $\mathscr{RF}(G)$ on $\mathbf{X}_G$ is transitive), it is clear that the number of directions is $\leq |\mathscr{RF}(G)|$. For a non-trivial group G, define

$$\mathscr{U} = \{f \in \mathscr{RF}(G) \mid L(f) = 1\}$$

and

$$\mathscr{V} = \{f|_{(0,1)} \mid f \in \mathscr{U}\}.$$

Let $\alpha = |\mathscr{RF}(G)|$, $\beta = |\mathscr{U}|$, and $\gamma = |\mathscr{V}|$.

The map $\varphi : \mathscr{U} \times \mathbb{R}^{>0} \to \mathscr{RF}(G)$ defined by $\varphi(f,a)(x) = f(a^{-1}x)$ $(x \in [0,a])$ is a bijection, so $2^{\aleph_0}\beta = \alpha$. The map $\mathscr{U} \to \mathscr{V} \times G \times G$, given by $f \mapsto (f|_{(0,1)}, f(0), f(1))$ is also bijective, so $\beta = \gamma|G|^2$. Hence $\alpha = \gamma|G|^2 2^{\aleph_0}$.

Proposition 3.34 *We have $\alpha = \gamma$.*

Proof The first step is to note that $\gamma \geq 2$. Choose $g \in G$ with $g \neq 1$. Define $f_1(x) = g$ for all $x \in (0,1)$. Then clearly $f_1 \in \mathscr{V}$. Also, define f_2 on $(0,1)$ by

$$f_2(x) = \begin{cases} 1_G & \text{if } x \text{ is irrational,} \\ g & \text{if } x \text{ is rational.} \end{cases}$$

To see that $f_2 \in \mathscr{V}$, suppose that $f_2(x) = 1_G$. Then x is irrational; let $\varepsilon > 0$ be such that $(x-\varepsilon, x+\varepsilon) \subseteq (0,1)$. Choose a rational $q \in (x-\varepsilon, x)$ and let $\delta = x - q$, so that $\delta < \varepsilon$ and δ is irrational. Then $x + \delta = q + 2\delta$ is irrational, so we have

$$f_2(x-\delta) = f_2(q) = g \neq 1_G = f_2(x+\delta).$$

[6] For the definition of *direction* at a vertex of a Λ-tree, see the discussion after Remark A.24.

It follows that f_2 is the restriction of a function in $\mathscr{U}$, that is, it is in $\mathscr{V}$. Clearly $f_1 \neq f_2$, establishing that $\gamma \geq 2$.

Next, let $\sigma = (f_n)_{n\geq 1}$ be a sequence of functions in $\mathscr{V}$. Define f_σ on $(0,1)$ by

$$\begin{cases} f_\sigma(x) = f_n(n(n+1)\left(x - \frac{1}{n+1}\right)), & x \in \left(\frac{1}{n+1}, \frac{1}{n}\right),\ n \geq 1, \\ f_\sigma(\frac{1}{n}) = g, & n \geq 2. \end{cases}$$

It is easy to see that $f_\sigma \in \mathscr{V}$, and if $\sigma \neq \tau$ then $f_\sigma \neq f_\tau$. Thus the mapping $\sigma \mapsto f_\sigma$ is injective, showing that $\gamma^{\aleph_0} \leq \gamma$; hence $\gamma^{\aleph_0} = \gamma$. Since $\gamma \geq 2$ we have $\gamma \geq 2^{\aleph_0}$ and thus $\alpha = \gamma|G|^2$.

For $h \in G \setminus \{1_G\}$, the function with constant value h on $(0,1)$ belongs to $\mathscr{V}$. Hence, if on the one hand G is infinite then $\gamma \geq |G \setminus \{1_G\}| = |G| = |G|^2$, so $\gamma|G|^2 = \gamma$. If on the other hand $|G|$ is finite then $\gamma|G|^2 = \gamma$, because γ is infinite. In either case, we have $\alpha = \gamma$. □

Now, for $f \in \mathscr{V}$, define $\tilde{f}$ on $[0,1]$ by

$$\tilde{f}(x) = \begin{cases} f(n(n+1)\left(x - \frac{1}{n+1}\right)), & x \in \left(\frac{1}{n+1}, \frac{1}{n}\right), \\ g, & x = \frac{1}{n},\ n \geq 1, \\ 1_G, & x = 0. \end{cases}$$

It is easy to see that $\tilde{f} \in \mathscr{U}$. Moreover, this gives $\gamma = \alpha$ different directions at x_0 in $\mathbf{X}_G$, defined by $[x_0, \tilde{f}x_0]$ for $f \in \mathscr{V}$, because if $g \in \mathscr{V}$, $g \neq f$, then $c(\tilde{f}, \tilde{g}) = 0$. Hence there are α directions at x_0, as claimed. (Note that $\tilde{f}$ is cyclically reduced, so its axis passes through x_0 by Lemma 3.20.)

Remark 3.35 We shall not be able to calculate $|\mathscr{RF}(G)|$ until the theory of test functions has been developed; see Corollary 10.4 as well as Section 10.4.

3.10 Exercises

3.1. Show that the segment $[x_0, \langle f, \alpha\rangle]$ spanned by the base-point x_0 and an arbitrary point $\langle f, \alpha\rangle$ of $\mathbf{X}_G$ is given by equation (3.2).

3.2. Use Lemma 3.15 to show that a group of the form

$$\Gamma_s = \langle a, b \,|\, a^s b^s = 1 \rangle, \quad s \geq 2,$$

does not embed as a hyperbolic subgroup in $\mathscr{RF}(G)$ for any group G.

3.3. Establish Lemma 3.22 by a direct cancellation argument. [Hint: if

$$f = t_1 g_1 t_1^{-1} t_2 g_2 t_2^{-1},$$

where $L(t_1^{-1} t_2) > 0$, consider $t_1^{-1} f t_1$ and apply part (ii) of Proposition 3.13.]

3.4. Let $f = g_1 s g_2 s^{-1} g_3$, where $g_1, g_2, g_3 \in G_0 - \{\mathbf{1}_G\}$ and $L(s) > 0$. Show that f is elliptic if and only if $g_1 g_3 = \mathbf{1}_G$.

3.5. Let $f \in \mathscr{RF}(G)$ be a hyperbolic element, and let $f = s f_1 s^{-1}$ where f_1 is cyclically reduced. In this situation, is f_1 necessarily the core of f?

4

Free $\mathbb{R}$-tree actions and universality

4.1 Introduction

The simplest case of Bass–Serre theory, which describes the structure of a group acting on a (simplicial) tree in terms of standard constructions of geometric group theory, occurs when the action is free. In Bass–Serre theory, one assumes that no group element is an inversion, that is, no group element interchanges the endpoints of an edge. This is needed to ensure that one can form the quotient graph for the action. From another viewpoint, if one forms the geometric realisation of the tree, so that each edge corresponds to a copy of the compact real interval $[0,1]$, the induced action will not be free if there are inversions, as an inversion will fix the midpoint of the corresponding interval. It is a consequence of Bass–Serre theory (although the result is much older, going back at least to Reidemeister and the early 1930s) that a group acts freely and without inversions on a tree if and only if it is a free group.

Given a simplicial tree T, there is a corresponding $\mathbb{Z}$-tree (X,d), where X is the set of vertices of T and $d(x,y)$ is the length of the unique reduced path in T from x to y; the length of a path being the number of edges in it. Moreover, all $\mathbb{Z}$-trees arise in this way (see Example A.8). It is easy to see that a graph automorphism of T is an inversion if and only if the corresponding isometry of (X,d) is an inversion in the sense of the previous chapter (see Definition A.48).

There appears at present to be no useful analogue of Bass–Serre theory for actions on Λ-trees, where Λ is an ordered abelian group different from $\mathbb{Z}$, even in the case where $\Lambda = \mathbb{R}$, a case of particular interest to us here. However, the analogy with Bass–Serre theory does at least suggest that one should begin by studying actions on Λ-trees that are free and without inversions. There has indeed been some progress for specific choices of Λ, and this is discussed in Section A.6.

Note that in an action on an $\mathbb{R}$-tree there are no inversions, and so we shall simply be considering free actions on $\mathbb{R}$-trees. See the remark after Definition A.48.

In this chapter we shall present two embedding results due to the authors. The first applies to Λ-trees for any Λ and says that if G acts freely and without inversions on a Λ-tree (X,d) then there are a group $\hat{G}$ and an action of $\hat{G}$ on a Λ-tree $(\widehat{X},\hat{d})$ that is free, without inversions, and additionally transitive together with a group embedding $G \to \hat{G}$ and an equivariant embedding $(X,d) \to (\widehat{X},\hat{d})$. This is an important result, used in our second embedding theorem, and it is likely to have other applications to the theory of Λ-trees. It is an improvement on Theorem 3.4 in Alperin and Moss [1]. The argument is inspired by a construction in [1] but we have adopted a quite different approach, involving string rewriting and Lyndon length functions. The reason is that important details are omitted in [1], and it appears difficult (if not outright impossible) to fill in these details.

The second embedding result states that if G is a group acting freely and transitively on an $\mathbb{R}$-tree (X,d) then G embeds in $\mathscr{RF}(H)$ for some group H, and there is an equivariant embedding $(X,d) \to \mathbf{X}_H$. Combining these two results, it emerges that $\mathscr{RF}$-groups and their associated $\mathbb{R}$-trees are in fact *universal*, with respect to inclusion, for free $\mathbb{R}$-tree actions. The proof of our second embedding result is based on the argument for Theorem 4.2 in [1], where again some details seem problematic. The motivation for our argument will eventually be explained, near the end of Chapter 11.

According to the classification of isometries noted in Section 3.5, an action by isometries of a group G on a Λ-tree $\mathbf{X} = (X,d)$ is free and without inversions if, for each $g \in G - \{1\}$, g acts as a hyperbolic isometry on $\mathbf{X}$. By Lemma A.56 this is equivalent to saying that, for any point $x \in X$, the displacement length L_x is a free length function, that is it satisfies

$$L_x(g^2) > L_x(g), \quad g \in G - \{1\}. \tag{4.1}$$

4.2 An embedding theorem

In this section we shall discuss the following important result, which is Theorem 5.4 from Chiswell and Müller [12].

Theorem 4.1 *Let G be a group acting freely and without inversions on a*

Λ-tree $\mathbf{X} = (X,d)$. Then there exists a group $\hat{G}$ acting freely, without inversions, and transitively on a Λ-tree $\widehat{\mathbf{X}} = (\widehat{X},\hat{d})$, together with a group embedding $\varphi : G \to \hat{G}$ and a G-equivariant isometry $\mu : \mathbf{X} \to \widehat{\mathbf{X}}$.

In what follows, we shall sketch the proof of Theorem 4.1; for more details see Sections 2–5 of [12].

4.2.1 Construction of $\hat{G}$

Let $B = \{b_i : i \in I\}$ be a set of representatives for the G-orbits of X. For convenience we suppose that $0 \in I$, and we shall take b_0 as the base-point of $\mathbf{X}$. Of course, we suppose that the map $i \mapsto b_i$ is bijective, and we shall assume that $X \cap I = \varnothing$.

We shall work with strings (or words) $\mathbf{x}$ over the alphabet $X \cup I$ of the form[1]

$$\mathbf{x} = (x_0, i_1, x_1, \ldots, i_n, x_n), \tag{4.2}$$

where $n \geq 0$, $x_j \in X$ and $i_k \in I$. Consider the following *rewrite rules*:

(1) $gb_i, i, x \longrightarrow gx$ for $g \in G$, $i \in I$, and $x \in X$;

(2) $i, b_i \longrightarrow \varepsilon$, where ε is the empty word, for all $i \in I$.

We shall write $\mathbf{x}_1 \longrightarrow \mathbf{x}_2$ or, more explicitly, $\mathbf{x}_1 \overset{(i)}{\longrightarrow} \mathbf{x}_2$ if $\mathbf{x}_2$ results from $\mathbf{x}_1$ by a move of type (i), where $i = 1,2$. Let S be the set of all expressions of the form (4.2). Clearly, S is closed under moves of types (1) or (2). We define an equivalence relation $\sim$ on S by

$$\begin{aligned} \mathbf{x} \sim \mathbf{y} \;:\Longleftrightarrow\; & \text{there exists a finite sequence } \mathbf{x} = \mathbf{x}_1, \mathbf{x}_2, \ldots, \mathbf{x}_m = \mathbf{y}, \\ & \text{where } m \geq 1,\ \mathbf{x}_j \in S, \text{ and, for } 1 \leq j < m, \\ & \text{either } \mathbf{x}_j \longrightarrow \mathbf{x}_{j+1} \text{ or } \mathbf{x}_{j+1} \longrightarrow \mathbf{x}_j. \end{aligned}$$

We denote by $[\mathbf{x}]$ the equivalence class of a word $\mathbf{x} \in S$. It is not hard to see that each such equivalence class contains exactly one *reduced* expression, that is, an expression to which none of the rewrite rules can be applied. Indeed, since the rewrite rules (1) and (2) both shorten the length of a word and S is closed under moves of type (1) or (2), each equivalence class $[\mathbf{x}]$ contains at least one reduced word and uniqueness follows from an application of Newman's

[1] We shall usually separate the letters of a word by commas to improve readability, and we shall often add parentheses at the beginning and end of a word.

'diamond lemma'; see Newman [41] or Section 1.5 of Cohn [15]. Next, we define a binary operation on S by

$$\mathbf{x}.\mathbf{y} := \text{the concatenation } \mathbf{x}, 0, \mathbf{y}.$$

This operation is clearly associative, and $\sim$ is a congruence on S, so that $S/\sim$ is a semigroup with respect to the operation $[\mathbf{x}][\mathbf{y}] := [\mathbf{x}.\mathbf{y}]$. Further, we have for $\mathbf{x} \in S$

$$\mathbf{x}.(b_0) = \mathbf{x}, 0, b_0 \xrightarrow{(2)} \mathbf{x} \xleftarrow{(1)} b_0, 0, \mathbf{x} = (b_0).\mathbf{x},$$

so that $[(b_0)]$ is a (two-sided) identity element for $S/\sim$. Moreover, for

$$\mathbf{x} = (g_0 b_{j_0}, i_1, g_1 b_{j_1}, i_2, \ldots, i_n, g_n b_{j_n}), \tag{4.3}$$

set

$$\bar{\mathbf{x}} := (b_0, j_n, g_n^{-1} b_{i_n}, j_{n-1}, g_{n-1}^{-1} b_{i_{n-1}}, \ldots, j_1, g_1^{-1} b_{i_1}, j_0, g_0^{-1} b_0).$$

Then a straightforward calculation, involving an alternating chain of moves, shows that

$$[\mathbf{x}][\bar{\mathbf{x}}] = [(b_0)] = [\bar{\mathbf{x}}][\mathbf{x}];$$

hence $S/\sim$ is a group, which we denote by $\hat{G}$. It is easy to see that $\hat{G}$ is independent, up to isomorphism, of the system of representatives for the G-orbits used in its definition.

Our next result computes the cardinality of $\hat{G}$, a result which we shall need later.

Lemma 4.2 *We have*

$$|\hat{G}| = \begin{cases} \aleph_0, & G = 1 \textit{ and } 1 < |I| < \infty, \\ \max\{|G|, |I|\} & \textit{otherwise.} \end{cases} \tag{4.4}$$

Proof As we have seen, the elements of $\hat{G}$ are in bijective correspondence with the reduced words in S. We claim that there are precisely

$$|X|(|I|-1)^n(|X|-1)^n \tag{4.5}$$

reduced words $(x_0, i_1, x_1, \ldots, x_n) \in S$. This is certainly true for $n = 0$. Suppose that our claim holds for some $n \geq 0$, and consider a reduced word

$$(x_0, i_1, x_1, \ldots, x_n, i_{n+1}, x_{n+1}) \in S. \tag{4.6}$$

By the induction hypothesis, the prefix $(x_0, i_1, \ldots, x_n)$ can be chosen in the number of ways given in (4.5) and, once this has been done, the index i_{n+1} may be chosen in $|I| - 1$ ways (we need to avoid only the index corresponding to the orbit of x_n, so as to prevent a type (1) move), and the point x_{n+1} can be chosen in $|X|-1$ many ways (we need to avoid only the orbit representative $b_{i_{n+1}}$, so as to prevent a type (2) move $i_{n+1}, b_{i_{n+1}} \longrightarrow \varepsilon$). Hence, there must be exactly

$$|X|(|I|-1)^{n+1}(|X|-1)^{n+1}$$

reduced words, as in (4.6), proving our claim. It follows that

$$|\hat{G}| = |X| \sum_{n\geq 0} (|I|-1)^n(|X|-1)^n. \tag{4.7}$$

If $|I|$ is infinite, then so is

$$|X| = |G|\,|I| = \max\{|G|,|I|\};$$

thus $(|I|-1)^n = |I|$ as well as $(|X|-1)^n = |X|$, in this case, and (4.7) gives

$$|\hat{G}| = \max\{|G|,|I|\}\,\aleph_0 \max\{|I|,\max\{|G|,|I|\}\} = \max\{|G|,|I|\},$$

in accordance with (4.4). Again, if $|I|$ is finite and G is infinite then $|X|$ is infinite; thus

$$(|I|-1)^n(|X|-1)^n = |X| = |G|,$$

so $|\hat{G}| = |G|$ in this case, again in accordance with (4.4). Finally, if $G=1$ and $|I|$ is finite then $|X| = |I|$ and (4.7) gives

$$|\hat{G}| = |I| \sum_{n\geq 0} (|I|-1)^{2n} = \begin{cases} 1, & |I| = 1, \\ \aleph_0, & |I| > 1. \end{cases}$$

□

4.2.2 The length function L and the tree $\widehat{\mathbf{X}}$

For a reduced expression $\mathbf{x} \in S$ as in (4.3), set

$$L([\mathbf{x}]) := d(b_0, g_0 b_{j_0}) + \sum_{k=1}^{n} d(b_{i_k}, g_k b_{j_k}).$$

Since each equivalence class of S contains a unique reduced word, this defines a map $L : \hat{G} \to \Lambda$. We shall show that L is a Lyndon length function.

Lemma 4.3 *For $[\mathbf{x}] \in \hat{G}$ we have $L([\mathbf{x}]) = L([\mathbf{x}]^{-1})$.*

Proof (sketch). For $\mathbf{x} \in S$ reduced and as in (4.3), the only possible type (1) move in $\bar{\mathbf{x}}$ is

$$b_0, j_n, g_n^{-1} b_{i_n} \longrightarrow g_n^{-1} b_{i_n}, \quad j_n = 0,$$

while the only possible type (2) move is

$$j_0, g_0^{-1} \longrightarrow \varepsilon, \quad j_0 = 0 \text{ and } g_0 = 1.$$

Moreover, carrying out these moves where possible, gives the reduced expression in the equivalence class $[\mathbf{x}]^{-1}$. The result follows now by a straightforward calculation. □

Our next result calculates the quantity

$$c([\mathbf{x}],[\mathbf{y}]) = \tfrac{1}{2}\big(L([\mathbf{x}]) + L([\mathbf{y}]) - L([\mathbf{x}]^{-1}[\mathbf{y}])\big)$$

for $[\mathbf{x}], [\mathbf{y}] \in \hat{G}$.

Lemma 4.4 *Let*

$$\mathbf{x} = (x_0, i_1, x_1, \ldots, i_m, x_m), \quad \mathbf{y} = (y_0, j_1, y_1, \ldots, j_n, y_n) \in S$$

be reduced expressions. Further, let $k \geq 0$ be maximal with respect to the condition

$$(x_0, i_1, x_1, \ldots, i_k) = (y_0, j_1, y_1, \ldots, j_k),$$

and set

$$\mathbf{w} = \mathbf{w}(\mathbf{x},\mathbf{y}) := \begin{cases} (x_0, i_1, x_1, \ldots, x_{k-1}), & k \geq 1, \\ \varepsilon, & k = 0. \end{cases}$$

Then we have

$$c([\mathbf{x}],[\mathbf{y}]) = L([\mathbf{w}(\mathbf{x},\mathbf{y})]) + (x_k \cdot y_k)_{b_{i_k}} \tag{4.8}$$

where $(x_k \cdot y_k)_{b_{i_k}}$ is the product defined at the beginning of Section A.2, the length $L([\mathbf{w}(\mathbf{x},\mathbf{y})])$ is to be interpreted as 0 if $\mathbf{w} = \varepsilon$, and where $i_0 := 0$. In particular, we have $c([\mathbf{x}],[\mathbf{y}]) \geq L([\mathbf{w}(\mathbf{x},\mathbf{y})])$ and $c([\mathbf{x}],[\mathbf{y}]) \in \Lambda$.

The proof of Lemma 4.4 is rather long and technical and will be omitted; see Lemma 4.2 in Chiswell and Müller [12] for details.

Proposition 4.5 *The mapping L is a Lyndon length function on $\hat{G}$ with $c(g,h) \in \Lambda$ for all $g,h \in \hat{G}$.*

Proof. In view of Lemmas 4.3 and 4.4, we have to show only that if

$$\mathbf{x} = (x_0, i_1, x_1, \ldots, i_\ell, x_\ell), \quad \mathbf{y} = (y_0, i'_1, y_1, \ldots, i'_m, y_m),$$

$$\mathbf{z} = (z_0, i''_1, z_1, \ldots, i''_n, z_n) \in S$$

are reduced words then at least two of the quantities

$$c([\mathbf{x}],[\mathbf{y}]), \quad c([\mathbf{y}],[\mathbf{z}]), \quad c([\mathbf{x}],[\mathbf{z}])$$

are equal and less than or equal to the third. Let $\mathbf{p}(\mathbf{x},\mathbf{y})$ denote the largest common prefix of the words $\mathbf{x}$ and $\mathbf{y}$, and let a similar definition apply to the pairs $\mathbf{y},\mathbf{z}$ and $\mathbf{x},\mathbf{z}$. Without loss of generality we may assume that $\mathbf{p}(\mathbf{x},\mathbf{y}) = \mathbf{p}(\mathbf{y},\mathbf{z})$ is a prefix of $\mathbf{p}(\mathbf{x},\mathbf{z})$. We distinguish two main cases according to whether $\mathbf{p}(\mathbf{x},\mathbf{y})$ ends in a point or an index.

Suppose first that $\mathbf{p}(\mathbf{x},\mathbf{y}) = x_0, i_1, x_1, \ldots, x_{\kappa-1}$ for some $\kappa \geq 1$, that is, $\mathbf{p}(\mathbf{x},\mathbf{y})$ is non-empty and ends in a point. Then, in the notation of Lemma 4.4, $k = \kappa - 1$ and so

$$\mathbf{w}(\mathbf{x},\mathbf{y}) = \begin{cases} (x_0, i_1, x_1, \ldots, x_{\kappa-2}), & \kappa \geq 2, \\ \varepsilon, & \kappa = 1, \end{cases}$$

and, by equation (4.8), we obtain

$$c([\mathbf{x}],[\mathbf{y}]) = L([\mathbf{p}(\mathbf{x},\mathbf{y})]) = c([\mathbf{y}],[\mathbf{z}]).$$

Moreover, if $i_\kappa \neq i''_\kappa$ then $\mathbf{p}(\mathbf{x},\mathbf{y}) = \mathbf{p}(\mathbf{x},\mathbf{z})$ and hence

$$c([\mathbf{x}],[\mathbf{z}]) = c([\mathbf{x}],[\mathbf{y}]) = c([\mathbf{y}],[\mathbf{z}]),$$

whereas, for $i_\kappa = i''_\kappa$, the string $\mathbf{p}(\mathbf{x},\mathbf{y})$ is a prefix of $\mathbf{z}(\mathbf{x},\mathbf{z})$ and so, by Lemma 4.4,

$$c([\mathbf{x}],[\mathbf{z}]) \geq L([\mathbf{w}(\mathbf{x},\mathbf{z})]) \geq L([\mathbf{p}(\mathbf{x},\mathbf{y})]) = c([\mathbf{x}],[\mathbf{y}]) = c([\mathbf{y}],[\mathbf{z}]).$$

Next, suppose that $\mathbf{p}(\mathbf{x},\mathbf{y}) = x_0, i_1, \ldots, x_{\kappa-1}, i_\kappa$ for some $\kappa \geq 0$, that is, $\mathbf{p}(\mathbf{x},\mathbf{y})$ is either empty or ends in an index. In this case, $\mathbf{w}(\mathbf{x},\mathbf{y}) = \mathbf{w}(\mathbf{y},\mathbf{z})$ is a prefix of $\mathbf{w}(\mathbf{x},\mathbf{z})$, and, by (4.8), we have

$$c([\mathbf{x}],[\mathbf{y}]) = L([\mathbf{w}(\mathbf{x},\mathbf{y})]) + (x_\kappa \cdot y_\kappa)_{b_{i_\kappa}}$$

as well as

$$c([\mathbf{y}],[\mathbf{z}]) = L([\mathbf{w}(\mathbf{x},\mathbf{y})]) + (y_\kappa \cdot z_\kappa)_{b_{i_\kappa}}.$$

Now we distinguish two subcases.

(a) Assume that $x_\kappa = z_\kappa$. Then $c([\mathbf{x}],[\mathbf{y}]) = c([\mathbf{y}],[\mathbf{z}])$ and

$$\begin{aligned} c([\mathbf{x}],[\mathbf{z}]) &\geq L([\mathbf{w}(\mathbf{x},\mathbf{y})]) + d(b_{i_\kappa}, x_\kappa) \\ &\geq L([\mathbf{w}(\mathbf{x},\mathbf{y})]) + (x_\kappa \cdot y_\kappa)_{b_{i_\kappa}} \\ &= c([\mathbf{x}],[\mathbf{y}]), \end{aligned}$$

since, by the triangle inequality,

$$\begin{aligned} (x_\kappa \cdot y_\kappa)_{b_{i_\kappa}} &= \tfrac{1}{2}\big(d(b_{i_\kappa}, x_\kappa) + d(b_{i_\kappa}, y_\kappa) - d(x_\kappa, y_\kappa)\big) \\ &\leq \tfrac{1}{2}\big(d(b_{i_\kappa}, x_\kappa) + d(b_{i_\kappa}, y_\kappa) - d(b_{i_\kappa}, y_\kappa) + d(b_{i_\kappa}, x_\kappa)\big) \\ &= d(b_{i_\kappa}, x_\kappa). \end{aligned}$$

(b) Assume that $x_\kappa \neq z_\kappa$. Then $\mathbf{p}(\mathbf{x},\mathbf{z}) = \mathbf{p}(\mathbf{x},\mathbf{y})$, so $\mathbf{w}(\mathbf{x},\mathbf{z}) = \mathbf{w}(\mathbf{x},\mathbf{y})$ and, by (4.8),

$$c([\mathbf{x}],[\mathbf{z}]) = L([\mathbf{w}(\mathbf{x},\mathbf{y})]) + (x_\kappa \cdot z_\kappa)_{b_{i_\kappa}}.$$

In this case, the desired result follows since $\mathbf{X}$ is 0-hyperbolic; thus two of the quantities

$$(x_\kappa \cdot y_\kappa)_{b_{i_\kappa}}, \quad (y_\kappa \cdot z_\kappa)_{b_{i_\kappa}}, \quad (x_\kappa \cdot z_\kappa)_{b_{i_\kappa}}$$

are equal and are less than or equal to the third. This completes the proof. □

With Proposition 4.5 at our disposal, it follows from Theorem A.29 that there exists an Λ-tree $\widehat{\mathbf{X}} = (\widehat{X}, \hat{d})$ on which $\hat{G}$ acts by isometries, with a base-point b such that $L = L_b$ is the displacement function with respect to b.

4.2.3 The action of $\hat{G}$ on $\widehat{\mathbf{X}}$

Lemma 4.6 *The action of $\hat{G}$ on $\widehat{\mathbf{X}}$ is free and without inversions.*

In order to establish Lemma 4.6, one needs to show that each non-trivial element $g = [\mathbf{x}] \in \hat{G}$ acts as a hyperbolic isometry on $\widehat{\mathbf{X}}$; thus, in view of (4.1) we need to prove that $L(g^2) > L(g)$. The latter assertion is shown by induction on the combinatorial length of g, that is, the length of the reduced word $\mathbf{x}$; see Lemma 5.1 in Chiswell and Müller [12] for details of the (rather technical) argument, or try Exercise 4.3.

Lemma 4.7 *The action of $\hat{G}$ on $\widehat{\mathbf{X}}$ is transitive.*

Proof By Theorem A.29, $\widehat{\mathbf{X}}$ is spanned by the orbit $\hat{G}b$. In view of Proposition A.37 it therefore suffices to show that $L = L_b$ is strongly regular.

Let $\mathbf{x}$ be a reduced word as in (4.3), and let $\gamma \in \Lambda$ be such that $0 \leq \gamma \leq L([\mathbf{x}])$. Set $i_0 := 0$, so that $L([\mathbf{x}]) = \sum_{\nu=0}^{n} d(b_{i_\nu}, g_\nu b_{j_\nu})$. Let $N \in \mathbb{N}_0$ be minimal, subject to the condition that $\gamma \leq \sum_{\nu=0}^{N} d(b_{i_\nu}, g_\nu b_{j_\nu})$, and let

$$\delta := \gamma - \sum_{\nu=0}^{N-1} d(b_{i_\nu}, g_\nu b_{j_\nu});$$

thus $0 \leq \delta \leq d(b_{i_N}, g_N b_{j_N})$ and $\delta > 0$ for $N \geq 1$.

Let z be the point on the segment $[b_{i_N}, g_N b_{j_N}]$ of $\mathbf{X}$ at distance δ from b_{i_N}, and define

$$\mathbf{y} := \begin{cases} (g_0 b_{j_0}, i_1, g_1 b_{j_1}, \ldots, i_{N-1}, g_{N-1} b_{j_{N-1}}, i_N, z), & N \geq 1, \\ (z), & N = 0. \end{cases}$$

Then $\mathbf{y}$ is reduced, and

$$\begin{aligned} L([\mathbf{y}]) &= \sum_{\nu=0}^{N-1} d(b_{i_\nu}, g_\nu b_{j_\nu}) + d(b_{i_N}, z) \\ &= \sum_{\nu=0}^{N-1} d(b_{i_\nu}, g_\nu b_{j_\nu}) + \delta \\ &= \gamma. \end{aligned}$$

Moreover, by Lemma 4.4, we have

$$\begin{aligned} c([\mathbf{x}], [\mathbf{y}]) &= L([(g_0 b_{j_0}, i_1, g_1 b_{j_1}, \ldots, i_{N-1}, g_{N-1} b_{j_{N-1}})]) + (g_N b_{j_N} \cdot z)_{b_{i_N}} \\ &= L([(g_0 b_{j_0}, i_1, g_1 b_{j_1}, \ldots, i_{N-1}, g_{N-1} b_{j_{N-1}})]) + d(b_{i_N}, z) \\ &= L([\mathbf{y}]). \end{aligned}$$

Letting $g := [\mathbf{x}]$, $g_1 := [\mathbf{y}]$, and $g_2 := g_1^{-1} g$, we find that $L(g_1) = \gamma$ and that

$$\begin{aligned} L(g_2) &= L([\mathbf{y}]^{-1} [\mathbf{x}]) \\ &= L([\mathbf{x}]^{-1} [\mathbf{y}]) \\ &= L([\mathbf{x}]) - L([\mathbf{y}]) \\ &= L(g) - L(g_1), \end{aligned}$$

so L is strongly regular, as required. □

4.2.4 Proof of Theorem 4.1

Define a map $\varphi : G \to \hat{G}$ by $\varphi(g) := [(gb_0)]$. Then φ is a group homomorphism since, for $g,h \in G$,

$$\begin{aligned}\varphi(g)\varphi(h) &= [(gb_0)][(hb_0)] \\ &= [(gb_0).(hb_0)] \\ &= [(gb_0,0,hb_0)] \\ &= [(ghb_0)] \\ &= \varphi(gh),\end{aligned}$$

where we have used the type (1) move $gb_0,0,hb_0 \longrightarrow ghb_0$ in step 4. Moreover, since the action of G on $\mathbf{X}$ is free we have, for $g \in G$, that

$$\varphi(g) = 1 \iff [(gb_0)] = [(b_0)] \iff gb_0 = b_0 \iff g = 1,$$

using the fact that (gb_0) and (b_0) are reduced words. Hence, φ is an embedding.

Since the action of $\hat{G}$ on $\widehat{\mathbf{X}}$ is transitive by Lemma 4.7, we have

$$\widehat{\mathbf{X}} = \left\{ [\mathbf{x}]b : [\mathbf{x}] \in \hat{G} \right\},$$

and the metric $\hat{d}$ on $\widehat{\mathbf{X}}$ is given by

$$\hat{d}([\mathbf{x}]b,[\mathbf{y}]b) = \hat{d}(b,[\mathbf{x}]^{-1}[\mathbf{y}]b) = L([\mathbf{x}]^{-1}[\mathbf{y}]),$$

since $L = L_b$. Define a map $\mu : X \to \widehat{X}$ via $\mu(gb_i) := [(gb_i)]b$. Then

$$\begin{aligned}\hat{d}(\mu(gb_i),\mu(hb_j)) &= \hat{d}([(gb_i)]b,[(hb_j)]b) \\ &= L([(gb_i)]^{-1}[(hb_j)]) \\ &= L([(b_0,i,g^{-1}b_0)][(hb_j)]) \\ &= L([(b_0,i,g^{-1}b_0,0,hb_j)]) \\ &= L([(b_0,i,g^{-1}hb_j)]) \\ &= d(b_i,g^{-1}hb_j) \text{ (whether or not } (b_0,i,g^{-1}hb_j) \text{ is reduced)} \\ &= d(gb_i,hb_j),\end{aligned}$$

so μ is an isometry. Finally, μ is G-equivariant, since

$$\begin{aligned}\mu(hgb_i) &= [(hgb_i)]b \\ &= [(hb_0)][(gb_i)]b \\ &= [(hb_0)]\mu(gb_i) \\ &= \varphi(h)\mu(gb_i).\end{aligned}$$

This completes the proof of the theorem.

Remark 4.8 The results of Myasnikov, Remeslennikov, and Serbin [40] motivate the question whether a group with a Lyndon length function L can always be embedded in a length-preserving way into a group with a regular Lyndon length function. Modulo an obvious necessary condition, Theorem 4.1 provides an affirmative answer to this question in the case when L is free; see Corollary 5.5 in Chiswell and Müller [12].

4.3 Universality of $\mathscr{RF}$-groups and their associated $\mathbb{R}$-trees

We shall discuss here a second embedding theorem, this time for $\mathbb{R}$-tree actions that are free and transitive.

Theorem 4.9 *Let G be a group acting freely and transitively on an $\mathbb{R}$-tree $\mathbf{X} = (X,d)$. Then there exists a group H, an injective group homomorphism $\psi : G \to \mathscr{RF}(H)$, and a G-equivariant isometry $\nu : \mathbf{X} \to \mathbf{X}_H$.*

Combining Theorem 4.9 with Theorem 4.1, we obtain a *universal property* of $\mathscr{RF}$-groups and their associated $\mathbb{R}$-trees.

Theorem 4.10 *Let G be a group acting freely on an $\mathbb{R}$-tree $\mathbf{X} = (X,d)$. Then there exist a group H, a group embedding $\chi : G \to \mathscr{RF}(H)$, and a G-equivariant isometry $\lambda : \mathbf{X} \to \mathbf{X}_H$ containing the canonical base-point x_0 of $\mathbf{X}_H$ in its image.*

A slightly improved version of Theorem 4.10 will be discussed in Chapter 6; see Theorem 6.4. In the remainder of this chapter we shall sketch the proof of Theorem 4.9; see Section 7 in Chiswell and Müller [12] for more details.

4.3.1 Construction of the group H

Let G be a group acting freely and transitively on an $\mathbb{R}$-tree $\mathbf{X} = (X, d)$, and choose a base-point $y_0 \in X$. We define an equivalence relation $\approx$ on $G - \{1_G\}$ via

$$g \approx h :\Longleftrightarrow c(g^{-1}, h^{-1}) = \tfrac{1}{2}\big(L_{y_0}(g) + L_{y_0}(h) - L_{y_0}(gh^{-1})\big) > 0$$

(see Exercise 4.4). Let s_g denote the equivalence class of g, set $s_{1_G} := \{1_G\}$, and let

$$S := \{s_g : g \in G\}.$$

We endow S with a group structure (which is arbitrary and need not be related to the group structure of G) and denote the resulting group by H.[2]

4.3.2 The functions F_g

Given $g \in G$, we define a function $F_g : [0, L_{y_0}(g)] \to H$ as follows. Let ξ be such that $0 \leq \xi \leq L_{y_0}(g)$. Then the point in $[y_0, gy_0]$ at distance ξ from y_0 has the form $g_\xi y_0$ for some unique $g_\xi \in G$ (since G acts freely and transitively), and we set

$$F_g(\xi) := s_{g_\xi}^{-1} s_{g^{-1}g_\xi}, \quad g \in G,\ \xi \in [0, L_{y_0}(g)].$$

By definition $F_g \in \mathscr{F}(H)$ and $L(F_g) = L_{y_0}(g)$. Moreover, it is not hard to see that, for $g \neq 1_G$, we have $F_g(\xi) \neq 1_H$ for every $\xi \in [0, L_{y_0}(g)]$ (see Lemma 7.1 in [12] for the details); in particular, F_g is reduced. Our aim is to show that the map $\psi : G \to \mathscr{RF}(H)$ given by $g \mapsto F_g$ is a homomorphism.

Lemma 4.11 *Let $g, h \in G$. For $0 \leq \xi < c(g,h)$ we have $F_g(\xi) = F_h(\xi)$, but $F_g(c(g,h)) \neq F_h(c(g,h))$ unless $g = h$.*

See Lemma 7.2 in Section 7 of [12] for the proof of the above important, if somewhat technical, observation. Next, we show the following.

Lemma 4.12 *For each $g \in G$ we have $F_{g^{-1}} = F_g^{-1}$; that is, the map ψ respects inverses.*

[2] This can always be done. In particular, if S is infinite then it can for instance be made into a free group. A somewhat less trivial observation is that an infinite set S may be endowed with the structure of an elementary-abelian 2-group; see, for example, Section IV.4 in Kertész [27].

Proof We have

$$L(F_{g^{-1}}) = L_{y_0}(g^{-1}) = L_{y_0}(g) = L(F_g) = L(F_g^{-1}).$$

Furthermore, since, for $0 \leq \xi \leq L_{y_0}(g)$, $[y_0, gy_0] = [y_0, g_\xi y_0, gy_0]$, we have

$$[g^{-1}y_0, y_0] = [g^{-1}y_0, g^{-1}g_\xi y_0, y_0]$$

and, by the definition of g_ξ, $d(g^{-1}y_0, g^{-1}g_\xi y_0) = \xi$. Hence

$$d(y_0, g^{-1}g_\xi y_0) = d(y_0, g^{-1}y_0) - \xi = L_{y_0}(g) - \xi,$$

so that $g^{-1}g_\xi y_0$ is the unique point on the segment $[y_0, g^{-1}y_0]$ at distance $L_{y_0}(g) - \xi$ from y_0. This gives

$$g^{-1}_{L_{y_0}(g)-\xi} = g^{-1}g_\xi, \quad (g \in G,\ \xi \in [0, L_{y_0}(g)]),$$

and we conclude that

$$\begin{aligned}
F_{g^{-1}}(L(F_g) - \xi) &= F_{g^{-1}}(L_{y_0}(g) - \xi) \\
&= s^{-1}_{g^{-1}_{L_{y_0}(g)-\xi}} s_{gg^{-1}_{L_{y_0}(g)-\xi}} \\
&= \left(s^{-1}_{g_\xi} s_{g^{-1}g_\xi}\right)^{-1} \\
&= (F_g(\xi))^{-1},
\end{aligned}$$

as claimed. □

Combining the last two lemmas, one easily deduces the following; see Corollary 7.4 in [12].

Corollary 4.13 *For all $g, h \in G$, we have $\varepsilon_0(F_g, F_h) = c(g^{-1}, h)$.*

4.3.3 The map ψ

We are now in a position to show the following crucial result.

Lemma 4.14 *The mapping $\psi : G \to \mathscr{RF}(H)$ given by $g \mapsto F_g$ is an injective group homomorphism.*

Proof By Corollary 4.13 we have $c(g^{-1}, h) = \varepsilon_0(F_g, F_h)$, and ψ is length-preserving by the definition of F_g, so that $L_{y_0}(g) = L(F_g)$; thus, by the definition of the function c,

$$\begin{aligned} L(F_{gh}) &= L_{y_0}(gh) \\ &= L_{y_0}(g) + L_{y_0}(h) - 2c(g^{-1}, h) \\ &= L(F_g) + L(F_h) - 2\varepsilon_0(F_g, F_h) \\ &= L(F_g F_h). \end{aligned}$$

Now let $g, h \in G$, and let $Y(y_0, g^{-1}y_0, hy_0) = ky_0$. Then we have

$$c(g^{-1}, h) = d(y_0, ky_0) = L_{y_0}(k).$$

Moreover, we have $gky_0 = Y(gy_0, y_0, ghy_0)$, so that

$$\begin{aligned} c(g, gh) &= d(y_0, gky_0) \\ &= d(g^{-1}y_0, ky_0) \\ &= d(g^{-1}y_0, y_0) - d(y_0, ky_0) \\ &= L_{y_0}(g) - c(g^{-1}, h), \end{aligned}$$

since $ky_0 \in [y_0, g^{-1}y_0]$. This is illustrated by the following figure and its translate by g.

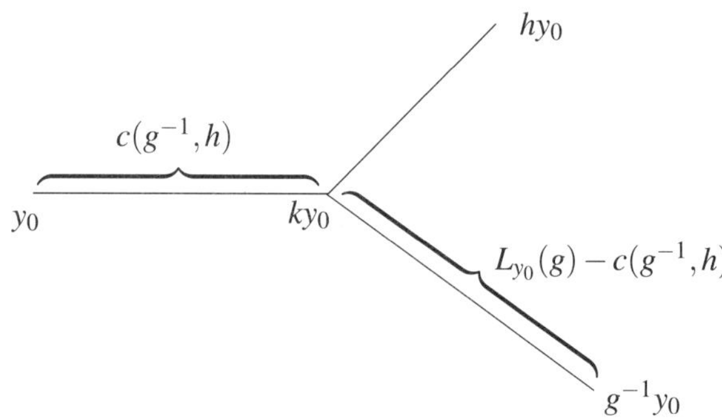

By Lemma 4.11,

$$F_{gh}(\xi) = F_g(\xi), \quad 0 \le \xi < L_{y_0}(g) - c(g^{-1}, h). \tag{4.9}$$

Next, suppose that $\xi = L_{y_0}(g) - c(g^{-1}, h)$, so that the point at distance ξ from y_0 on $[y_0, ghy_0]$ and on $[y_0, gy_0]$ is gky_0. Thus, $(gh)_\xi = gk = g_\xi$, and so $F_{gh}(\xi) = s_{gk}^{-1} s_{h^{-1}k}$ as well as $F_g(\xi) = s_{gk}^{-1} s_k$. Also, ky_0 is the point on $[y_0, hy_0]$ at distance $c(g^{-1}, h)$ from y_0, thus

$$F_h(c(g^{-1}, h)) = s_k^{-1} s_{h^{-1}k}.$$

From the last three equations we conclude that

$$F_g(L_{y_0}(g)-c(g^{-1},h))\,F_h(c(g^{-1},h)) = F_{gh}(L_{y_0}(g)-c(g^{-1},h)). \qquad (4.10)$$

Finally, suppose that $L_{y_0}(g)-c(g^{-1},h) < \xi \leq L_{y_0}(gh)$. Then the point p at distance ξ from y_0 on $[y_0, ghy_0]$ is at distance

$$\xi - d(y_0, gky_0) = \xi - L_{y_0}(g) + c(g^{-1},h) > 0$$

from gky_0. Further, $[gy_0, ghy_0] = [gy_0, gky_0, p, ghy_0]$; therefore

$$[y_0, hy_0] = [y_0, ky_0, g^{-1}p, hy_0]$$

and

$$\begin{aligned} d(y_0, g^{-1}p) &= d(y_0, ky_0) + d(ky_0, g^{-1}p) \\ &= c(g^{-1},h) + d(gky_0, p) \\ &= \xi - L_{y_0}(g) + 2c(g^{-1},h). \end{aligned}$$

Thus, setting $\xi' := \xi - L_{y_0}(g) + 2c(g^{-1},h)$, we have $g^{-1}p = h_{\xi'}y_0$, hence $p = gh_{\xi'}y_0$ and so $(gh)_\xi = gh_{\xi'}$. It follows that $F_{gh}(\xi) = s_{gh_{\xi'}}^{-1} s_{h^{-1}h_{\xi'}}$. It is easy to see that $ky_0 = Y(y_0, h_{\xi'}y_0, g^{-1}y_0)$, so we have

$$h_{\xi'}^{-1}ky_0 = Y(h_{\xi'}^{-1}y_0, y_0, h_{\xi'}^{-1}g^{-1}y_0)$$

and hence

$$c(h_{\xi'}^{-1}, h_{\xi'}^{-1}g^{-1}) = d(y_0, h_{\xi'}^{-1}ky_0) = d(p, gky_0) > 0.$$

Since clearly $h_{\xi'}, gh_{\xi'} \neq 1_G$, it follows that $s_{gh_{\xi'}} = s_{h_{\xi'}}$ and so

$$F_h(\xi') = s_{h_{\xi'}}^{-1} s_{h^{-1}h_{\xi'}} = F_{gh}(\xi).$$

Summarising, we have shown that

$$F_{gh}(\xi) = F_h(\xi - L_{y_0}(g) + 2c(g^{-1},h)), \quad L_{y_0}(g) - c(g^{-1},h) < \xi \leq L_{y_0}(gh). \qquad (4.11)$$

Combining equations (4.9), (4.10), and (4.11) we obtain, abbreviating $\varepsilon_0(F_g, F_h)$ to ε_0,

$$F_{gh}(\xi) = \left\{ \begin{array}{ll} F_g(\xi), & 0 \leq \xi < L(F_g) - \varepsilon_0 \\ F_g(L(F_g)-\varepsilon_0)\,F_h(\varepsilon_0), & \xi = L(F_g) - \varepsilon_0 \\ F_h(\xi - L(F_g) + 2\varepsilon_0), & L(F_g) - \varepsilon_0 < \xi \leq L(F_{gh}) \end{array} \right\} = (F_g F_h)(\xi),$$

for $0 \leq \xi \leq L(F_{gh})$. Thus $F_{gh} = F_g F_h$, showing that ψ is a group homomorphism. Also, since ψ is length-preserving it has a trivial kernel and so is injective. □

4.3.4 Proof of Theorem 4.9

Given a base-point $y_0 \in X$, we know already how to construct a group H and an injective homomorphism $\psi : G \to \mathscr{RF}(H)$. Define a map $\nu : \mathbf{X} \to \mathbf{X}_H$ by

$$\nu(gy_0) := \psi(g)x_0 = F_g x_0,$$

making use of the fact that the action of G is free and transitive. Then we have, for $g, h \in G$,

$$\begin{aligned} d(\nu(gy_0), \nu(hy_0)) &= d(F_g x_0, F_h x_0) \\ &= d(x_0, F_g^{-1} F_h x_0) \\ &= L(F_g^{-1} F_h) \\ &= L(F_{g^{-1}h}) \\ &= L_{y_0}(g^{-1}h) \\ &= d(gy_0, hy_0); \end{aligned}$$

that is, ν is an isometry. Also, ν is G-equivariant since

$$\nu(hgy_0) = \psi(hg)x_0 = \psi(h)\psi(g)x_0 = \psi(h)\nu(gy_0),$$

and the proof of the theorem is complete.

4.4 Exercises

4.1. Fill in the details in the proof of Lemma 4.3.

4.2. Prove Lemma 4.4.

4.3. Establish Lemma 4.6. [Hint: use induction on the length of the reduced word $\mathbf{x}$ to show that each non-trivial element $g = [\mathbf{x}] \in \hat{G}$ satisfies the inequality $L(g^2) > L(g)$.]

4.4. Let G be a group acting freely on an $\mathbb{R}$-tree $\mathbf{X} = (X,d)$, and let $x_0 \in X$ be a base-point. Show that the relation $\approx$ introduced on the set $G - \{1_G\}$ via

$$g \approx h :\Longleftrightarrow c(g^{-1},h^{-1}) = \tfrac{1}{2}\big(L_{x_0}(g) + L_{x_0}(h) - L_{x_0}(gh^{-1})\big) > 0$$

is an equivalence relation (for the transitivity of $\approx$ see axiom (iii) for a length function at the start of Section A.3).

4.5. Show that every infinite set may be endowed with the structure of an elementary-abelian 2-group.

4.6. Provide proofs of Lemma 4.11 and Corollary 4.13.

5
Exponent sums

5.1 Introduction

A principal problem faced when studying the groups $\mathscr{RF}(G)$ is the apparent lack of known homomorphisms involving them, a phenomenon due in no small part to the fact that no useful generating system nor defining relations are known for these groups. In this chapter we are going to construct, for each element $g \in G$, a certain homomorphism

$$e_g : \mathscr{RF}(G) \to \mathbb{R}$$

via Lebesgue measure theory. These homomorphisms e_g may, to some extent, be viewed as continuous analogues of exponent sum maps for free groups formed with respect to the elements of a basis, for which reason we shall call them exponent sums as well.

The construction of these exponent sums also bears on another, related, problem: as we saw in Lemma 3.22, the product of two elliptic elements is usually hyperbolic and hence $E(G)$, the subgroup of $\mathscr{RF}(G)$ generated by its elliptic elements, contains a vast amount of hyperbolic elements. Could it be that for some non-trivial group G we have $E(G) = \mathscr{RF}(G)$? That is, can it happen that $\mathscr{RF}(G)$ is generated by its elliptic elements?

As we shall see, the map e_g is surjective whenever $g \neq g^{-1}$, that is, whenever g is not an involution, and is trivial otherwise. We shall also observe that, as a consequence of the construction, we have

$$e_g(E(G)) = 0, \quad g \in G.$$

Thus, if on the one hand G contains a non-involution then we can conclude that

$$(\mathscr{RF}(G) : E(G)[\mathscr{RF}(G), \mathscr{RF}(G)]) \geq 2^{\aleph_0};$$

in particular, $\mathscr{RF}(G)$ is not generated by its elliptic elements. If, on the other hand, G is an elementary abelian 2-group, then all exponent sums are trivial, and the question whether $\mathscr{RF}(G)$ is generated by its elliptic elements in this case has to remain open at this stage. It will be resolved in Chapter 9 by the theory of test functions; see Section 9.4.

5.2 Some measure theory

We shall use some ideas and results from measure theory. Let l denote the length of an interval, with the convention that the empty interval is allowed and has length 0, and let $\mathscr{B}(\mathbb{R})$ (the Boolean of $\mathbb{R}$) be the set of all subsets of $\mathbb{R}$. Let m^* denote Lebesgue outer measure on $\mathbb{R}$, that is,

$$m^*(A) = \inf_{A \subseteq \bigcup I_n} \sum l(I_n), \qquad A \in \mathscr{B}(\mathbb{R}),$$

where the infimum is taken over all countable collections $\{I_n\}_{n\geq 1}$ of open intervals covering A, that is, such that $A \subseteq \bigcup_{n\geq 1} I_n$. The following properties of m^* are well known and are discussed in many text books; see, for instance, Chapter 3.2 in Royden [43]:

(i) m^* is *translation invariant*, that is,

$$m^*(A) = m^*(A + x), \qquad x \in \mathbb{R}, A \in \mathscr{B}(\mathbb{R});$$

(ii) m^* is *countably subadditive*, that is,

$$m^*\left(\bigcup_{n=1}^{\infty} A_n\right) \leq \sum_{n=1}^{\infty} m^*(A_n)$$

for each countable collection $\{A_n\}_{n\geq 1}$ of sets of real numbers;

(iii) $m^*(A) = 0$ if A is countable;

(iv) m^* is *monotone*, that is, $A \subseteq B$ implies $m^*(A) \leq m^*(B)$;

(v) if I is an interval then $m^*(I) = l(I)$.

We shall need some further, less well-known, properties of m^*.

Lemma 5.1 *Let $\alpha,\beta \geq 0$ be real numbers, let A be a subset of $[0,\alpha]$ and B a subset of $[\alpha,\alpha+\beta]$. Then $m^*(A\cup B) = m^*(A)+m^*(B)$.*

Proof Let $\{I_n\}$ be a countable collection of open intervals such that

$$A\cup B \subseteq \bigcup_{n=1}^{\infty} I_n.$$

Furthermore, set

$$I_n' := I_n \cap [0,\alpha] \quad \text{and} \quad I_n'' := I_n \cap [\alpha,\alpha+\beta].$$

Then the sets I_n' and I_n'' are bounded intervals, possibly empty. Let $\varepsilon > 0$ be given. If a_n, b_n are the endpoints of I_n', with $a_n \leq b_n$, let

$$J_n' := \begin{cases} \left(a_n - \varepsilon/2^{n+2}, b_n + \varepsilon/2^{n+2}\right), & I_n' \neq \varnothing, \\ \left(-\varepsilon/2^{n+2}, \varepsilon/2^{n+2}\right), & I_n' = \varnothing. \end{cases}$$

Then we have

$$l(J_n') = l(I_n') + \varepsilon/2^{n+1} \quad \text{and} \quad I_n' \subseteq J_n'.$$

Similarly, we can find open intervals J_n'' such that

$$l(J_n'') = l(I_n'') + \varepsilon/2^{n+1} \quad \text{and} \quad I_n'' \subseteq J_n''.$$

Now

$$A \subseteq \left(\bigcup_{n=1}^{\infty} I_n\right) \cap [0,\alpha] = \bigcup_{n=1}^{\infty} I_n' \subseteq \bigcup_{n=1}^{\infty} J_n'$$

and similarly

$$B \subseteq \bigcup_{n=1}^{\infty} J_n'',$$

implying that

$$m^*(A) \leq \sum_{n=1}^{\infty} l(J_n') \quad \text{and} \quad m^*(B) \leq \sum_{n=1}^{\infty} l(J_n'')$$

by definition of Lebesgue outer measure. Moreover, since

$$I_n' \cup I_n'' \subseteq I_n \quad \text{and} \quad I_n' \cap I_n'' \subseteq \{\alpha\}$$

we have

$$l(I_n') + l(I_n'') \leq l(I_n).$$

Therefore

$$\begin{aligned}\sum_{n=1}^{\infty} l(I_n) &\geq \sum_{n=1}^{\infty} l(I_n') + \sum_{n=1}^{\infty} l(I_n'') \\ &= \sum_{n=1}^{\infty} l(J_n') + \sum_{n=1}^{\infty} l(J_n'') - \sum_{n=1}^{\infty} \varepsilon/2^n \\ &\geq m^*(A) + m^*(B) - \varepsilon.\end{aligned}$$

Since this is true for each $\varepsilon > 0$, we must have

$$\sum_{n=1}^{\infty} l(I_n) \geq m^*(A) + m^*(B)$$

for every countable collection $\{I_n\}_{n\geq 1}$ of open intervals covering $A \cup B$. It follows that

$$m^*(A \cup B) \geq m^*(A) + m^*(B),$$

and the reverse inequality comes from property (ii) of m^* (that is, its countable subadditivity), finishing the proof of the lemma. □

Lemma 5.2 *Lebesgue outer measure is reflection invariant; that is,*

$$m^*(A) = m^*(-A), \quad A \in \mathscr{B}(\mathbb{R}).$$

Proof If $\{I_n\}_{n\geq 1}$ is a countable collection of open intervals covering A, that is,

$$A \subseteq \bigcup_{n\geq 1} I_n,$$

then

$$-A \subseteq \bigcup_{n\geq 1} (-I_n),$$

so that $\{-I_n\}_{n\geq 1}$ is a countable family of open intervals covering $-A$. Also, if $I_n = (a_n, b_n)$ then $-I_n = (-b_n, -a_n)$, so that

$$l(-I_n) = b_n - a_n = l(I_n),$$

with the usual conventions for dealing with $\pm\infty$. It follows that

$$m^*(-A) \leq \sum_{n=1}^{\infty} l(-I_n) = \sum_{n=1}^{\infty} l(I_n)$$

for every such collection $\{I_n\}_{n\geq 1}$ of open intervals covering A. We conclude that $m^*(-A) \leq m^*(A)$, and replacing A by $-A$ gives the reverse inequality. □

Before introducing and investigating the maps e_g we shall need one further piece of preparation.

Lemma 5.3 *For each $x \in \mathbb{R}$ and every $A \in \mathscr{B}(\mathbb{R})$, we have*

$$m^*(A - \{x\}) = m^*(A) = m^*(A \cup \{x\}).$$

Proof Since

$$A - \{x\} \subseteq A \subseteq A \cup \{x\},$$

the monotonicity of m^* (that is, property (iv) above) gives

$$m^*(A - \{x\}) \leq m^*(A) \leq m^*(A \cup \{x\}).$$

Thus, it suffices to show that

$$m^*(A \cup \{x\}) \leq m^*(A - \{x\}).$$

However, $A \cup \{x\} = (A - \{x\}) \cup \{x\}$ so, by properties (ii) and (iii),

$$m^*(A \cup \{x\}) \leq m^*(A - \{x\}) + m^*(\{x\}) = m^*(A - \{x\}),$$

as required. □

5.3 The maps μ_g

Definition 5.4 Let G be a group. For $g \in G$ and $f \in \mathscr{RF}(G)$, we set

$$\mu_g(f) := m^*(\{\xi \in [0, L(f)] : f(\xi) = g\}).$$

We note that, by properties (iv) and (v) of the outer measure m^*, we have

$$\mu_g(f) \leq L(f), \quad f \in \mathscr{RF}(G);$$

in particular, $\mu_g(f)$ is always finite. Our next result records two properties of the maps $\mu_g : \mathscr{RF}(G) \to \mathbb{R}$ that will be important in dealing with the exponent sum maps to be introduced in the next section.

Lemma 5.5 *For $g \in G$, and $f, f_1, f_2 \in \mathscr{RF}(G)$ with $\varepsilon_0(f_1, f_2) = 0$, we have:*

(i) $\mu_g(f^{-1}) = \mu_{g^{-1}}(f)$;

(ii) $\mu_g(f_1 \circ f_2) = \mu_g(f_1) + \mu_g(f_2)$.

Proof (i) Let

$$A := \{\xi \in [0, L(f)] : f^{-1}(\xi) = g\}$$

and

$$B := \{\xi \in [0, L(f)] : f(\xi) = g^{-1}\},$$

so that

$$m^*(A) = \mu_g(f^{-1}) \quad \text{and} \quad m^*(B) = \mu_{g^{-1}}(f).$$

Then

$$\begin{aligned} \xi \in B &\iff 0 \le \xi \le L(f) \text{ and } f(\xi) = g^{-1} \\ &\iff 0 \le L(f) - \xi \le L(f) \text{ and } f^{-1}(L(f) - \xi) = g \\ &\iff L(f) - \xi \in A, \end{aligned}$$

that is,

$$B = L(f) - A.$$

By Lemma 5.2 (reflection invariance of m^*) and property (i) (translation invariance of m^*), it follows that

$$\begin{aligned} \mu_g(f^{-1}) &= m^*(A) \\ &= m^*(-A) \\ &= m^*(-A + L(f)) \\ &= m^*(B) \\ &= \mu_{g^{-1}}(f), \end{aligned}$$

as claimed.

(ii) Let

$$A_1 := \{\xi \in [0, L(f_1)] : f_1(\xi) = g\}$$

and

$$A_2 := \{\xi \in [0, L(f_2)] : f_2(\xi) = g\},$$

so that

$$m^*(A_1) = \mu_g(f_1) \quad \text{and} \quad m^*(A_2) = \mu_g(f_2),$$

and let

$$A_2' := L(f_1) + A_2.$$

Then the set

$$\{\xi \in [0, L(f_1 \circ f_2)] : (f_1 \circ f_2)(\xi) = g\}$$

must be either $A_1 \cup A_2'$, $(A_1 \cup A_2') - \{L(f_1)\}$, or $(A_1 \cup A_2') \cup \{L(f_1)\}$. It follows that

$$\begin{aligned}\mu_g(f_1 \circ f_2) &= m^*(A_1 \cup A_2') \\ &= m^*(A_1) + m^*(A_2') \\ &= m^*(A_1) + m^*(A_2) \\ &= \mu_g(f_1) + \mu_g(f_2).\end{aligned}$$

Here we have used Lemma 5.3 to obtain the first equality, Lemma 5.1 for the second, property (i) (the translation invariance of m^*) in step 3, and the definition of the map μ_g in the last step. □

5.4 The maps e_g

Definition 5.6 Given $g \in G$, we define a mapping $e_g : \mathscr{RF}(G) \to \mathbb{R}$, called the *exponent sum* of $\mathscr{RF}(G)$ *relative to* g, by

$$e_g(f) := \mu_g(f) - \mu_{g^{-1}}(f).$$

Clearly, the definition of the map e_g is inspired by that of the exponent sum relative to a free generator in a free group, whence the name.

Proposition 5.7 *For each $g \in G$, the mapping e_g is a group homomorphism.*

Proof Let $f_1, f_2 \in \mathscr{RF}(G)$. According to Lemma 2.15 we can write $f_1 = f_1' \circ u$, $f_2 = u^{-1} \circ f_2'$, with $f_1 f_2 = f_1' \circ f_2'$. By the second part of Lemma 5.5, we have

$$\begin{aligned}\mu_g(f_1 f_2) &= \mu_g(f_1' \circ f_2') \\ &= \mu_g(f_1') + \mu_g(f_2') \\ &= \mu_g(f_1) - \mu_g(u) + \mu_g(f_2) - \mu_g(u^{-1}).\end{aligned}$$

Since this applies to every $g \in G$, replacing g by g^{-1} gives

$$\mu_{g^{-1}}(f_1 f_2) = \mu_{g^{-1}}(f_1) - \mu_{g^{-1}}(u) + \mu_{g^{-1}}(f_2) - \mu_{g^{-1}}(u^{-1}).$$

Subtracting, we get

$$\begin{aligned} e_g(f_1 f_2) &= \big(\mu_g(f_1) - \mu_{g^{-1}}(f_1)\big) + \big(\mu_g(f_2) - \mu_{g^{-1}}(f_2)\big) \\ &\quad + \big(\mu_{g^{-1}}(u^{-1}) - \mu_g(u)\big) + \big(\mu_{g^{-1}}(u) - \mu_g(u^{-1})\big) \\ &= \big(\mu_g(f_1) - \mu_{g^{-1}}(f_1)\big) + \big(\mu_g(f_2) - \mu_{g^{-1}}(f_2)\big) \\ &= e_g(f_1) + e_g(f_2), \end{aligned}$$

where we have used the first part of Lemma 5.5 in the second step. Hence, e_g is indeed a homomorphism, as claimed. □

Now let

$$K_g = \ker(e_g) = \Big\{ f \in \mathscr{R}\mathscr{F}(G) : \mu_g(f) = \mu_{g^{-1}}(f) \Big\}.$$

Since $\mathscr{R}\mathscr{F}(G)/K_g$ is abelian,

$$[\mathscr{R}\mathscr{F}(G), \mathscr{R}\mathscr{F}(G)] \leq K_g.$$

Moreover, if $f \in G_0$, then the preimage of g, and that of g^{-1}, under the map f is either empty or consists of one point; hence $\mu_g(f) = 0 = \mu_{g^{-1}}(f)$, so $f \in K_g$ and therefore $G_0 \leq K_g$. It follows that $E(G) \leq K_g$ and thus

$$E(G)\,[\mathscr{R}\mathscr{F}(G), \mathscr{R}\mathscr{F}(G)] \leq K_g. \tag{5.1}$$

Also, if $g = g^{-1}$ then clearly $\mu_g = \mu_{g^{-1}}$; thus $\mathrm{Im}(e_g) = \{0\}$ and $K_g = \mathscr{R}\mathscr{F}(G)$. Our final lemma considers the case where $g \neq g^{-1}$.

Lemma 5.8 *If $g \neq g^{-1}$ then* $\mathrm{Im}(e_g) = \mathbb{R}$.

Proof Let $r \in \mathbb{R}$, $r \geq 0$, and let f be the constant function with value g on $[0, r]$. Then, by property (v) of m^*, we have $\mu_g(f) = r$ and $\mu_{g^{-1}}(f) = 0$, so $e_g(f) = r$. Also $e_g(f^{-1}) = -r$, hence e_g is surjective, as claimed. □

As an application of Proposition 5.7 and Lemma 5.8, we now have the following.

Corollary 5.9 *If G is not an elementary abelian 2-group then we have*

$$(\mathscr{R}\mathscr{F}(G) : E(G)[\mathscr{R}\mathscr{F}(G), \mathscr{R}\mathscr{F}(G)]) \geq 2^{\aleph_0}. \tag{5.2}$$

Proof Since G is not an elementary abelian 2-group we can find an element $g \in G$ such that $g \neq g^{-1}$. Put $\mathscr{H} = E(G)[\mathscr{RF}(G), \mathscr{RF}(G)]$. By (5.1) and Lemma 5.8,

$$(\mathscr{RF}(G):\mathscr{H}) = (\mathscr{RF}(G):K_g)(K_g:\mathscr{H}) \geq (\mathscr{RF}(G):K_g) = 2^{\aleph_0};$$

whence (5.2). □

In the case where G is an elementary abelian 2-group, all exponent sums are trivial and nothing can be deduced by the technique developed in this chapter concerning the index of $E(G)$ in $\mathscr{RF}(G)$. For a solution to this problem see Chapter 9, in particular Section 9.4.

5.5 The map e_G

Let

$$\mathrm{Inv}(G) := \{x \in G : x^2 = 1_G\}$$

be the set of proper involutions of G together with 1_G, and let R_G be a system of representatives for the equivalence relation $\sim$ on $G - \mathrm{Inv}(G)$ given by

$$x \sim y :\Longleftrightarrow x = y \text{ or } x = y^{-1};$$

that is, we choose an element from each pair $\{g, g^{-1}\}$ of inverses in $G - \mathrm{Inv}(G)$. Assuming that G is not an elementary abelian 2-group (that is, $R_G \neq \varnothing$), we can define a map $e_G : \mathscr{RF}(G) \to \prod_{g \in R_G} \mathbb{R}$ via

$$e_G(f) := \{e_g(f)\}_{g \in R_G}, \quad f \in \mathscr{RF}(G).$$

By what we have shown above, e_G is a homomorphism whose kernel $K_G = \bigcap_{g \in R_G} K_g$ contains the commutator subgroup $[\mathscr{RF}(G), \mathscr{RF}(G)]$ as well as the subgroup $E(G)$ of $\mathscr{RF}(G)$ generated by the elliptic elements.

It appears difficult to pin down the image of the map e_G precisely; however, we can at least show that this image is 'large' in some suitable sense. For

$$\mathbf{x} = \{x_g\}_{g \in R_G} \in \prod_{g \in R_G} \mathbb{R},$$

let

$$\mathrm{supp}(\mathbf{x}) := \{g \in R_G : x_g \neq 0\}$$

be the support of $\mathbf{x}$, and define subspaces $\ell^1(R_G)$ and $\bar{\ell}^1(R_G)$ of $\prod_{g\in R_G}\mathbb{R}$ as follows:

$$\ell^1(R_G) := \left\{\mathbf{x} \in \prod_{g\in R_G}\mathbb{R} : \mathbf{x} \text{ has countable support and } \sum_{g\in R_G}|x_g| < \infty\right\},$$

and

$$\bar{\ell}^1(R_G) := \left\{\mathbf{x} \in \prod_{g\in R_G}\mathbb{R} : \sum_{g\in S}|x_g| < \infty \text{ for every countable set } S \subseteq R_G\right\}.$$

Proposition 5.10 *We have*

$$\ell^1(R_G) = \bar{\ell}^1(R_G) \subseteq e_G(\mathscr{R}\mathscr{F}(G)).$$

For the proof, we shall need the following auxiliary result; see also Problem 19 in Section 4, Chapter 2 of Royden [43].

Lemma 5.11 *Let $A \subseteq \mathbb{R}$ be a set of positive real numbers such that*

$$\sum_{b\in B} b < \infty \tag{5.3}$$

for every countable subset B of A. Then A itself is countable.

Proof Suppose that on the contrary A is uncountable. If the set

$$A_\varepsilon := \{a \in A : a \geq \varepsilon\}$$

were countable for every real number $\varepsilon > 0$ then

$$A = \bigcup_{n\geq 1} A_{1/n}$$

would be countable. Thus, since A is assumed to be uncountable, there must exist $\varepsilon_1 > 0$ such that the set A_{ε_1} is uncountable. But then, choosing a countably infinite subset B of A_{ε_1}, we would have

$$\sum_{b\in B} b \geq n\varepsilon_1, \quad n \in \mathbb{N};$$

that is, the sum $\sum_{b\in B} b$ would be infinite, contradicting (5.3). Hence, A is countable as claimed. □

Proof of Proposition 5.10 Let $\mathbf{x} = \{x_g\}_{g\in R_G} \in \ell^1(R_G)$ be given, and enumerate the support of $\mathbf{x}$ as

$$\operatorname{supp}(\mathbf{x}) = \{g_1, g_2, \ldots, g_n, \ldots\} \subseteq R_G.$$

Define a function $f_{\mathbf{x}} \in \mathscr{RF}(G)$ of length $s := \sum_{g\in R_G} |x_g|$, via

$$f_{\mathbf{x}}(\xi) := \left\{ \begin{array}{ll} g_j^{\operatorname{sgn}(x_{g_j})}, & \sum_{i=1}^{j-1} |x_{g_i}| < \xi < \sum_{i=1}^{j} |x_{g_i}| \ (j \geq 1) \\ 1_G, & \text{otherwise} \end{array} \right\} \qquad (0 \leq \xi \leq s).$$

By construction, for $j \geq 1$ and $0 < \eta < \min\{|x_{g_j}|, |x_{g_{j+1}}|\}$,

$$f_{\mathbf{x}}\left(\sum_{i=1}^{j} |x_{g_i}| - \eta\right) f_{\mathbf{x}}\left(\sum_{i=1}^{j} |x_{g_i}| + \eta\right) = g_j^{\operatorname{sgn}(x_{g_j})} g_{j+1}^{\operatorname{sgn}(x_{g_{j+1}})} \neq 1_G,$$

since g_j and g_{j+1} are different (and hence non-equivalent) elements of R_G. Thus $f_{\mathbf{x}}$ is indeed reduced, as claimed. Also, by definition of the function $f_{\mathbf{x}}$,

$$\mu_g(f_{\mathbf{x}}) = \begin{cases} |x_{g_j}|, & g = g_j^{\operatorname{sgn}(x_{g_j})}, \\ 0, & \text{otherwise,} \end{cases}$$

and so

$$e_g(f_{\mathbf{x}}) = \left\{ \begin{array}{ll} x_{g_j}, & g = g_j \\ 0, & g \notin \operatorname{supp}(\mathbf{x}) \end{array} \right\} \qquad (g \in R_G).$$

It follows that

$$e_G(f_{\mathbf{x}}) = \{x_g\}_{g\in R_G} = \mathbf{x}$$

and hence

$$\ell^1(R_G) \subseteq e_G(\mathscr{RF}(G)),$$

as claimed.

It remains to show that $\ell^1(R_G) = \bar{\ell}^1(R_G)$. By definition we have $\ell^1(R_G) \subseteq \bar{\ell}^1(R_G)$. Conversely, let $\mathbf{x} = \{x_g\}_{g\in R_G} \in \bar{\ell}^1(R_G)$. Then the set

$$A_{\mathbf{x}} := \{|x_g| : g \in \operatorname{supp}(\mathbf{x})\}$$

satisfies the hypotheses of Lemma 5.11. Indeed, if B is a countable subset of $A_{\mathbf{x}}$ then we can find a countable set $S \subseteq R_G$ such that

$$B = \{|x_g| : g \in S\},$$

and so

$$\sum_{b \in B} b = \sum_{g \in S} |x_g| < \infty$$

by the definition of $\bar{\ell}^1(R_G)$. It follows now from Lemma 5.11 that $A_{\mathbf{x}}$ is countable, and, given any $a \in A_{\mathbf{x}}$, the set

$$X_a := \{g \in \operatorname{supp}(\mathbf{x}) : |x_g| = a\}$$

must be finite, again by definition of $\bar{\ell}^1(R_G)$. Hence

$$\operatorname{supp}(\mathbf{x}) = \bigcup_{a \in A_{\mathbf{x}}} X_a$$

is countable, so $\mathbf{x} \in \ell^1(R_G)$ and the proof is complete. □

6
Functoriality

6.1 Introduction

We are going to work with three categories of groups:

> **Groups** – the category of groups and group homomorphisms,
>
> $\widetilde{\textbf{Groups}}$ – the category of groups and embeddings (that is, injective homomorphisms),
>
> $\widehat{\textbf{Groups}}$ – the category whose objects are groups and whose morphisms $\varphi \in \mathrm{Mor}(G, H)$ are injective maps $\varphi : G \to H$ satisfying $\varphi(1_G) = 1_H$ and $\varphi(x^{-1}) = \varphi(x)^{-1}$ for $x \in G$.

We have

$$\widehat{\textbf{Groups}} \geq \widetilde{\textbf{Groups}} \leq \textbf{Groups}.$$

We shall show that the $\mathscr{R}\mathscr{F}$-construction may be viewed as a covariant functor

$$\widehat{\mathscr{R}\mathscr{F}}(-) : \widehat{\textbf{Groups}} \to \widehat{\textbf{Groups}},$$

which restricts to a functor

$$\widetilde{\mathscr{R}\mathscr{F}}(-) : \widetilde{\textbf{Groups}} \to \widetilde{\textbf{Groups}}.$$

Given a suitable (and fairly natural) definition of induced morphisms, these facts are rather straightforward to establish, although not completely trivial; nevertheless, some noteworthy conclusions may be drawn from them.

First, the existence of the functor $\widetilde{\mathscr{R}\mathscr{F}}(-)$ provides a certain supply of (injective) homomorphisms involving $\mathscr{R}\mathscr{F}$-groups and in particular yields the conclusion that

$$G \cong H \implies \mathscr{RF}(G) \cong \mathscr{RF}(H). \tag{6.1}$$

Of course, (6.1) is to be expected (to the extent that the construction itself would be compromised if this fact could not be established), but it is interesting to note that the proof of (6.1) (that is, the functoriality of $\widetilde{\mathscr{RF}}(-)$) is not trivial. Incidentally, the question to what extent the converse of (6.1) holds, that is, whether $\mathscr{RF}(G) \cong \mathscr{RF}(H)$ implies $G \cong H$, turns out to be difficult and only partial results are known at present; see Müller [37].

Second, the existence of the functor $\widehat{\mathscr{RF}}(-)$ implies the following. *If two groups G and H satisfy both* $|\mathrm{Inv}(G)| = |\mathrm{Inv}(H)|$ *and* $|G - \mathrm{Inv}(G)| = |H - \mathrm{Inv}(H)|$ *then we have* $|\mathscr{RF}(G)| = |\mathscr{RF}(H)|$. Somewhat loosely, this last statement may be rephrased by saying that *the cardinality of $\mathscr{RF}(G)$ depends only on the cardinality of the subset* $\mathrm{Inv}(G)$ *of G and that of its complement.* This is a non-trivial result, even though it does not tell us what the cardinality of $\mathscr{RF}(G)$ actually is; see Corollary 10.4 for the computation of $|\mathscr{RF}(G)|$ by means of test function theory.

Third, the functorial property of $\widetilde{\mathscr{RF}}(-)$ also allows us to extend an automorphism α of G to an automorphism $\hat{\alpha}$ of $\mathscr{RF}(G)$ in such a way that the assignment $\alpha \mapsto \hat{\alpha}$ gives an embedding of $\mathrm{Aut}(G)$ into $\mathrm{Aut}(\mathscr{RF}(G))$.

The main result of this chapter, however, lies considerably deeper; namely, we will show that $\widehat{\mathscr{RF}}(-)$ induces in a natural way a functor

$$\widehat{\mathscr{RF}}_0(-) : \widehat{\mathbf{Groups}} \to \mathbf{Groups},$$

whose action on objects is given by

$$\widehat{\mathscr{RF}}_0(G) = \mathscr{RF}(G)/E(G).$$

As an consequence, we obtain the following rather surprising result.

Rigidity theorem *Suppose that G and H are groups such that* $|\mathrm{Inv}(G)| = |\mathrm{Inv}(H)|$ *and* $|G - \mathrm{Inv}(G)| = |H - \mathrm{Inv}(H)|$. *Then we have*

$$\mathscr{RF}(G)/E(G) \cong \mathscr{RF}(H)/E(H).$$

Hence, again somewhat loosely, we may say that *the isomorphism type of the quotient group $\mathscr{RF}(G)/E(G)$ depends only on the cardinality of the set of involutions of G and that of its complement.*

We are, of course, well aware of the somewhat different connotation of the term 'rigidity' in the theory of semisimple Lie groups; see Chapter VII in Margulis [32]. However, here, as well as there, this term refers to a certain extension property of morphisms so our choice of terminology appears at least to some extent justified.

6.2 The functor $\widehat{\mathscr{R}\mathscr{F}}(-)$

Let G and H be groups, and let $\varphi : G \to H$ be a map. We shall assume the following:

(i) φ is injective;

(ii) $\varphi(1_G) = 1_H$;

(iii) $\varphi(x^{-1}) = (\varphi(x))^{-1}, \quad x \in G$;

that is, φ is a morphism in the category $\widehat{\textbf{Groups}}$. Let $\hat{\varphi} : \mathscr{F}(G) \to \mathscr{F}(H)$ be the map induced by φ via

$$\hat{\varphi}(f) = \varphi \circ f, \quad f \in \mathscr{F}(G).$$

By construction $\hat{\varphi}$ is length-preserving; that is, we have

(I) $L(\hat{\varphi}(f)) = L(f), \quad f \in \mathscr{F}(G)$.

We also have the analogues of properties (i)–(iii) for the map $\hat{\varphi}$, as follows.

(II) *The map* $\hat{\varphi} : \mathscr{F}(G) \to \mathscr{F}(H)$ *is injective.*

Proof If $\hat{\varphi}(f_1) = \hat{\varphi}(f_2)$ for some $f_1, f_2 \in \mathscr{F}(G)$ then, by (I),

$$L(f_1) = L(\hat{\varphi}(f_1)) = L(\hat{\varphi}(f_2)) = L(f_2) =: \alpha$$

and we have

$$\varphi(f_1(\xi)) = \varphi(f_2(\xi)), \quad 0 \leq \xi \leq \alpha,$$

implying $f_1 = f_2$ by the injectivity of φ. □

(III) *We have* $\hat{\varphi}(\mathbf{1}_G) = \mathbf{1}_H$.

Proof The function $\hat{\varphi}(\mathbf{1}_G)$ has length 0 by (I); and, by property (ii) of φ,

$$\hat{\varphi}(\mathbf{1}_G)(0) = \varphi(\mathbf{1}_G(0)) = \varphi(1_G) = 1_H = \mathbf{1}_H(0),$$

whence our claim. □

(IV) *The map* $\hat{\varphi}$ *respects inverses; that is,*

$$\hat{\varphi}(f^{-1}) = (\hat{\varphi}(f))^{-1}, \quad f \in \mathscr{F}(G).$$

Proof By assertion (I), both sides of the above relation have length equal to $L(f)$. Moreover, for $0 \le \xi \le L(f)$ we have

$$\begin{aligned}\hat{\varphi}(f^{-1})(\xi) &= \varphi(f^{-1}(\xi)) \\ &= \varphi((f(L(f) - \xi))^{-1}) \\ &= (\varphi(f(L(f) - \xi)))^{-1} \\ &= (\hat{\varphi}(f)(L(f) - \xi))^{-1} \\ &= (\hat{\varphi}(f))^{-1}(\xi),\end{aligned}$$

where again we have made use of property (iii). □

Next, we claim the following.

(V) *An element* $f \in \mathscr{F}(G)$ *is reduced if and only if* $\hat{\varphi}(f) \in \mathscr{F}(H)$ *is reduced; in particular,* $\hat{\varphi}$ *restricts to a map* $\hat{\varphi}|_{\mathscr{RF}(G)} : \mathscr{RF}(G) \to \mathscr{RF}(H)$.

Proof Suppose that f is reduced and that there exists (a sufficiently small) $\varepsilon > 0$ such that for some point $\xi_0 \in (0, L(f))$ we have

$$\hat{\varphi}(f)(\xi_0 - \eta) = (\hat{\varphi}(f)(\xi_0 + \eta))^{-1}, \quad 0 \le \eta \le \varepsilon,$$

as well as

$$\hat{\varphi}(f)(\xi_0) = 1_H.$$

Since φ respects inverses, we have

$$(\hat{\varphi}(f)(\xi_0 + \eta))^{-1} = (\varphi(f(\xi_0 + \eta)))^{-1} = \varphi((f(\xi_0 + \eta))^{-1});$$

hence, by properties (i) and (ii) of φ (see the start of this section), we find that

$$f(\xi_0 - \eta) = (f(\xi_0 + \eta))^{-1}, \quad 0 \leq \eta \leq \varepsilon,$$

and that

$$f(\xi_0) = 1_G.$$

We conclude that $[\xi_0 - \varepsilon, \xi_0 + \varepsilon]$ is a cancelling neighbourhood for f around the interior point ξ_0, a contradiction, since f is assumed to be reduced.

Conversely, suppose that $\hat{\varphi}(f)$ is reduced and that there exists $\varepsilon > 0$ such that, for $\xi_0 \in (0, L(f))$,

$$f(\xi_0 - \eta) = (f(\xi_0 + \eta))^{-1}, \quad 0 \leq \eta \leq \varepsilon,$$

and

$$f(\xi_0) = 1_G.$$

Applying φ to these equations and using properties (ii) and (iii) of φ, we obtain

$$\hat{\varphi}(f)(\xi_0 - \eta) = (\hat{\varphi}(f)(\xi_0 + \eta))^{-1}, \quad 0 \leq \eta \leq \varepsilon,$$

and

$$\hat{\varphi}(f)(\xi_0) = 1_H,$$

contradicting our assumption that $\hat{\varphi}(f)$ is reduced. □

We are now in a position to explain the first result mentioned in the introduction to this chapter.

Proposition 6.1 *Setting*

$$\widehat{\mathscr{R}\mathscr{F}}(G) := \mathscr{R}\mathscr{F}(G), \quad G \in |\widehat{\mathbf{Groups}}|,$$

and

$$\widehat{\mathscr{R}\mathscr{F}}(\varphi) := \hat{\varphi}|_{\mathscr{R}\mathscr{F}(G)}, \quad \varphi \in \mathrm{Mor}_{\widehat{\mathbf{Groups}}}(G, H),$$

we obtain a covariant functor

$$\widehat{\mathscr{R}\mathscr{F}}(-) : \widehat{\mathbf{Groups}} \to \widehat{\mathbf{Groups}}.$$

Proof By assertions (II)–(V), the map $\hat{\varphi}|_{\mathscr{R}\mathscr{F}(G)}$ is a morphism in the category $\widehat{\mathbf{Groups}}$ connecting $\widehat{\mathscr{R}\mathscr{F}}(G)$ to $\widehat{\mathscr{R}\mathscr{F}}(H)$. One needs to check that

$$\widehat{\mathrm{id}_G}|_{\mathscr{R}\mathscr{F}(G)} = \mathrm{id}_{\mathscr{R}\mathscr{F}(G)}, \quad G \in |\widehat{\mathbf{Groups}}|$$

and that, for morphisms $\varphi : G \to H$ and $\psi : H \to K$, we have

$$(\widehat{\psi \circ \varphi})|_{\mathscr{RF}(G)} = \hat{\psi}|_{\mathscr{RF}(H)} \circ \hat{\varphi}|_{\mathscr{RF}(G)};$$

these equations, however, are immediate from the definition of $\hat{\varphi}$. □

We continue with some observations concerning the category $\widehat{\textbf{Groups}}$.

(VI)

(a) *A morphism* $\varphi \in \mathrm{Mor}_{\widehat{\textbf{Groups}}}(G,H)$ *is an isomorphism if and only if* φ *is bijective as a map.*

(b) *Assuming the axiom of choice, two objects* $G,H \in |\widehat{\textbf{Groups}}|$ *are isomorphic if and only if* $|\mathrm{Inv}(G)| = |\mathrm{Inv}(H)|$ *and* $|G - \mathrm{Inv}(G)| = |H - \mathrm{Inv}(H)|$.

Proof (a) If φ is an isomorphism then there exists $\psi \in \mathrm{Mor}_{\widehat{\textbf{Groups}}}(H,G)$ such that $\psi \circ \varphi = \mathrm{id}_G$ and $\varphi \circ \psi = \mathrm{id}_H$; in particular, φ is invertible as a map and hence is a bijection. Conversely, suppose that φ is bijective. Then the inverse map $\psi : H \to G$ exists and is a bijection, so it only remains to establish properties (ii) and (iii) for ψ in order to be able to conclude that φ is an isomorphism. However, using properties (ii) and (iii) of φ, we clearly have

$$(\varphi(\psi(1_H)) = 1_H = \varphi(1_G)$$

and

$$\varphi(\psi(x^{-1})) = x^{-1} = \varphi((\psi(x))^{-1}), \quad x \in H;$$

the desired conclusion follows now from the injectivity of φ.

(b) Suppose that two groups G and H are isomorphic in the category $\widehat{\textbf{Groups}}$. By part (a) this means that there exists a bijection $\varphi : G \to H$ sending 1_G to 1_H and respecting inverses. It follows that φ sends involutions of G to involutions of H and non-involutions to non-involutions; hence it induces bijections $\varphi_1 : \mathrm{Inv}(G) \to \mathrm{Inv}(H)$ and $\varphi_2 : G - \mathrm{Inv}(G) \to H - \mathrm{Inv}(H)$ and, consequently, $|\mathrm{Inv}(G)| = |\mathrm{Inv}(H)|$ and $|G - \mathrm{Inv}(G)| = |H - \mathrm{Inv}(H)|$.

Conversely, suppose that the last two equations hold, and let R_G, R_H be defined as in Section 5.5. By cardinal arithmetic (using the axiom of choice), we have

$$\begin{aligned} |\mathrm{Inv}(G) - \{1_G\}| &= |\mathrm{Inv}(G)| - 1 \\ &= |\mathrm{Inv}(H)| - 1 \\ &= |\mathrm{Inv}(H) - \{1_H\}| \end{aligned}$$

and

$$
\begin{aligned}
|R_G| + |R_G| &= |G - \mathrm{Inv}(G)| \\
&= |H - \mathrm{Inv}(H)| \\
&= |R_H| + |R_H|,
\end{aligned}
$$

the last computation implying that $|R_G| = |R_H|$. Indeed, this is clear if $|R_G|$ is finite while for $|R_G|$ infinite we have

$$|R_G| + |R_G| = \max\left\{|R_G|, |R_G|\right\} = |R_G|;$$

moreover, in this situation, $|R_H|$ must also be infinite and the same argument gives that

$$|R_H| + |R_H| = |R_H|,$$

whence the result. Consequently, there exist bijections

$$\varphi_1 : \mathrm{Inv}(G) - \{1_G\} \to \mathrm{Inv}(H) - \{1_H\}$$

and

$$\varphi_2 : R_G \to R_H.$$

We now define a map $\varphi : G \to H$ by

$$
\varphi(x) := \left\{
\begin{array}{ll}
1_H, & x = 1_G \\
\varphi_1(x), & x \in \mathrm{Inv}(G) - \{1_G\} \\
\varphi_2(x), & x \in R_G \\
(\varphi_2(x^{-1}))^{-1}, & x \in (G - \mathrm{Inv}(G)) - R_G
\end{array}
\right\} \quad (x \in G).
$$

Then φ satisfies properties (ii) and (iii) by definition and is clearly a bijection, hence an isomorphism; so G and H are isomorphic in the category $\widehat{\mathbf{Groups}}$. □

Combining Proposition 6.1 with assertion (VI), we obtain the following.

Corollary 6.2 *Let G and H be groups such that*

$$|\mathrm{Inv}(G)| = |\mathrm{Inv}(H)| \quad \textit{and} \quad |G - \mathrm{Inv}(G)| = |H - \mathrm{Inv}(H)|.$$

Then we have $|\mathscr{RF}(G)| = |\mathscr{RF}(H)|$.

Proof By part (b) of assertion (VI), our hypothesis is equivalent to that of the existence of an isomorphism $\varphi : G \to H$ in the category $\widehat{\mathbf{Groups}}$. By Proposition 6.1, that is, the functoriality of $\widehat{\mathscr{RF}}(-)$, we have that $\widehat{\mathscr{RF}}(\varphi)$ is an isomorphism in the same category, connecting $\mathscr{RF}(G)$ to $\mathscr{RF}(H)$; in particular, applying part (a) of assertion (VI) we have $|\mathscr{RF}(G)| = |\mathscr{RF}(H)|$. □

The proof of Corollary 6.2 may essentially be rephrased by saying that if $\varphi \in \mathrm{Mor}_{\widehat{\mathbf{Groups}}}(G,H)$ is surjective then $\hat{\varphi}|_{\mathscr{RF}(G)}$ is surjective. Concerning the surjectivity of the maps φ, $\hat{\varphi}$, and $\hat{\varphi}|_{\mathscr{RF}(G)}$, we can say a little more.

(VII) *Let* $\varphi : G \to H$ *be a morphism in the category* $\widehat{\mathbf{Groups}}$. *Then the following assertions are equivalent:*

(a) *the map* $\varphi : G \to H$ *is surjective;*

(b) *the map* $\hat{\varphi} : \mathscr{F}(G) \to \mathscr{F}(H)$ *is surjective;*

(c) *the map* $\hat{\varphi}|_{\mathscr{RF}(G)} : \mathscr{RF}(G) \to \mathscr{RF}(H)$ *is surjective.*

Proof (a) $\Rightarrow$ (b). Let $\hat{f} \in \mathscr{F}(H)$ be an arbitrary element. Using the axiom of choice, define a function $f \in \mathscr{F}(G)$ with $L(f) = L(\hat{f})$ in such a way that

$$f(\xi) \in \varphi^{-1}(\hat{f}(\xi)), \quad 0 \leq \xi \leq L(f).$$

Then $\hat{\varphi}(f) = \hat{f}$, so $\hat{\varphi}$ is surjective.

(b) $\Rightarrow$ (c). Let $\hat{f} \in \mathscr{RF}(H)$ be arbitrary. Since $\hat{\varphi}$ is surjective there exists $f \in \mathscr{F}(G)$ such that $\hat{\varphi}(f) = \hat{f}$. Moreover, by the reverse implication in assertion (V), f is reduced since $\hat{f}$ is reduced; hence $\hat{\varphi}_{|\mathscr{RF}(G)}$ is surjective.

(c) $\Rightarrow$ (a). Since $\hat{\varphi}_{|\mathscr{RF}(G)}$ preserves length and is surjective, the subgroup G_0 of $\mathscr{RF}(G)$ is mapped under $\hat{\varphi}_{|\mathscr{RF}(G)}$ onto the subgroup H_0 of $\mathscr{RF}(H)$. Identifying G_0 with G and H_0 with H in the canonical way, we obtain, by the definition of $\hat{\varphi}$, a commutative square

$$\begin{array}{ccc} G_0 & \xrightarrow{\hat{\varphi}|_{G_0}} & H_0 \\ {\scriptstyle\cong}\downarrow & & \downarrow{\scriptstyle\cong} \\ G & \xrightarrow[\varphi]{} & H \end{array}$$

Hence, φ is surjective, since the map $\hat{\varphi}|_{G_0}$ is surjective. □

6.3 The functor $\widetilde{\mathscr{RF}}(-)$

We shall need some further observations concerning the hat construction.

(VIII) *For* $\varphi \in \mathrm{Mor}_{\widehat{\mathbf{Groups}}}(G,H)$ *and* $f_1, f_2 \in \mathscr{F}(G)$, *we have*

$$\varepsilon_0(f_1, f_2) = \varepsilon_0(\hat{\varphi}(f_1), \hat{\varphi}(f_2)).$$

Proof By properties (i) and (iii) of φ,

$$f_1(L(f_1)) = (f_2(0))^{-1} \iff \hat{\varphi}(f_1)(L(\hat{\varphi}(f_1))) = (\hat{\varphi}(f_2)(0))^{-1}.$$

Hence, if $f_1(L(f_1)) \neq (f_2(0))^{-1}$ then $\hat{\varphi}(f_1)(L(\hat{\varphi}(f_1))) \neq (\hat{\varphi}(f_2)(0))^{-1}$, so

$$\varepsilon_0(f_1, f_2) = 0 = \varepsilon_0(\hat{\varphi}(f_1), \hat{\varphi}(f_2))$$

in this case. If, however, $f_1(L(f_1)) = (f_2(0))^{-1}$ then $\hat{\varphi}(f_1)(L(\hat{\varphi}(f_1))) = (\hat{\varphi}(f_2)(0))^{-1}$, hence

$$\varepsilon_0(f_1, f_2) = \sup \mathscr{E}(f_1, f_2)$$

and

$$\varepsilon_0(\hat{\varphi}(f_1), \hat{\varphi}(f_2)) = \sup \mathscr{E}(\hat{\varphi}(f_1), \hat{\varphi}(f_2)).$$

Moreover, for $\varepsilon \in [0, \min\{L(f_1), L(f_2)\}]$,

$$\begin{aligned}
\varepsilon \in \mathscr{E}(f_1, f_2) &\iff f_1(L(f_1) - \eta) = (f_2(\eta))^{-1} \text{ for } 0 \leq \eta \leq \varepsilon \\
&\iff \hat{\varphi}(f_1)(L(\hat{\varphi}(f_1)) - \eta) = (\hat{\varphi}(f_2)(\eta))^{-1} \text{ for } 0 \leq \eta \leq \varepsilon \\
&\iff \varepsilon \in \mathscr{E}(\hat{\varphi}(f_1), \hat{\varphi}(f_2)).
\end{aligned}$$

Hence

$$\mathscr{E}(f_1, f_2) = \mathscr{E}(\hat{\varphi}(f_1), \hat{\varphi}(f_2)),$$

and again our claim holds. □

(IX) *Let* G, H *be groups, and let* $\varphi : G \to H$ *be an injective homomorphism. Then the induced map* $\hat{\varphi} : \mathscr{F}(G) \to \mathscr{F}(H)$ *respects reduced multiplication, that is, we have*

$$\hat{\varphi}(f_1 f_2) = \hat{\varphi}(f_1)\hat{\varphi}(f_2), \quad f_1, f_2 \in \mathscr{F}(G). \tag{6.2}$$

In particular, the restriction $\hat{\varphi}|_{\mathscr{RF}(G)} : \mathscr{RF}(G) \to \mathscr{RF}(H)$ *is an injective group homomorphism.*

Proof By assertions (I) and (VIII) we have

$$\begin{aligned}
L(\hat{\varphi}(f_1 f_2)) &= L(f_1 f_2) \\
&= L(f_1) + L(f_2) - \varepsilon_0(f_1, f_2) \\
&= L(\hat{\varphi}(f_1)) + L(\hat{\varphi}(f_2)) - \varepsilon_0(\hat{\varphi}(f_1), \hat{\varphi}(f_2)) \\
&= L(\hat{\varphi}(f_1)\hat{\varphi}(f_2)).
\end{aligned}$$

Moreover, making use of (VIII) and the fact that φ is a homomorphism, we have, for $0 \leq \xi \leq L(f_1 f_2)$,

$$(\hat{\varphi}(f_1)\hat{\varphi}(f_2))(\xi)$$

$$= \begin{cases} \hat{\varphi}(f_1)(\xi), & 0 \leq \xi < L(f_1) - \varepsilon_0(f_1, f_2), \\ \hat{\varphi}(f_1)(L(f_1) - \varepsilon_0(f_1, f_2))\hat{\varphi}(f_2)(\varepsilon_0(f_1, f_2)), & \xi = L(f_1) - \varepsilon_0(f_1, f_2), \\ \hat{\varphi}(f_2)(\xi - L(f_1) + 2\varepsilon_0(f_1, f_2)), & L(f_1) - \varepsilon_0(f_1, f_2) < \xi \leq L(f_1 f_2), \end{cases}$$

$$= \begin{cases} \varphi(f_1(\xi)), & 0 \leq \xi < L(f_1) - \varepsilon_0(f_1, f_2), \\ \varphi\big(f_1(L(f_1) - \varepsilon_0(f_1, f_2))f_2(\varepsilon_0(f_1, f_2))\big), & \xi = L(f_1) - \varepsilon_0(f_1, f_2), \\ \varphi\big(f_2(\xi - L(f_1) + 2\varepsilon_0(f_1, f_2))\big), & L(f_1) - \varepsilon_0(f_1, f_2) < \xi \leq L(f_1 f_2), \end{cases}$$

$$= \varphi((f_1 f_2)(\xi))$$

$$= \hat{\varphi}(f_1 f_2)(\xi),$$

whence (6.2). The particular statement at the end of the assertion follows from this and assertion (II). □

We can now state a second result.

Proposition 6.3 *Setting*

$$\widetilde{\mathscr{RF}}(-) := \text{ the restriction of } \widehat{\mathscr{RF}}(-) \text{ to } \widetilde{\widehat{\mathbf{Groups}}},$$

we obtain a covariant functor $\widetilde{\mathscr{RF}}(-)$ on the category $\widetilde{\widehat{\mathbf{Groups}}}$ of groups and embeddings; in particular, $G \cong H$ implies $\mathscr{RF}(G) \cong \mathscr{RF}(H)$. Moreover, every automorphism α of G extends to an automorphism $\hat{\alpha}$ of $\mathscr{RF}(G)$ (identifying G with G_0) in such a way that mapping α to $\hat{\alpha}$ gives an embedding of $\mathrm{Aut}(G)$ *into* $\mathrm{Aut}(\mathscr{RF}(G))$.

Proof The first statement is a consequence of (IX) and the particular statement follows immediately from the functoriality of $\widetilde{\mathscr{RF}}(-)$. Further, if α is an automorphism of G then $\hat{\alpha} := \widetilde{\mathscr{RF}}(\alpha)$ is an automorphism of $\mathscr{RF}(G)$ such that $\hat{\alpha}|_{G_0} = \alpha$. The mapping given by $\alpha \mapsto \hat{\alpha}$ is a homomorphism from $\mathrm{Aut}(G)$ to $\mathrm{Aut}(\mathscr{RF}(G))$ since $\widetilde{\mathscr{RF}}(-)$ is a covariant functor, and the fact that $\hat{\alpha}$ extends α shows that this homomorphism is an embedding. □

At this point we wish to return briefly to the universality property of $\mathscr{RF}$-groups and their associated $\mathbb{R}$-trees, that is, to Theorem 4.10. If $\iota : G \to H$ is a group embedding then, according to Proposition 6.3, $\widetilde{\mathscr{RF}}(\iota)$ is a group embedding of $\mathscr{RF}(G)$ into $\mathscr{RF}(H)$. Moreover, $\widetilde{\mathscr{RF}}(\iota)$ is length-preserving; hence, by Theorem 4.6 in Chapter 2 of Chiswell [10], there is a unique $\mathscr{RF}(G)$-equivariant isometry $\mu : \mathbf{X}_G \to \mathbf{X}_H$ and μ maps the base-point of $\mathbf{X}_G$ to the base-point of $\mathbf{X}_H$.

Now suppose that we are given a group G acting freely on an $\mathbb{R}$-tree $\mathbf{X}$ by means of a homomorphism $\varphi : G \to \mathrm{Isom}(\mathbf{X})$. Then we can form first the group $\hat{G}$ and then the quotient set $S = \hat{G}/\approx$, as explained in Chapter 4. Let us denote the cardinality $|\hat{G}/\approx|$ of the set S, which at the same time is the (cardinal) number of directions of the $\mathbb{R}$-tree $\widehat{\mathbf{X}}$ (that is, the number of directions of $\widehat{\mathbf{X}}$ at any given point), by $\kappa_{\mathbf{X},\varphi}$. Note that Lemma 4.2 provides an upper bound for $\kappa_{\mathbf{X},\varphi}$ in terms of the cardinality of G and that of the number of orbits of G on $\mathbf{X}$.

Consider a group K containing a subgroup H_0 of cardinality $\kappa_{\mathbf{X},\varphi}$ and let $\psi : H_0 \to S$ be any bijection transporting the structure of H_0 to S, thus turning S into a group H. By Theorem 4.10 we have a group embedding $\chi : G \to \mathscr{RF}(H)$ and a G-equivariant isometry $\lambda : \mathbf{X} \to \mathbf{X}_H$ mapping a given base-point of $\mathbf{X}$ to the canonical base-point of $\mathbf{X}_H$. Further, by the discussion above, we also have a group embedding $\chi' : \mathscr{RF}(H) \to \mathscr{RF}(K)$ and an $\mathscr{RF}(H)$-equivariant isometry $\lambda' : \mathbf{X}_H \to \mathbf{X}_K$ mapping base-point to base-point. Combining these maps we obtain a group embedding $\tilde{\chi} = \chi' \circ \chi : G \to \mathscr{RF}(K)$ and a G-equivariant isometry $\tilde{\lambda} = \lambda' \circ \lambda : \mathbf{X} \to \mathbf{X}_K$ mapping a given base-point of $\mathbf{X}$ to the canonical base-point of $\mathbf{X}_K$. Thus, we have established the following *strong version* of the universality principle for $\mathscr{RF}$-groups and their associated $\mathbb{R}$-trees.

Theorem 6.4 *Let G be a group acting freely on an $\mathbb{R}$-tree $\mathbf{X}$ by means of a homomorphism $\varphi : G \to \mathrm{Isom}(\mathbf{X})$, and let $\kappa_{\mathbf{X},\varphi} := |\hat{G}/\approx|$ be the cardinal number of directions of the tree $\widehat{\mathbf{X}}$. Moreover, let $\{b_i : i \in I\}$ be a set of representatives for the G-orbits on $\mathbf{X}$. Then we have the following.*

(i) $$\kappa_{\mathbf{X},\varphi} \leq \begin{cases} \aleph_0, & G = 1 \text{ and } 1 < |I| < \infty, \\ \max\{|G|, |I|\}, & \text{otherwise.} \end{cases}$$

(ii) *For every group H containing a subgroup of cardinality $\kappa_{\mathbf{X},\varphi}$, there exists an injective group homomorphism $\chi : G \to \mathscr{RF}(H)$ and a G-equivariant isometry $\lambda : \mathbf{X} \to \mathbf{X}_H$ whose image contains the base-point of $\mathbf{X}_H$.*

6.4 The functor $\widehat{\mathscr{RF}}_0(-)$

The purpose of this section is to construct the functor $\widehat{\mathscr{RF}}_0(-)$ mentioned in the introduction, thus in particular establishing the rigidity theorem explained there. For this, we shall need to continue our analysis of the hat notation.

(X) *For $\varphi \in \mathrm{Mor}_{\widehat{\mathbf{Groups}}}(G,H)$ and $f_1, f_2 \in \mathscr{F}(G)$ we have*

$$\hat{\varphi}(f_1 * f_2) = \hat{\varphi}(f_1) * h_{f_1,f_2} * \hat{\varphi}(f_2), \tag{6.3}$$

where $L(h_{f_1,f_2}) = 0$ and

$$h_{f_1,f_2}(0) = \left(\varphi(f_1(L(f_1)))\right)^{-1} \varphi\left(f_1(L(f_1))f_2(0)\right) \left(\varphi(f_2(0))\right)^{-1}.$$

In particular we have

$$\hat{\varphi}(f_1 * f_2) = \hat{\varphi}(f_1) * \hat{\varphi}(f_2),$$

provided that $f_1(L(f_1)) = 1_G$ or $f_2(0) = 1_G$.

Proof By assertion (I), both sides of equation (6.3) have length equal to $L(f_1) + L(f_2)$. Moreover, for $0 \le \xi \le L(f_1) + L(f_2)$ we have

$$\hat{\varphi}(f_1 * f_2)(\xi) = \begin{cases} \varphi(f_1(\xi)), & 0 \le \xi < L(f_1), \\ \varphi(f_1(L(f_1))f_2(0)), & \xi = L(f_1), \\ \varphi(f_2(\xi - L(f_1))), & L(f_1) < \xi \le L(f_1) + L(f_2), \end{cases}$$

$$= \begin{cases} \hat{\varphi}(f_1)(\xi), & 0 \le \xi < L(f_1), \\ \varphi(f_1(L(f_1))f_2(0)), & \xi = L(f_1), \\ \hat{\varphi}(f_2)(\xi - L(\hat{\varphi}(f_1))), & L(f_1) < \xi \le L(f_1) + L(f_2), \end{cases}$$

$$= \left(\hat{\varphi}(f_1) * h_{f_1,f_2} * \hat{\varphi}(f_2)\right)(\xi),$$

where h_{f_1,f_2} is as defined above. □

(XI) *For* $\varphi \in \mathrm{Mor}_{\widehat{\mathbf{Groups}}}(G,H)$ *and* $f_1, f_2 \in \mathscr{F}(G)$ *with* $\varepsilon_0(f_1, f_2) = 0$, *we have*

$$\hat{\varphi}(f_1 \circ f_2) = \hat{\varphi}(f_1) \circ h_{f_1,f_2} \circ \hat{\varphi}(f_2),$$

where h_{f_1,f_2} *is defined as in* (X); *in particular, if* $f_1(L(f_1)) = 1_G$ *or* $f_2(0) = 1_G$ *then*

$$\hat{\varphi}(f_1 \circ f_2) = \hat{\varphi}(f_1) \circ \hat{\varphi}(f_2).$$

Proof By assertion (X) we have

$$\hat{\varphi}(f_1 \circ f_2) = \hat{\varphi}(f_1) * h_{f_1,f_2} * \hat{\varphi}(f_2).$$

Next, $\varepsilon_0(\hat{\varphi}(f_1), h_{f_1,f_2}) = 0$ since $L(h_{f_1,f_2}) = 0$. We claim that also

$$\varepsilon_0(\hat{\varphi}(f_1) \circ h_{f_1,f_2}, \hat{\varphi}(f_2)) = 0, \tag{6.4}$$

which would finish the proof of (XI). This is certainly true if

$$\hat{\varphi}(f_1)(L(f_1))h_{f_1,f_2}(0) \neq \left(\hat{\varphi}(f_2)(0)\right)^{-1};$$

thus we may suppose that

$$\hat{\varphi}(f_1)(L(f_1))h_{f_1,f_2}(0)\hat{\varphi}(f_2)(0) = 1_H, \tag{6.5}$$

so that

$$\varepsilon_0(\hat{\varphi}(f_1) \circ h_{f_1,f_2}, \hat{\varphi}(f_2)) = \sup \mathscr{E}(\hat{\varphi}(f_1) \circ h_{f_1,f_2}, \hat{\varphi}(f_2)).$$

By the definition of h_{f_1,f_2}, equation (6.5) yields

$$\varphi(f_1(L(f_1))f_2(0)) = 1_H,$$

implying that

$$f_1(L(f_1))f_2(0) = 1_G$$

by properties (i) and (ii) of φ, and so $\varepsilon_0(f_1, f_2) = \sup \mathscr{E}(f_1, f_2)$. Since $\varepsilon_0(f_1, f_2) = 0$ by assumption, we conclude that, for every ε with $0 < \varepsilon \leq \min\{L(f_1), L(f_2)\}$, there exists some $\eta = \eta(\varepsilon)$ such that $0 < \eta \leq \varepsilon$ and such that

$$f_1(L(f_1) - \eta) \neq (f_2(\eta))^{-1}.$$

By properties (i) and (iii) of φ and the definition of the circle operation, this implies that

$$\left(\hat{\varphi}(f_1) \circ h_{f_1,f_2}\right)(L(f_1) - \eta) = \hat{\varphi}(f_1)(L(f_1) - \eta) \neq \left(\hat{\varphi}(f_2)(\eta)\right)^{-1}$$

for this $\eta(\varepsilon)$, whence (6.4). The particular statement follows from the definition of the element h_{f_1,f_2}. □

(XII) *$f \in \mathscr{RF}(G)$ is elliptic $\Longleftrightarrow$ $\hat{\varphi}(f) \in \mathscr{RF}(H)$ is elliptic.*

Proof Write $f = t \circ f_1 \circ t^{-1}$ with $t, f_1 \in \mathscr{RF}(G)$ and f_1 cyclically reduced according to Lemma 3.7. Applying Corollary 2.18 (associativity of the circle product) together with assertions (IV) and (XI), we find that

$$\begin{aligned}\hat{\varphi}(f) &= \hat{\varphi}((t \circ f_1) \circ t^{-1})\\ &= \hat{\varphi}(t \circ f_1) \circ h_{t\circ f_1, t^{-1}} \circ \hat{\varphi}(t^{-1})\\ &= \hat{\varphi}(t) \circ \left(h_{t,f_1} \circ \hat{\varphi}(f_1) \circ h_{t\circ f_1, t^{-1}}\right) \circ (\hat{\varphi}(t))^{-1}.\end{aligned}$$

If f is elliptic then $L(f_1) = 0$ by Proposition 3.13; hence, by assertion (I),

$$L(h_{t,f_1} \circ \hat{\varphi}(f_1) \circ h_{t\circ f_1, t^{-1}}) = 0$$

and thus $\hat{\varphi}(f)$ is elliptic, again by Proposition 3.13. Now suppose that f is hyperbolic; hence $L(f_1) > 0$ by Proposition 3.13. We claim that the element

$$h_{t,f_1} \circ \hat{\varphi}(f_1) \circ h_{t\circ f_1, t^{-1}}$$

of $\mathscr{RF}(H)$ is again cyclically reduced, so that Proposition 3.13 allows us to conclude that $\hat{\varphi}(f)$ is hyperbolic as required. Our claim is certainly true if

$$\hat{\varphi}(f_1)(L(f_1))\, h_{t\circ f_1, t^{-1}}(0) \neq \left(h_{t,f_1}(0)\, \hat{\varphi}(f_1)(0)\right)^{-1};$$

hence we may suppose that

$$\hat{\varphi}(f_1)(L(f_1))\, h_{t\circ f_1, t^{-1}}(0)\, h_{t,f_1}(0)\, \hat{\varphi}(f_1)(0) = 1_H, \tag{6.6}$$

so that

$$\begin{aligned}&\varepsilon_0(h_{t,f_1} \circ \hat{\varphi}(f_1) \circ h_{t\circ f_1, t^{-1}}, h_{t,f_1} \circ \hat{\varphi}(f_1) \circ h_{t\circ f_1, t^{-1}})\\ &\qquad\qquad = \sup \mathscr{E}(h_{t,f_1} \circ \hat{\varphi}(f_1) \circ h_{t\circ f_1, t^{-1}}, h_{t,f_1} \circ \hat{\varphi}(f_1) \circ h_{t\circ f_1, t^{-1}}).\end{aligned}$$

By definition of the elements h_{t,f_1} and $h_{t\circ f_1, t^{-1}}$ and property (iii) of φ, we have

$$h_{t,f_1}(0) = \left(\varphi(t(L(t)))\right)^{-1} \varphi(t(L(t)) f_1(0)) \left(\varphi(f_1(0))\right)^{-1}$$

and

$$\begin{aligned}h_{t\circ f_1, t^{-1}}(0) &= \left(\varphi((t\circ f_1)(L(t\circ f_1)))\right)^{-1} \varphi((t\circ f_1)(L(t\circ f_1)) t^{-1}(0)) \left(\varphi(t^{-1}(0))\right)^{-1}\\ &= \left(\varphi(f_1(L(f_1)))\right)^{-1} \varphi(f_1(L(f_1))(t(L(t)))^{-1})\, \varphi(t(L(t))).\end{aligned}$$

Putting the right-hand sides of the last two equations back into (6.6), we find after some simplification that

$$\varphi(f_1(L(f_1))(t(L(t)))^{-1})\,\varphi(t(L(t))f_1(0)) = 1_H.$$

Rewriting the last equation as

$$\varphi(f_1(L(f_1))(t(L(t)))^{-1}) = \varphi((t(L(t))f_1(0))^{-1})$$

by means of property (iii), the injectivity of φ yields

$$f_1(L(f_1))f_1(0) = 1_G,$$

so that $\varepsilon_0(f_1,f_1) = \sup\mathscr{E}(f_1,f_1)$. Just as in the proof of (XI), the fact that $\varepsilon_0(f_1,f_1) = 0$ provides, for each ε with $0 < \varepsilon < L(f_1)$, a real number $\eta = \eta(\varepsilon)$ such that $0 < \eta \leq \varepsilon$ and such that

$$f_1(L(f_1)-\eta) \neq (f_1(\eta))^{-1}.$$

Applying φ to the last equation and using properties (i) and (iii), we find that

$$\hat{\varphi}(f_1)(L(\hat{\varphi}(f_1))-\eta) \neq (\hat{\varphi}(f_1)(\eta))^{-1},$$

and so

$$(h_{t,f_1}\circ\hat{\varphi}(f_1)\circ h_{t\circ f_1,t^{-1}})(L(\hat{\varphi}(f_1)-\eta)) \neq ((h_{t,f_1}\circ\hat{\varphi}(f_1)\circ h_{t\circ f_1,t^{-1}})(\eta))^{-1}.$$

It follows that indeed

$$\varepsilon_0(h_{t,f_1}\circ\hat{\varphi}(f_1)\circ h_{t\circ f_1,t^{-1}}, h_{t,f_1}\circ\hat{\varphi}(f_1)\circ h_{t\circ f_1,t^{-1}}) = 0,$$

and the proof is complete. □

As another application of assertion (XI), we can show the crucial fact that $\hat{\varphi}|_{\mathscr{RF}(G)}$ always respects reduced multiplication 'up to an elliptic element'.

(XIII) *For $\varphi \in \mathrm{Mor}_{\widehat{\mathbf{Groups}}}(G,H)$ and $f_1, f_2 \in \mathscr{RF}(G)$, we have*

$$\hat{\varphi}(f_1 f_2) = \hat{\varphi}(f_1)\,e_H(f_1,f_2)\,\hat{\varphi}(f_2),$$

where $e_H(f_1,f_2)$ is an elliptic element of $\mathscr{RF}(H)$.

Proof Using Lemma 2.15, there are decompositions $f_1 = f_1'\circ u$ and $f_2 = u^{-1}\circ f_2'$ with $f_1', f_2', u \in \mathscr{RF}(G)$, such that $L(u) = \varepsilon_0(f_1,f_2)$ and $f_1 f_2 = f_1'\circ f_2'$. Adapting $f_1'(L(f_1'))$ and $f_2'(0)$ if necessary, we can ensure that $u(0) = 1_G$ while maintaining the equations $f_1 = f_1'\circ u$ and $f_2 = u^{-1}\circ f_2'$, and Lemma 2.16

applies to show that we still have $f_1 f_2 = f_1' \circ f_2'$. By (IV) and the particular statement in (XI),

$$\begin{aligned}\hat{\varphi}(f_1) &= \hat{\varphi}(f_1' \circ u)\\ &= \hat{\varphi}(f_1') \circ \hat{\varphi}(u)\\ &= \hat{\varphi}(f_1')\hat{\varphi}(u)\end{aligned}$$

and

$$\begin{aligned}\hat{\varphi}(f_2) &= \hat{\varphi}(u^{-1} \circ f_2')\\ &= \hat{\varphi}(u^{-1}) \circ \hat{\varphi}(f_2')\\ &= (\hat{\varphi}(u))^{-1} \circ \hat{\varphi}(f_2')\\ &= (\hat{\varphi}(u))^{-1}\hat{\varphi}(f_2').\end{aligned}$$

Consequently, by the general statement in (XI),

$$\begin{aligned}\hat{\varphi}(f_1 f_2) &= \hat{\varphi}(f_1' \circ f_2')\\ &= \hat{\varphi}(f_1') \circ h_{f_1', f_2'} \circ \hat{\varphi}(f_2')\\ &= \hat{\varphi}(f_1')\hat{\varphi}(u)(\hat{\varphi}(u))^{-1} h_{f_1', f_2'} \hat{\varphi}(u)(\hat{\varphi}(u))^{-1}\hat{\varphi}(f_2')\\ &= \hat{\varphi}(f_1)\, e_H(f_1, f_2)\, \hat{\varphi}(f_2),\end{aligned}$$

where

$$e_H(f_1, f_2) := (\hat{\varphi}(u))^{-1} h_{f_1', f_2'} \hat{\varphi}(u)$$

is an elliptic element of $\mathscr{RF}(H)$, since $h_{f_1', f_2'} \in H_0$. □

We shall need two further auxiliary results before we are ready for the main result of this chapter.

(XIV) *Let* $\varphi \in \mathrm{Mor}_{\widehat{\mathbf{Groups}}}(G, H)$, *let* f, t *be elements of* $\mathscr{RF}(G)$, *and suppose that* $\hat{\varphi}(f) \in E(H)$. *Then* $\hat{\varphi}(tft^{-1}) \in E(H)$.

Proof Applying (XIII) twice and using (IV), we find that

$$\begin{aligned}\hat{\varphi}(tft^{-1}) &= \hat{\varphi}((tf)t^{-1})\\ &= \hat{\varphi}(tf)\, e_H(tf, t^{-1})\hat{\varphi}(t^{-1})\\ &= \hat{\varphi}(t)\, e_H(t, f)\, \hat{\varphi}(f)\, e_H(tf, t^{-1})(\hat{\varphi}(t))^{-1},\end{aligned}$$

which is contained in $E(H)$, since $e_H(t,f), e_H(tf,t^{-1}), \hat{\varphi}(f) \in E(H)$ and since $E(H)$ is normal in $\mathscr{RF}(H)$. $\square$

By definition, every element of the subgroup $E(G)$ is a product of finitely many elliptic elements; we define the *complexity* $\chi(f)$ of $f \in E(G)$ to be the minimal number of elliptic elements required to represent f as such a product and, for a non-negative integer k, let

$$E(G)_k := \{f \in E(G) : \chi(f) = k\}.$$

Clearly, each set $E(G)_k$ is normal in $\mathscr{RF}(G)$, and we have

$$E(G) = \bigcup_{k \geq 0} E(G)_k.$$

(XV) *For* $\varphi \in \mathrm{Mor}_{\widehat{\mathbf{Groups}}}(G,H)$, *we have* $\hat{\varphi}(E(G)) \subseteq E(H)$.

Proof We use induction on k to show that, for every non-negative integer k,

$$\hat{\varphi}(E(G)_k) \subseteq E(H). \tag{6.7}$$

For $k = 0$, this is true in view of (III), while (XII) in particular ensures the validity of (6.7) for $k = 1$. Let $K \geq 2$, suppose inductively that assertion (6.7) holds with k replaced by $K - 1$, and let $f \in E(G)_K$. Write $f = f_1 t g t^{-1}$, where $f_1 \in E(G)_{K-1}$ and $g \in G_0 \setminus \{\mathbf{1}_G\}$. Then $\hat{\varphi}(f_1) \in E(H)$ by the inductive hypothesis, and $\hat{\varphi}(t^{-1} f_1 t) \in E(H)$, arguing either by the inductive hypothesis plus the normality of $E(G)_{K-1}$ or by (XIV). Hence

$$\hat{\varphi}(t^{-1} f_1 t g) \in E(H),$$

since the left-hand side differs from $\hat{\varphi}(t^{-1} f_1 t)$ only at the endpoint, so that

$$\hat{\varphi}(t^{-1} f_1 t g) \in E(H) H_0 = E(H);$$

finally,

$$\hat{\varphi}(f) = \hat{\varphi}(t \cdot t^{-1} f_1 t g \cdot t^{-1}) \in E(H)$$

by (XIV). Since f is an arbitrary element of $E(G)_K$ we conclude that

$$\hat{\varphi}(E(G)_K) \subseteq E(H),$$

so that (6.7) holds for every $k \in \mathbb{N}_0$, whence our claim. $\square$

We are now in a position to establish the main result of this chapter.

Theorem 6.5

(i) *Defining*

$$\underline{\hat{\varphi}}(fE(G)) := \hat{\varphi}(f)E(H), \quad \varphi \in \mathrm{Mor}_{\widehat{\mathbf{Groups}}}(G,H),$$

we obtain a well-defined group homomorphism

$$\underline{\hat{\varphi}} : \mathscr{RF}(G)/E(G) \to \mathscr{RF}(H)/E(H).$$

(ii) *Setting*

$$\widehat{\mathscr{RF}}_0(G) := \mathscr{RF}(G)/E(G), \quad G \in |\widehat{\mathbf{Groups}}|$$

and

$$\widehat{\mathscr{RF}}_0(\varphi) := \underline{\hat{\varphi}}, \quad \varphi \in \mathrm{Mor}_{\widehat{\mathbf{Groups}}}(G,H),$$

we obtain a covariant functor

$$\widehat{\mathscr{RF}}_0(-) : \widehat{\mathbf{Groups}} \to \mathbf{Groups}.$$

Proof (i) For $e \in E(G)$, we have

$$\hat{\varphi}(fe) = \hat{\varphi}(f)e_H(f,e)\hat{\varphi}(e)$$

by assertion (XIII); thus

$$\hat{\varphi}(fe)E(H) = \hat{\varphi}(f)E(H),$$

since $\hat{\varphi}(e) \in E(H)$ by assertion (XV). Hence, the map $\underline{\hat{\varphi}}$ is well defined.

Next, let $f_1, f_2 \in \mathscr{RF}(G)$ be given. Then, using assertion (XIII), we have

$$\begin{aligned}
\underline{\hat{\varphi}}(f_1E(G)\cdot f_2E(G)) &= \underline{\hat{\varphi}}(f_1f_2E(G)) \\
&= \hat{\varphi}(f_1f_2)E(H) \\
&= \hat{\varphi}(f_1)e_H(f_1,f_2)\hat{\varphi}(f_2)E(H) \\
&= \hat{\varphi}(f_1)e_H(f_1,f_2)E(H)\cdot\hat{\varphi}(f_2)E(H) \\
&= \hat{\varphi}(f_1)E(H)\cdot\hat{\varphi}(f_2)E(H) \\
&= \underline{\hat{\varphi}}(f_1E(G))\cdot\underline{\hat{\varphi}}(f_2E(G));
\end{aligned}$$

that is, $\underline{\hat{\varphi}}$ is a group homomorphism, as claimed.

(ii) In view of part (i), in order to show our claim concerning the functoriality of $\widehat{\mathscr{RF}}_0(-)$ it remains only to check that

$$\underline{\widehat{\mathrm{id}}_G} = \mathrm{id}_{\mathscr{RF}(G)/E(G)}$$

and that, for morphisms $\varphi : G \to H$ and $\psi : H \to K$ in the category $\widehat{\mathbf{Groups}}$, we have

$$\underline{\widehat{\psi \circ \varphi}} = \underline{\hat{\varphi}} \circ \underline{\hat{\varphi}};$$

however, these equations are immediate from the definition of the map $\underline{\hat{\varphi}}$ given in part (i). □

Combining Theorem 6.5 with assertion (VI)(b), we obtain the following important consequence.[1]

Corollary 6.6 **(The rigidity theorem)** *Let G and H be groups such that*

$$|\mathrm{Inv}(G)| = |\mathrm{Inv}(H)| \quad \textit{and} \quad |G - \mathrm{Inv}(G)| = |H - \mathrm{Inv}(H)|.$$

Then we have a group isomorphism

$$\mathscr{RF}(G)/E(G) \cong \mathscr{RF}(H)/E(H).$$

6.5 A remark concerning the automorphism group of $\mathscr{RF}(G)/E(G)$

By Corollary 5.9 the group $\mathscr{RF}(G)/E(G)$ is 'usually' large, and Corollary 6.6, while not actually determining the structure of this quotient, makes a non-trivial assertion concerning the isomorphism type of $\mathscr{RF}(G)/E(G)$ (namely that it depends only on the two cardinal numbers $|\mathrm{Inv}(G)|$ and $|G - \mathrm{Inv}(G)|$ and not on the structure of G itself). Our next result exhibits a certain, usually non-trivial, subgroup of the automorphism group of $\mathscr{RF}(G)/E(G)$.

For a group G, denote by $S^+(G)$ the set of all permutations φ on G such that

$$\varphi(1_G) = 1_G$$

and

$$\varphi(x^{-1}) = (\varphi(x))^{-1}, \quad x \in G.$$

[1] This result was first conjectured by J.-C. Schlage-Puchta and was subsequently proved by the second-named author of the present text.

Clearly, $S^+(G)$ forms a group under composition; in fact, it is the automorphism group of the group G in the category $\widehat{\mathbf{Groups}}$. As before, let R_G be a system of representatives for the equivalence relation $\sim$ on $G - \operatorname{Inv}(G)$ given by

$$x \sim y :\Longleftrightarrow x = y \text{ or } x = y^{-1};$$

that is, we choose an element from each pair $\{x, x^{-1}\}$ of inverses in $G - \operatorname{Inv}(G)$. Our final result in this chapter in particular determines the structure of the group $S^+(G)$.

Proposition 6.7 *Let G be a group.*

(i) *We have a group isomorphism*[2]

$$S^+(G) \cong \operatorname{Sym}\big(\operatorname{Inv}(G) - \{1_G\}\big) \times \big(C_2 \wr \operatorname{Sym}(R_G)\big). \tag{6.8}$$

(ii) *Sending φ to $\underline{\hat{\varphi}}$, the group $S^+(G)$ is mapped homomorphically into $\operatorname{Aut}(\mathscr{RF}(G)/E(G))$. Moreover, if $\underline{\hat{\varphi}} = \mathrm{id}_{\mathscr{RF}(G)/E(G)}$ then φ fixes every non-involution of G.*

Proof (i) A map $\varphi \in S^+(G)$ gives rise to a triple $(\sigma, \varepsilon, \pi)$, where σ is a permutation of $\operatorname{Inv}(G) - \{1_G\}$ (the restriction of φ to the proper involutions of G), π is a permutation of R_G (describing the assignment of inverse pairs in $G - \operatorname{Inv}(G)$ effected by φ), and $\varepsilon : R_G \to \{1, -1\}$ determines to which of the two possible images in $\{\pi(x), (\pi(x))^{-1}\}$ a representative $x \in R$ is sent under φ; more precisely,

$$\varepsilon(x) = \left\{ \begin{array}{ll} +1, & \varphi(x) \in R_G \\ -1, & \varphi(x) \notin R_G \end{array} \right\} \quad (x \in R_G).$$

Conversely, given such a triple $(\sigma, \varepsilon, \pi)$, we obtain a map $\varphi_{(\sigma,\varepsilon,\pi)} \in S^+(G)$ by setting

$$\varphi_{(\sigma,\varepsilon,\pi)}(x) := \left\{ \begin{array}{ll} 1_G, & x = 1_G \\ \sigma(x), & x \in \operatorname{Inv}(G) - \{1_G\} \\ (\pi(x))^{\varepsilon(x)}, & x \in R_G \\ (\pi(x^{-1}))^{-\varepsilon(x^{-1})}, & x \in R_G^{-1} \end{array} \right\} \quad (x \in G).$$

[2] The wreath product on the right-hand side of (6.8) is formed using the natural permutation representation of $\operatorname{Sym}(R_G)$.

Clearly, the maps given by $\varphi \mapsto (\sigma, \varepsilon, \pi)$ and $(\sigma, \varepsilon, \pi) \mapsto \varphi_{(\sigma,\varepsilon,\pi)}$ are inverses of each other, hence are bijections.

Next, we study the effect of composition on the triples $(\sigma, \varepsilon, \pi)$. We have, for $x \in G$,

$$\varphi_{(\sigma_2,\varepsilon_2,\pi_2)}\big(\varphi_{(\sigma_1,\varepsilon_1,\pi_1)}(x)\big) = \varphi_{(\sigma_2,\varepsilon_2,\pi_2)}\left(\begin{cases} 1_G, & x = 1_G \\ \sigma_1(x), & x \in \mathrm{Inv}(G) - \{1_G\} \\ (\pi_1(x))^{\varepsilon_1(x)}, & x \in R_G \\ (\pi_1(x^{-1}))^{-\varepsilon_1(x^{-1})}, & x \in R_G^{-1} \end{cases}\right)$$

$$= \begin{cases} 1_G, & x = 1_G, \\ \sigma_2(\sigma_1(x)), & x \in \mathrm{Inv}(G) - \{1_G\}, \\ (\pi_2(\pi_1(x)))^{\varepsilon_2(\pi_1(x))}, & x \in R_G,\ \varepsilon_1(x) = 1, \\ (\pi_2(\pi_1(x)))^{-\varepsilon_2(\pi_1(x))}, & x \in R_G,\ \varepsilon_1(x) = -1, \\ (\pi_2(\pi_1(x^{-1})))^{-\varepsilon_2(\pi_1(x^{-1}))}, & x \in R_G^{-1},\ \varepsilon_1(x^{-1}) = 1, \\ (\pi_2(\pi_1(x^{-1})))^{\varepsilon_2(\pi_1(x^{-1}))}, & x \in R_G^{-1},\ \varepsilon_1(x^{-1}) = -1, \end{cases}$$

$$= \begin{cases} 1_G, & x = 1_G, \\ (\sigma_2 \circ \sigma_1)(x), & x \in \mathrm{Inv}(G) - \{1_G\}, \\ ((\pi_2 \circ \pi_1)(x))^{\varepsilon_1(x)\varepsilon_2(\pi_1(x))}, & x \in R_G, \\ ((\pi_2 \circ \pi_1)(x^{-1}))^{-\varepsilon_1(x^{-1})\varepsilon_2(\pi_1(x^{-1}))}, & x \in R_G^{-1}, \end{cases}$$

$$= \varphi_{(\sigma_2\circ\sigma_1,\varepsilon,\pi_2\circ\pi_1)}(x),$$

where

$$\varepsilon(x) := \varepsilon_1(x)\varepsilon_2(\pi_1(x)), \quad x \in R_G.$$

Hence, the assignment $(\sigma, \varepsilon, \pi) \mapsto \varphi_{(\sigma,\varepsilon,\pi)}$, when interpreted as a map

$$\mathrm{Sym}(\mathrm{Inv}(G) - \{1_G\}) \times \big(C_2 \wr \mathrm{Sym}(R_G)\big) \to S^+(G),$$

is an isomorphism (in the category **Groups**), whence (6.8).

(ii) The fact that the map

$$\Psi_G : S^+(G) \to \mathrm{Aut}(\mathscr{RF}(G)/E(G))$$

given by

$$\Psi_G(\varphi) := \widehat{\mathscr{RF}}_0(\varphi) = \underline{\hat{\varphi}}, \quad \varphi \in S^+(G),$$

is a homomorphism is immediate from the assertion of part (ii) of Theorem 6.5, that is, the functoriality of $\widehat{\mathscr{RF}}_0(-)$.

Finally, suppose that

$$\Psi_G(\varphi) = \mathrm{id}_{\mathscr{RF}(G)/E(G)},$$

that is,

$$f^{-1}\hat{\varphi}(f) \in E(G), \quad f \in \mathscr{RF}(G). \tag{6.9}$$

If $x \in G - \mathrm{Inv}(G)$ and $\varphi(x) \neq x$, define an element $f \in \mathscr{RF}(G)$ with $L(f) = 1$ via

$$f(\xi) = x, \quad 0 \le \xi \le 1.$$

Then $\hat{\varphi}(f)$ has length 1 and satisfies

$$\hat{\varphi}(f)(\xi) = \varphi(x), \quad 0 \le \xi \le 1,$$

and we have

$$\varepsilon_0(f^{-1}, \hat{\varphi}(f)) = 0$$

since

$$f^{-1}(1)\hat{\varphi}(f)(0) = x^{-1}\varphi(x) \neq 1_G.$$

We now distinguish two cases.

(1) $\varphi(x) = x^{-1}$. Then, by part (ii) of Lemma 5.5,

$$\begin{aligned}\mu_{x^{-1}}(f^{-1}\hat{\varphi}(f)) &= \mu_{x^{-1}}(f^{-1} \circ \hat{\varphi}(f)) \\ &= \mu_{x^{-1}}(f^{-1}) + \mu_{x^{-1}}(\hat{\varphi}(f)) \\ &= 2,\end{aligned}$$

while

$$\begin{aligned}\mu_x(f^{-1}\hat{\varphi}(f)) &= \mu_x(f^{-1} \circ \hat{\varphi}(f)) \\ &= \mu_x(f^{-1}) + \mu_x(\hat{\varphi}(f)) \\ &= 0.\end{aligned}$$

It follows that

$$e_x(f^{-1}\hat{\varphi}(f)) = -2.$$

(2) $\varphi(x) \not\sim x$. Then we still have

$$\mu_x(f^{-1}\hat{\varphi}(f)) = 0,$$

while

$$\mu_{x^{-1}}(f^{-1}\hat{\varphi}(f)) = 1,$$

so that

$$e_x(f^{-1}\hat{\varphi}(f)) = -1.$$

In both cases, we have found that $e_x(f^{-1}\hat{\varphi}(f)) \neq 0$, and, since $e_x(E(G)) = 0$, we conclude that

$$f^{-1}\hat{\varphi}(f) \notin E(G),$$

contradicting (6.9). We have thus shown that $\varphi \in \ker(\Psi_G)$ fixes non-involutions of G pointwise, as claimed. □

From part (ii) of Proposition 6.7, we obtain the following.

Corollary 6.8 *Suppose that the group G does not have non-trivial involutions; that is, that* $\mathrm{Inv}(G) = \{1_G\}$. *Then $S^+(G)$ embeds into* $\mathrm{Aut}(\mathscr{RF}(G)/E(G))$ *via the map Ψ_G.*

6.6 Exercises

6.1. Show that the functor $\widetilde{\mathscr{RF}}(-)$ is faithful, that is, that the maps

$$\widetilde{\mathscr{RF}}(-)_{G,H} : \mathrm{Mor}_{\widetilde{\mathbf{Groups}}}(G,H) \longrightarrow \mathrm{Mor}_{\widetilde{\mathbf{Groups}}}(\mathscr{RF}(G), \mathscr{RF}(H))$$

are injective.

7

Conjugacy of hyperbolic elements

7.1 Introduction

Two elliptic elements $a = sgs^{-1}$ and $b = tht^{-1}$, with $g,h \in G_0 - \{\mathbf{1}_G\}$, are conjugate in $\mathscr{RF}(G)$ if and only if g and h are conjugate in G_0; see Exercise 7.1. Hence, nothing further can be said in general about the conjugacy of the elliptic elements in $\mathscr{RF}(G)$ without restricting or specifying the group G. For the hyperbolic elements, however, the situation is different and much more interesting.

Let us begin by recalling the solution of the corresponding (conjugacy) problem for free groups, following the account in Section 1.4 of Magnus, Karrass, and Solitar [31]. Let F be a free group with basis $X = \{x_1, x_2, \ldots, x_n\}$. The first step is to introduce a specific process σ for cyclically reducing an arbitrary word w in the free generators x_i. Roughly speaking, σ cyclically reduces w by first freely reducing it and then cancelling the first and last symbols, if necessary. For instance,

$$\begin{aligned}
\sigma(x_1x_2x_3x_3^{-1}x_2x_1x_2^{-1}x_1^{-1}) &= \sigma(x_1x_2^2x_1x_2^{-1}x_1^{-1}) \\
&= \sigma(x_2^2x_1x_2^{-1}) \\
&= \sigma(x_2x_1) \\
&= x_2x_1.
\end{aligned}$$

More precisely, one first introduces a process ρ which freely reduces a given word w in the generators x_i by going through the word w from left to right and deleting every inverse pair of the form $x_i^{\varepsilon}x_i^{-\varepsilon}$ when we first encounter it. For instance, in order to compute the word

$$\rho(x_1x_2^{-1}x_3x_3^{-1}x_2x_2^{-1}),$$

one successively computes

$$\begin{aligned}
\rho(x_1) &= x_1,\\
\rho(x_1x_2^{-1}) &= x_1x_2^{-1},\\
\rho(x_1x_2^{-1}x_3) &= x_1x_2^{-1}x_3,\\
\rho(x_1x_2^{-1}x_3x_3^{-1}) &= x_1x_2^{-1},\\
\rho(x_1x_2^{-1}x_3x_3^{-1}x_2) &= x_1,\\
\rho(x_1x_2^{-1}x_3x_3^{-1}x_2x_2^{-1}) &= x_1x_2^{-1}.
\end{aligned}$$

In general, ρ is defined by induction on the word length via

$$\rho(w) = \begin{cases} 1, & w = 1,\\ x_i^{\varepsilon}, & w = x_i^{\varepsilon},\\ x_{i_1}^{\varepsilon_1}\dots x_{i_r}^{\varepsilon_r}x_i^{\varepsilon}, & w = vx_i^{\varepsilon}, \rho(v) = x_{i_1}^{\varepsilon_1}\dots x_{i_r}^{\varepsilon_r}, \text{ and } (i_r \neq i \text{ or } \varepsilon = \varepsilon_r),\\ x_{i_1}^{\varepsilon_1}\dots x_{i_{r-1}}^{\varepsilon_{r-1}}, & w = vx_i^{\varepsilon}, \rho(v) = x_{i_1}^{\varepsilon_1}\dots x_{i_r}^{\varepsilon_r}, i_r = i, \text{ and } \varepsilon = -\varepsilon_r. \end{cases}$$

Here $i_1,\dots,i_r,i \in [n]$, and $\varepsilon_1,\dots,\varepsilon_r,\varepsilon \in \{1,-1\}$.[1] One then defines σ for an arbitrary word w in the generators x_i by

$$\sigma(w) := \sigma(\rho(w)),$$

where σ is defined inductively for freely reduced words by

$$\sigma(w) = \begin{cases} 1, & w = 1,\\ x_i^{\varepsilon}, & w = x_i^{\varepsilon},\\ x_i^{\varepsilon}vx_j^{\eta}, & w = x_i^{\varepsilon}vx_j^{\eta} \text{ with } i \neq j \text{ or } \varepsilon = \eta,\\ \sigma(v), & w = x_i^{\varepsilon}vx_j^{\eta} \text{ with } i = j \text{ and } \varepsilon = -\eta. \end{cases}$$

Here, $i,j \in [n]$ and $\varepsilon,\eta \in \{1,-1\}$.

[1] As usual, $[n]$ denotes the standard set of cardinality n; that is, $[n] = \{1,2,\dots,n\}$.

The next step, usually not carried out in this degree of formality, consists in defining an equivalence relation τ on F via

$$w_1 \tau w_2 \;:\Longleftrightarrow\; w_1 \equiv uv \text{ and } w_2 \equiv vu.$$

Here the elements of F are viewed as reduced words, and $\equiv$ means identical as words (with concatenation as the binary operation). The reflexivity and symmetry of τ are clear, so we may focus on its transitivity. Suppose that we have $w_1 \tau w_2$ and $w_2 \tau w_3$. Then we have word identities

$$w_1 \equiv uv,$$

$$w_2 \equiv vu \equiv u'v',$$

$$w_3 \equiv v'u'.$$

Assume that $|v| \leq |u'|$, so that $u' \equiv vu'_1$ and $u \equiv u'_1 v'$. Then $w_1 \equiv u'_1(v'v)$ and $w_3 \equiv (v'v)u'_1$, hence $w_1 \tau w_3$ holds as claimed. The case where $|v| > |u'|$ is treated similarly.

Definition 7.1 If $w_1 \tau w_2$ holds for elements $w_1, w_2 \in F$ then w_2 is called a *cyclic permutation* of w_1.

We now have the following classical result, effectively resolving the conjugacy problem in a finitely generated free group with a specified basis.[2]

Proposition 7.2 *If F is the free group on the free generators $x_1, x_2, \ldots, x_n$, then two words w_1 and w_2 in the generators x_i define conjugate elements of F if and only if we have $\sigma(w_1)\tau\sigma(w_2)$.*

As we shall see, the conjugacy of hyperbolic elements in $\mathscr{R}\mathscr{F}(G)$ is governed by a result which (apart from being necessarily not effective) is an almost complete analogue of Proposition 7.2; cf. Theorem 7.5 in Section 7.2.

However, the analogy does not quite stop here.[3] Suppose that $w \in F$ is a non-trivial element of the free group F (again considered as a reduced word in a basis of F), and that $u \in F$ normalises the subgroup $\langle w \rangle$ generated by w. Since free groups are torsion-free, w has infinite order and thus $\langle w \rangle \cong C_\infty$; consequently, $w^{\pm 1}$ are the only two generators of $\langle w \rangle$. Since u normalises $\langle w \rangle$,

[2] See Theorem 1.3 in Section 1.4 of Magnus, Karrass, and Solitar [31].

[3] With suitable modification, we shall follow the argument of Proposition 2.19 in Chapter I of Lyndon and Schupp [30].

it follows that

$$u^{-1}wu = w^{\pm 1}.$$

But w is not conjugate to its inverse w^{-1}. Indeed, if it were, then, by Proposition 7.2, we would have word identities

$$\sigma(w) \equiv uv$$

and

$$\sigma(w^{-1}) \equiv vu.$$

Moreover, since $\sigma(w^{-1}) = (\sigma(w))^{-1}$ these identities in turn imply that

$$v^{-1}u^{-1} \equiv vu,$$

from which we conclude that $u \equiv u^{-1}$ and $v \equiv v^{-1}$ and, in particular, that $u^2 = v^2 = 1$. Since F is torsion-free we must have $u = v = 1$, thus $\sigma(w) = 1$, and so $w = 1$, a contradiction. It follows that

$$N_F(\langle w \rangle) \subseteq C_F(w)$$

and, since the reverse inclusion is trivial, that

$$N_F(\langle w \rangle) = C_F(w) \tag{7.1}$$

holds for all $w \in F$, which in turn allows us to conclude that normalisers of infinite cyclic subgroups in a free group are themselves cyclic. As we shall see in Section 7.3, a conclusion completely analogous to (7.1) can be deduced from our conjugacy result, Theorem 7.5, in the case when G does not have elements of order 2 (see Corollary 7.9), albeit with some more work.

7.2 The equivalence relation τ_G and the conjugacy theorem

As was explained in Section 2.6, there is in general no analogue, finite or transfinite, for the process of reduction, that is, for the map ρ defined above. Instead we shall have to work with the elements of the group $\mathscr{RF}(G)$ directly. There exists, however, a kind of analogue for the process σ of cyclic reduction; this is given by Lemma 3.7, which establishes the existence and uniqueness of the (cyclically reduced) core $c(f)$ of a reduced function $f \in \mathscr{RF}(G)$. Allowing ourselves to be guided by the situation for free groups, we now define an equivalence relation on $\mathscr{RF}(G)$ analogous to the relation τ on a free group discussed above.

Definition 7.3 Given a group G, we define a binary relation τ_G on $\mathscr{RF}(G)$ by

$$f_1 \, \tau_G \, f_2 \; :\Longleftrightarrow \; f_1 = p \circ q \text{ and } f_2 = q \circ p$$

$$\text{for some } p, q \in \mathscr{RF}(G) \quad (f_1, f_2 \in \mathscr{RF}(G)).$$

If $f_1 \tau_G f_2$, we say that f_2 is a *cyclic permutation* of f_1.

Next, we show that τ_G is indeed an equivalence relation on $\mathscr{RF}(G)$. Characteristically this task, while certainly rather straightforward, is not as easy as establishing the corresponding observation for free groups.

Lemma 7.4 *Relation τ_G is an equivalence relation on $\mathscr{RF}(G)$.*

Proof Symmetry is clear by the definition of τ_G, and reflexivity holds since we may write a given element $f \in \mathscr{RF}(G)$ as $f = f \circ \mathbf{1}_G = \mathbf{1}_G \circ f$. Hence, it only remains to establish the transitivity of τ_G. If $f_1 \tau_G f_2$ and $f_2 \tau_G f_3$ hold then there exist elements $p_1, p_2, q_1, q_2 \in \mathscr{RF}(G)$ such that

$$f_1 = p_1 \circ q_1,$$

$$f_2 = q_1 \circ p_1 = p_2 \circ q_2,$$

$$f_3 = q_2 \circ p_2.$$

Suppose first that $L(q_1) \leq L(p_2)$. Then we can apply Lemma 2.14 (dissection of reduced functions) to write $p_2 = q_1 \circ u$ for some $u \in \mathscr{RF}(G)$. Hence, applying Corollary 2.18 (associativity of the circle product), we find that

$$f_2 = q_1 \circ p_1 = (q_1 \circ u) \circ q_2 = q_1 \circ (u \circ q_2),$$

and Proposition 2.1 gives $p_1 = u \circ q_2$. Applying Corollary 2.18 again, we now obtain

$$f_1 = (u \circ q_2) \circ q_1 = u \circ (q_2 \circ q_1)$$

and

$$f_3 = q_2 \circ (q_1 \circ u) = (q_2 \circ q_1) \circ u,$$

so $f_1 \tau_G f_3$ holds as desired. An analogous argument serves in the case where $L(p_2) < L(q_1)$. □

We can now state the main result of this chapter, whose proof will occupy Sections 7.4 and 7.5.

Theorem 7.5 **(The conjugacy theorem for hyperbolic elements)** *Let $f_1, f_2 \in \mathscr{RF}(G)$ be hyperbolic elements. Then f_1 is conjugate to f_2 in $\mathscr{RF}(G)$ if and only if $c(f_1)\tau_G c(f_2)$.*

Remark 7.6 We note that if $f_1 = gf_2g^{-1}$ with $g \in G_0$ then $f_1 = p \circ q$ and $f_2 = q \circ p$, where $p := g$ and $q := f_2g^{-1}$; in other words, conjugating by a G_0-element does not change the τ_G-class of a reduced function. In particular, in view of Lemma 3.7 the core $c(f)$ of a reduced function $f \in \mathscr{RF}(G)$ is well determined, modulo the equivalence relation τ_G, so the criterion stated in Theorem 7.5 for the conjugacy of two hyperbolic elements actually makes sense.

7.3 Normalisers of infinite cyclic hyperbolic subgroups

Since our conjugacy result, Theorem 7.5, applies only to hyperbolic elements and since the argument leading to equation (7.1) repeatedly uses the fact that free groups are torsion-free (which need not be the case for the groups $\mathscr{RF}(G)$, see Corollary 3.26), we can expect only a somewhat limited analogue of (7.1) to hold in our context. The main point is to understand when a hyperbolic element is conjugate to its inverse. This is cleared up in our next result.

Lemma 7.7 *Let $f \in \mathscr{RF}(G)$ be a hyperbolic element. Then the following are equivalent.*

(i) *The element f is conjugate in $\mathscr{RF}(G)$ to its inverse f^{-1}.*

(ii) *The element f is the product of two involutions lying in different conjugates of G_0.*

Proof (i) $\Rightarrow$ (ii). Suppose that f is conjugate to f^{-1}, and write $f = t \circ f_1 \circ t^{-1}$ with f_1 cyclically reduced according to Lemma 3.7. Then f^{-1} is hyperbolic as well and, by part (i) of Lemma 3.15, we have $f^{-1} = t \circ f_1^{-1} \circ t^{-1}$, where f_1^{-1} is again cyclically reduced by part (ii) of Lemma 3.6. Hence $c(f) = f_1$ and $c(f^{-1}) = f_1^{-1}$ and Theorem 7.5 tells us that $f_1 = p \circ q$ and $f_1^{-1} = q \circ p$ for some $p, q \in \mathscr{RF}(G)$. Applying Lemma 2.12 (inversion of star products) yields

$$p \circ q = f_1 = p^{-1} \circ q^{-1}. \tag{7.2}$$

Comparing values of the left-hand and right-hand sides of (7.2), we find that, for $0 \le \xi < L(p)$,

$$\begin{aligned} p(\xi) &= (p \circ q)(\xi) \\ &= (p^{-1} \circ q^{-1})(\xi) \\ &= p^{-1}(\xi) \end{aligned}$$

whereas, for $0 < \xi \le L(q)$,

$$\begin{aligned} q(\xi) &= (p \circ q)(L(p) + \xi) \\ &= (p^{-1} \circ q^{-1})(L(p) + \xi) \\ &= q^{-1}(\xi). \end{aligned}$$

Since $L(f_1) > 0$ by part (ii) of Proposition 3.13, we must have $L(p) > 0$ or $L(q) > 0$. If on the one hand $L(q) > 0$ then

$$\begin{aligned} q(0) &= (q \circ p)(0) \\ &= (q^{-1} \circ p^{-1})(0) \\ &= q^{-1}(0); \end{aligned}$$

thus $q = q^{-1}$, implying $p = p^{-1}$. If on the other hand $L(q) = 0$ then we must have $L(p) > 0$, and a calculation similar to that above yields $p = p^{-1}$, again implying $q = q^{-1}$. In both cases we have found that

$$p^2 = \mathbf{1}_G = q^2,$$

so that f_1, and hence also f itself, is a product of two involutions. Moreover these two involutions cannot lie in the same conjugate of G_0 since their product, f, would then be contained in the same G_0-conjugate and hence would be elliptic, contradicting our hypothesis that f is hyperbolic.

(ii) $\Rightarrow$ (i). If f is the product of two involutions, say $f = pq$ with $p^2 = \mathbf{1}_G = q^2$, then

$$p^{-1} f p = qp = q^{-1} p^{-1} = f^{-1},$$

so that f is indeed conjugate to f^{-1}. □

We are now in a position to compute the normaliser of an arbitrary infinite cyclic hyperbolic subgroup of $\mathscr{RF}(G)$, thereby in particular obtaining our analogue to Equation (7.1).

Proposition 7.8 *Let $f \in \mathscr{RF}(G)$ be a hyperbolic element.*

(i) *If f is not a product of two involutions then we have*

$$N_{\mathscr{RF}(G)}(\langle f \rangle) = C_{\mathscr{RF}(G)}(f). \tag{7.3}$$

(ii) *If f is a product of two involutions, $f = pq$ with $p^2 = q^2 = \mathbf{1}_G$, then the centraliser $C_{\mathscr{RF}(G)}(f)$ has index 2 in the normaliser $N_{\mathscr{RF}(G)}(\langle f \rangle)$, the non-trivial coset being generated by p; that is, we have*

$$N_{\mathscr{RF}(G)}(\langle f \rangle) = C_{\mathscr{RF}(G)}(f) \cup pC_{\mathscr{RF}(G)}(f). \tag{7.4}$$

Proof Since f, being hyperbolic, has infinite order, $\langle f \rangle \cong C_\infty$ admits only two generators, namely $f^{\pm 1}$. Consequently, if $g \in \mathscr{RF}(G)$ normalises $\langle f \rangle$ then we must have

$$g^{-1}fg = f^{\pm 1}.$$

If on the one hand f is not a product of two involutions then, according to Lemma 7.7, f is not conjugate to its inverse so that we must have $g \in C_{\mathscr{RF}(G)}(f)$. This gives

$$N_{\mathscr{RF}(G)}(\langle f \rangle) \subseteq C_{\mathscr{RF}(G)}(f),$$

and, the reverse inclusion being trivial, equation (7.3) follows in this case, whence part (i).

If on the other hand $f = pq$ is the product of two involutions p and q then p conjugates f into f^{-1}. Let x be an arbitrary element of $\mathscr{RF}(G)$ and put $y = p^{-1}x$. Then

$$\begin{aligned} x^{-1}fx = f^{-1} &\iff y^{-1}f^{-1}y = f^{-1} \\ &\iff y^{-1}fy = f \\ &\iff y \in C_{\mathscr{RF}(G)}(f), \end{aligned}$$

whence (7.4), and part (ii) is proved as well. □

Corollary 7.9 *Let $f \in \mathscr{RF}(G)$ be a hyperbolic element, and suppose that* $\mathrm{Inv}(G) = \{1_G\}$, *that is, G does not contain proper involutions. Then we have*

$$N_{\mathscr{RF}(G)}(\langle f \rangle) = C_{\mathscr{RF}(G)}(f).$$

Proof By assumption, G does not contain proper involutions; hence, by Corollary 3.26, neither does $\mathscr{RF}(G)$. Consequently, part (i) of Proposition 7.8 applies and the result follows. □

7.4 The main lemma

The main step in the proof of Theorem 7.5 consists of the following.

Lemma 7.10 *Let $f_1, f_2 \in \mathscr{RF}(G)$ be cyclically reduced and of positive length, and let $f_1 = tf_2t^{-1}$ for some $t \in \mathscr{RF}(G)$. Then f_1 and f_2 are cyclic permutations of each other.*

Proof Suppose first that we have both $\varepsilon_0(t, f_2) > 0$ and $\varepsilon_0(f_2, t^{-1}) > 0$. Then there exists $\varepsilon > 0$ such that

$$t(L(t)-\eta)f_2(\eta) = 1_G = f_2(L(f_2)-\eta)t^{-1}(\eta), \quad 0 \leq \eta \leq \varepsilon,$$

implying that

$$f_2(L(f_2)-\eta)f_2(\eta) = 1_G, \quad 0 \leq \eta \leq \varepsilon,$$

and so

$$\varepsilon_0(f_2, f_2) \geq \varepsilon > 0,$$

contradicting our hypothesis that f_2 is cyclically reduced. Hence, at least one of $\varepsilon_0(t, f_2)$ and $\varepsilon_0(f_2, t^{-1})$ must vanish. We now distinguish three cases.

Case 1: $\varepsilon_0(t, f_2) = \varepsilon_0(f_2, t^{-1}) = 0$. Then, by Lemma 2.17(i) applied to the functions t, f_2, t^{-1} plus the fact that $L(f_2) > 0$, we have $\varepsilon_0(tf_2, t^{-1}) = 0$, so that $f_1 = t \circ f_2 \circ t^{-1}$. Applying Lemma 3.7(ii) together with the fact that f_1 and f_2 are cyclically reduced, we find that f_1 and f_2 are conjugate via an element from G_0; in particular $f_1 \tau_G f_2$ holds, by Remark 7.6.

Case 2: $\varepsilon_0(t, f_2) = 0$ and $\varepsilon_0(f_2, t^{-1}) > 0$. Then, in particular, $L(t) > 0$. Suppose that $\varepsilon_0(tf_2, t^{-1}) < L(t)$, and fix a real number ε such that $0 < \varepsilon < L(t) - \varepsilon_0(tf_2, t^{-1})$. Then, for $0 \leq \eta \leq \varepsilon$, we have

$$L(t) - \eta \geq L(t) - \varepsilon > \varepsilon_0(tf_2, t^{-1})$$

and hence

$$L(f_1) - \eta = L(tf_2) + L(t^{-1}) - 2\varepsilon_0(tf_2, t^{-1}) - \eta > L(tf_2) - \varepsilon_0(tf_2, t^{-1}).$$

Thus, for η in this range,

$$\begin{aligned} f_1(L(f_1)-\eta) &= t^{-1}(L(f_1) - \eta - L(tf_2) + 2\varepsilon_0(tf_2, t^{-1})) \\ &= t^{-1}(L(t)-\eta) \\ &= \big(t(\eta)\big)^{-1} \end{aligned}$$

as well as

$$f_1(\eta) = (t \circ f_2)(\eta) = t(\eta),$$

where we have used the facts that $\varepsilon_0(t, f_2) = 0$ according to our case assumption and that $\eta < L(t)$. We obtain

$$f_1(L(f_1) - \eta) f_1(\eta) = 1_G, \quad 0 \leq \eta \leq \varepsilon,$$

so that

$$\varepsilon_0(f_1, f_1) \geq \varepsilon > 0,$$

contradicting our hypothesis that f_1 is cyclically reduced. Hence, we must have

$$\varepsilon_0(t f_2, t^{-1}) = L(t); \tag{7.5}$$

in particular,

$$\begin{aligned} L(f_1) &= L(t f_2) + L(t^{-1}) - 2\varepsilon_0(t f_2, t^{-1}) \\ &= L(t \circ f_2) + L(t) - 2L(t) \\ &= L(f_2). \end{aligned}$$

Now there are two subcases.

Case 2(i): $L(f_2) > L(t)$. Then we can use Lemma 2.14 (dissection of reduced functions) to write $f_2 = x \circ u$, where $x, u \in \mathscr{RF}(G)$, $L(x) > 0$, and $L(u) = L(t)$. We have, on the one hand,

$$(t f_2)(L(t f_2) - \eta) = t(L(t) - \eta), \quad 0 \leq \eta < L(t),$$

from (7.5). On the other hand, for $0 \leq \eta < L(f_2)$,

$$\begin{aligned} (t f_2)(L(t f_2) - \eta) &= (t \circ f_2)(L(t f_2) - \eta) \\ &= f_2(L(f_2) - \eta), \end{aligned}$$

and, for $0 \leq \eta < L(t)$,

$$\begin{aligned} f_2(L(f_2) - \eta) &= (x \circ u)(L(f_2) - \eta) \\ &= u(L(u) - \eta). \end{aligned}$$

These three calculations, when taken together, imply that

$$u(\xi) = t(\xi), \quad 0 < \xi \leq L(t),$$

and, adjusting the value of $u(0)$ so that $u(0) = t(0)$ (as we are free to do, according to Lemma 2.14), we conclude that $u = t$. But then $f_2 = x \circ t$ and $f_1 = tx$. Moreover, applying Lemma 2.17(ii) to the functions t, x, t and using the facts that $\varepsilon_0(x,t) = \varepsilon_0(t,xt) = 0$ (the latter coming from the assumption of case 2) and that $L(x) > 0$, we find that $\varepsilon_0(t,x) = 0$ so $f_1 = t \circ x$ and hence $f_1 \tau_G f_2$, as desired.

Case 2(ii): $L(f_2) \leq L(t)$. Now we use Lemma 2.14 to write $t = t_1 \circ v$ with $t_1, v \in \mathscr{RF}(G)$ and $L(v) = L(f_2)$. Then, for $0 \leq \eta < L(f_2)$,

$$\begin{aligned} f_2(L(f_2) - \eta) &= (tf_2)(L(tf_2) - \eta) \\ &= t(L(t) - \eta) \\ &= v(L(v) - \eta). \end{aligned}$$

Here, we have used in step 1 the assumption of case 2 that $\varepsilon_0(t, f_2) = 0$, in step 2 equation (7.5), and in the last step the fact that $t = t_1 \circ v$. We conclude that

$$v(\xi) = f_2(\xi), \quad 0 < \xi \leq L(f_2),$$

and, adjusting the value of $v(0)$, according to Lemma 2.14, so that $v(0) = f_2(0)$, we find that $v = f_2$; hence $t = t_1 \circ f_2$ and $f_1 = t_1 f_2 t_1^{-1}$. At this stage, we are faced with the following possibilities:

(a) $L(t_1) = 0$; thus $f_1 \tau_G f_2$ by Remark 7.6, as required.

(b) $0 < L(t_1) < L(f_2)$; in this case, $\varepsilon_0(f_2, t_1^{-1}) > 0$ since

$$\varepsilon_0(t_1, f_2) = 0 = \varepsilon_0(f_2, t_1^{-1})$$

implies $\varepsilon_0(t, t_1^{-1}) = 0$ by part (i) of Lemma 2.17, applied to the functions t_1, f_2, t_1^{-1}; consequently, making use of the case 2(ii) assumption, we have

$$\begin{aligned} L(f_1) &= L(tt_1^{-1}) \\ &= L(t \circ t_1^{-1}) \\ &= L(t) + L(t_1) \\ &\geq L(f_2) + L(t_1) \\ &> L(f_2), \end{aligned}$$

contradicting the fact that $L(f_1) = L(f_2)$ in case 2. Hence, replacing t by t_1, the argument of case 2(i) can be applied to show that $f_1 \tau_G f_2$ in this situation as well.

(c) $L(t_1) \geq L(f_2)$; then, just as in (b), we see that $\varepsilon_0(f_2, t_1^{-1}) > 0$, so (replacing t by t_1) we are again in a case 2(ii) situation. Repeating the argument of that case finitely many times at most, we reach a situation where $f_1 = t_0 f_2 t_0^{-1}$, $\varepsilon_0(t_0, f_2) = 0$, and either $L(t_0) = 0$, or $\varepsilon_0(f_2, t_0^{-1}) > 0$ and $L(t_0) < L(f_2)$. The fact that $f_1 \tau_G f_2$ holds follows now either from Remark 7.6 or by another application of the argument in case 2(i).

Case 3: $\varepsilon_0(t, f_2) > 0$ and $\varepsilon_0(f_2, t^{-1}) = 0$. Then, in view of Lemma 2.12 (inversion of star products) and part (ii) of Lemma 3.6, we can apply case 2 to f_1^{-1}, f_2^{-1} to conclude that $f_1^{-1} \tau_G f_2^{-1}$, from which it follows that $f_1 \tau_G f_2$ holds by another application of Lemma 2.12. □

7.5 Proof of Theorem 7.5

We are now in a position to establish Theorem 7.5. Let $f_1, f_2 \in \mathscr{RF}(G)$ be hyperbolic elements, f_1 conjugate to f_2, and write $f_1 = s \circ f_1' \circ s^{-1}$ and $f_2 = t \circ f_2' \circ t^{-1}$ with f_1', f_2' cyclically reduced according to Lemma 3.7(i). Then f_1' and f_2' are conjugate and, in view of Proposition 3.13, have positive length. By Lemma 7.10, f_1' is a cyclic permutation of f_2'; also, $f_i' = c(f_i)$ for $i = 1, 2$ and thus $c(f_1) \tau_G c(f_2)$ holds, as claimed.

Conversely, if $c(f_1) \tau_G c(f_2)$ holds then $f_1' \tau_G f_2'$ holds with f_1', f_2' as above, so $f_1' = p \circ q$ and $f_2' = q \circ p$ for some $p, q \in \mathscr{RF}(G)$. Then

$$p f_2' p^{-1} = pq = f_1',$$

that is, f_1' and f_2' are conjugate in $\mathscr{RF}(G)$, hence so are f_1 and f_2, and Theorem 7.5 is proved.

7.6 Exercises

7.1. Show that two non-trivial elliptic elements $a = sgs^{-1}$ and $b = tht^{-1}$ with $g, h \in G_0$ are conjugate in $\mathscr{RF}(G)$ if and only if g and h are conjugate in G_0.

7.2. Complete the proof, given in Section 7.1, that the relation τ introduced there on a (finitely generated) free group F is an equivalence relation on F.

7.3. Complete the proof of Lemma 7.4; that is, give the arguments establishing the transitivity of τ_G in the case when $L(p_2) < L(q_1)$.

7.4. Take another look at Exercise 3.5 in the light of the present chapter.

8
The centralisers of hyperbolic elements

8.1 Introduction

By part (i) of Proposition 2.20, the centralisers of elliptic elements of $\mathscr{RF}(G)$ are determined, up to isomorphism, by the isomorphism types of centralisers in the group G itself; hence, without restricting the structure of G, nothing more can be said here.

The situation is very different, and much more interesting, for the centralisers of hyperbolic elements and it is this story that will be developed in the present, rather long and technical, chapter.

Among other things we shall obtain a criterion deciding, given a hyperbolic function f, whether

$$\mathfrak{C}_f := C_{\mathscr{RF}(G)}(f)$$

is infinite cyclic. Moreover, we shall obtain considerable insight into the structure of $\mathfrak{C}_f$ in the general case; in particular, we will see that the centralisers of hyperbolic elements always embed into the additive group of the real numbers and hence are abelian and relatively 'small' in some sense. We shall also obtain a presentation for the centraliser $\mathfrak{C}_f$ of an arbitrary hyperbolic function $f \in \mathscr{RF}(G)$. This is remarkable, since test function theory allows us to show that every non-trivial subgroup of the additive reals is in fact realised as the centraliser of some hyperbolic function (see part (ii) of Theorem 10.10); hence, for each torsion-free abelian group of rank at most $2^{\aleph_0}$, we obtain a presentation in terms of a generating system exhibiting considerable internal structure. Only the future will tell whether this information may be used to classify these abelian groups, which so far have resisted any such attempt.

As applications of the main result of this chapter, Theorem 8.16, we will show that $\mathscr{RF}(G)$ does not contain non-trivial soluble normal subgroups and enjoys an analogue of the centraliser partition property of free groups; see Propositions 8.18 and 8.23. See part (v) of Theorem 10.10 for a different proof of the former result.

8.2 A preliminary lemma

Definition 8.1 A function $f \in \mathscr{RF}(G)$ is called *normalised* if it satisfies $f(0) = 1_G$.

Our first result states that, up to an inner automorphism, we may replace a hyperbolic element of $\mathscr{RF}(G)$ by a cyclically reduced and normalised function of positive length. Hence, Theorem 8.16 will describe the centralisers of arbitrary hyperbolic elements up to conjugation.

Lemma 8.2 *Let $f \in \mathscr{RF}(G)$ be hyperbolic. Then f is conjugate to a normalised and cyclically reduced function of positive length.*

Proof By part (i) of Lemma 3.7 and part (ii) of Proposition 3.13, the core $f_1 := c(f)$ of a hyperbolic function f is cyclically reduced, conjugate to f, and of positive length. Let $\mathbf{x} \in G_0$ be such that $\mathbf{x}(0) = f_1(0)$, and set

$$f_1' := \mathbf{x}^{-1} f_1 \mathbf{x}.$$

Then

$$f_1'(0) = (\mathbf{x}^{-1} \circ f_1 \circ \mathbf{x})(0) = \mathbf{x}^{-1}(0) f_1(0) = \big(f_1(0)\big)^{-1} f_1(0) = 1_G$$

and $L(f_1') = L(f_1) > 0$. Moreover, f_1' is still cyclically reduced. Indeed, we have

$$f_1'(L(f_1'))f_1'(0) = f_1(L(f_1))f_1(0)$$

and, in particular,

$$f_1'(L(f_1'))f_1'(0) = 1_G \iff f_1(L(f_1))f_1(0) = 1_G.$$

Thus, either $f_1(L(f_1))f_1(0) \neq 1_G$, in which case $f_1'(L(f_1'))f_1'(0) \neq 1_G$ and so $\varepsilon_0(f_1', f_1') = 0$, or $f_1(L(f_1))f_1(0) = 1_G$, in which case $f_1'(L(f_1'))f_1'(0) = 1_G$ and, for every small $\varepsilon > 0$, there exists $\eta = \eta(\varepsilon)$ such that $0 < \eta \leq \varepsilon$ and such that

$$f_1(L(f_1) - \eta)f_1(\eta) \neq 1_G.$$

It follows that

$$f_1'(L(f_1') - \eta)f_1'(\eta) = f_1(L(f) - \eta)f_1(\eta) \neq 1_G$$

and so

$$\varepsilon_0(f_1', f_1') = \sup \mathscr{E}(f_1', f_1') = 0.$$

□

8.3 The periods of a hyperbolic function

As it turns out, the centraliser of a hyperbolic function f is determined by the set of strong periods of f, a concept which will be introduced next.

Definition 8.3 Let $f \in \mathscr{RF}(G)$ be an element of length $L(f) = \alpha > 0$.

(i) The points $\omega \in [0, \alpha]$ satisfying

$$\forall \gamma, \delta \in (0, \alpha] : |\gamma - \delta| = \omega \to f(\gamma) = f(\delta)$$

are called *periods* of f. The set of all periods of f is denoted Ω_f.

(ii) The elements of the set

$$\Omega_f^0 = \left\{ \omega \in \Omega_f : \alpha - \omega \in \Omega_f \right\}$$

are termed *strong periods* of f.

Remark 8.4 We note that, according to Definition 8.3, the numbers 0 and α are always strong periods of f; we call them the *trivial periods*.

Our next result collects together some useful properties of periods and strong periods.

Lemma 8.5 *Let $f \in \mathscr{RF}(G)$ be a function of length $\alpha > 0$.*

(i) *If $\omega_1, \omega_2 \in \Omega_f$ and $\omega_1 + \omega_2 \in [0, \alpha]$ then $\omega_1 + \omega_2 \in \Omega_f$.*

(ii) *If $\omega_1, \ldots, \omega_r \in \Omega_f^0$ for some $r \geq 1$ and $\omega_1 + \cdots + \omega_r \in [0, \alpha]$ then we have $\omega_1 + \cdots + \omega_r \in \Omega_f^0$.*

(iii) *If $\omega_1, \omega_2 \in \Omega_f^0$ and $\omega_1 - \omega_2 \in [0, \alpha]$ then $\omega_1 - \omega_2 \in \Omega_f^0$.*

(iv) *Let* $\omega_1, \ldots, \omega_r \in \Omega_f^0$, *where* $r \geq 1$. *Then*

$$\omega_1 + \cdots + \omega_r = k\alpha + \omega$$

for some non-negative integer k *and* $\omega \in \Omega_f^0 - \{\alpha\}$.

(v) *Denote by* $\langle \Omega_f^0 \rangle$ *the subgroup of* $(\mathbb{R}, +)$ *generated by the set* Ω_f^0. *Then*

$$\langle \Omega_f^0 \rangle \cap [0, \alpha] = \Omega_f^0. \tag{8.1}$$

Proof (i) Consider points $\gamma, \delta \in (0, \alpha]$ such that $|\gamma - \delta| = \omega_1 + \omega_2$; to fix our ideas, let us take $\gamma \leq \delta$. Then

$$0 < \gamma \leq \gamma + \omega_1 \leq \gamma + \omega_1 + \omega_2 = \delta \leq \alpha,$$

so that $\gamma + \omega_1 \in (0, \alpha]$. Moreover, we have

$$f(\gamma) = f(\gamma + \omega_1),$$

since ω_1 is a period of f, and

$$f(\gamma + \omega_1) = f(\delta),$$

since ω_2 is also a period of f. It follows that $f(\gamma) = f(\delta)$, so $\omega_1 + \omega_2$ is again a period of f.

(ii) Consider first the special case where $r = 2$. By definition $\omega_1, \omega_2 \in \Omega_f$ and we have $\omega_1 + \omega_2 \in [0, \alpha]$ by assumption; hence, $\omega_1 + \omega_2 \in \Omega_f$ by part (i). It remains to show that $\omega_1 + \omega_2$ is a strong period of f. Let $\gamma, \delta \in (0, \alpha]$ be points such that $|\gamma - \delta| = \alpha - (\omega_1 + \omega_2)$ and, without loss of generality, suppose that $\gamma \leq \delta$. If on the one hand $\gamma > \omega_1$ then $\gamma - \omega_1 \in (0, \alpha]$ and

$$f(\delta) = f(\gamma - \omega_1 + \alpha - \omega_2) = f(\gamma - \omega_1) = f(\gamma),$$

since $\omega_1, \alpha - \omega_2 \in \Omega_f$ by assumption. If on the other hand $\gamma \leq \omega_1$ then

$$\delta = \gamma + \alpha - \omega_1 - \omega_2 \leq \alpha - \omega_2;$$

hence

$$0 < \gamma + \alpha - \omega_1 = \delta + \omega_2 \leq \alpha$$

that is, $\delta + \omega_2 \in (0, \alpha]$, and

$$f(\gamma) = f(\gamma + \alpha - \omega_1) = f(\delta + \omega_2) = f(\delta)$$

since $\omega_2, \alpha - \omega_1 \in \Omega_f$. In both cases it follows that $\alpha - (\omega_1 + \omega_2) \in \Omega_f$, and so $\omega_1 + \omega_2 \in \Omega_f^0$ as claimed.

A straightforward induction on r, making use of the special case $r = 2$ just established, now finishes the proof of part (ii).

(iii) We have

$$\alpha - (\omega_1 - \omega_2) = (\alpha - \omega_1) + \omega_2,$$

so $\alpha - (\omega_1 - \omega_2)$ is a sum of two periods of f, since $\omega_2, \alpha - \omega_1 \in \Omega_f$ by assumption. Since also $\alpha - (\omega_1 - \omega_2) \in [0, \alpha]$, we conclude that $\alpha - (\omega_1 - \omega_2) \in \Omega_f$ by part (i). It remains to show that $\omega_1 - \omega_2$ is also a period of f. Let $\gamma, \delta \in (0, \alpha]$ be such that $|\gamma - \delta| = \omega_1 - \omega_2$, and suppose without loss of generality that $\gamma \leq \delta$. If on the one hand $\gamma > \omega_2$ then $\gamma - \omega_2 \in (0, \alpha]$ and

$$f(\gamma) = f(\gamma - \omega_2) = f(\gamma - \omega_2 + \omega_1) = f(\delta),$$

since by assumption ω_1, ω_2 are periods of f. If on the other hand $\gamma \leq \omega_2$ then $\delta \leq \omega_1$; hence $\delta + \alpha - \omega_1 \in (0, \alpha]$ and

$$f(\delta) = f(\delta + \alpha - \omega_1) = f(\gamma + \alpha - \omega_2) = f(\gamma)$$

since $\alpha - \omega_1, \alpha - \omega_2 \in \Omega_f$. It follows that $\omega_1 - \omega_2 \in \Omega_f$, and so $\omega_1 - \omega_2$ is a strong period of f, as claimed.

(iv) Set

$$s_\rho := \sum_{1 \leq j \leq \rho} \omega_j, \quad 0 \leq \rho \leq r.$$

We claim that

$$s_\rho = k_\rho \alpha + \omega_\rho^*, \quad (k_\rho, \omega_\rho^*) \in \mathbb{N}_0 \times (\Omega_f^0 - \{\alpha\})$$

for all $\rho = 0, 1, \ldots, r$. For $\rho = 0$, this holds for $k_\rho = \omega_\rho^* := 0$. Suppose that $s_{\rho_0} = k_{\rho_0} \alpha + \omega_{\rho_0}^*$ for some ρ_0 with $0 \leq \rho_0 < r$, where $k_{\rho_0} \in \mathbb{N}_0$ and $\omega_{\rho_0}^* \in \Omega_f^0 - \{\alpha\}$. Then

$$s_{\rho_0+1} = k_{\rho_0} \alpha + \omega_{\rho_0}^* + \omega_{\rho_0+1}.$$

Now there are three possibilities. If $\omega_{\rho_0}^* + \omega_{\rho_0+1} = \alpha$ then $s_{\rho_0+1} = (k_{\rho_0} + 1)\alpha$, which is of the required form. If $\omega_{\rho_0}^* + \omega_{\rho_0+1} < \alpha$ then

$$\omega' := \omega_{\rho_0}^* + \omega_{\rho_0+1} \in \Omega_f^0 - \{\alpha\}$$

by part (ii), and $s_{\rho_0+1} = k_{\rho_0} \alpha + \omega'$ is again of the required form. Finally, suppose that

$$\alpha < \omega_{\rho_0}^* + \omega_{\rho_0+1} < 2\alpha.$$

Then

$$0 < \omega' := \omega_{\rho_0}^* - (\alpha - \omega_{\rho_0+1}) < \alpha,$$

so $\omega' \in \Omega_f^0 - \{\alpha\}$ by part (iii), and so $s_{\rho_0+1} = (k_{\rho_0}+1)\alpha + \omega'$ is of the required form. This completes the inductive step and hence the proof of part (iv).

(v) Let $\xi \in \langle \Omega_f^0 \rangle \cap [0, \alpha]$. By part (iv) we have

$$\xi = k\alpha + (\omega_1 - \omega_2)$$

with $k \in \{0,1\}$ and $\omega_1, \omega_2 \in \Omega_f^0 - \{\alpha\}$. If on the one hand $k = 0$ then $\xi = \omega_1 - \omega_2 \in \Omega_f^0$ by part (iii). If on the other hand $k = 1$ then $\xi = \omega_1 + (\alpha - \omega_2) \in \Omega_f^0$ by part (ii). This shows that the left-hand side of Equation (8.1) is contained in the right-hand side, and the reverse inclusion holds trivially. □

Corollary 8.6 *Every element ξ of the subgroup $\langle \Omega_f^0 \rangle$ of $(\mathbb{R}, +)$ generated by the set Ω_f^0 of strong periods can be written in the form*

$$\xi = \sigma(k\alpha + \omega), \quad (k, \omega, \sigma) \in \mathbb{N}_0 \times (\Omega_f^0 - \{\alpha\}) \times \{1, -1\}. \tag{8.2}$$

Moreover, with the convention that 0 is written as $(+1)(0 \cdot \alpha + 0)$ and not as $(-1)(0 \cdot \alpha + 0)$, representation (8.2) is unique.

Proof Let $\xi \in \langle \Omega_f^0 \rangle$, and set $k := [|\xi|/\alpha]$.[1] Then $k \in \mathbb{N}_0$ and $\omega := |\xi| - k\alpha$ satisfies $0 \leq \omega < \alpha$; thus $\omega \in \Omega_f^0 - \{\alpha\}$ by part (v) of Lemma 8.5. Hence ξ can be written in the desired form (10.16). Next, if $\xi \neq 0$ and

$$\xi = \sigma_1(k_1\alpha + \omega_1) = \sigma_2(k_2\alpha + \omega_2)$$

then $k_1\alpha + \omega_1, k_2\alpha + \omega_2 > 0$, so $\sigma_1 = \sigma_2$. It follows that

$$|k_1 - k_2|\alpha = |\omega_2 - \omega_1| < \alpha$$

and therefore that $k_1 = k_2$ and $\omega_1 = \omega_2$. Finally, if $\xi = \sigma(k\alpha + \omega) = 0$ then $k\alpha + \omega = 0$; thus $k = \omega = 0$ and, by our convention, $\sigma = +1$. □

[1] As usual $[x]$, the Gauß bracket of the real number x, denotes the largest integer less than or equal to x.

Our next result provides two characterisations of the strong periods among the set of all periods of a given function f.

Lemma 8.7 *Let* $f \in \mathscr{RF}(G)$ *be a reduced function of length* $\alpha > 0$, *and let* ω *be a period of* f. *Then the following assertions are equivalent.*

(i) $\omega \in \Omega_f^0$.

(ii) *For every strong period* ω' *of* f *such that* $\omega' \geq \omega$, *the real number* $\omega' - \omega$ *is also a period of* f.

(iii) *There exists a period* ω' *and a strong period* ω'' *such that* $\omega + \omega' = \omega''$.

Proof If ω, ω' are strong periods of f, so is $\omega' - \omega$ by part (iii) of Lemma 8.5, provided that $\omega' - \omega \in [0, \alpha]$, which in turn is guaranteed by the assumption that $\omega' \geq \omega$; in particular, $\omega' - \omega \in \Omega_f$. Thus (i) implies (ii). The implication (ii) $\Rightarrow$ (iii) is trivial since α is a strong period of f. Now suppose that ω is a period and that there exist periods ω', ω'' as in (iii). Then we have

$$\alpha - \omega = \alpha - (\omega'' - \omega') = (\alpha - \omega'') + \omega'.$$

Since ω'' is a strong period, $\alpha - \omega''$ is also a period of f; furthermore, since the sum of two periods is a period (provided that it fits into the interval $[0, \alpha]$) by part (i) of Lemma 8.5, we conclude that $\alpha - \omega$ is also a period. Consequently ω is a strong period of f, as required. □

8.4 The subset C_f^- of $\mathfrak{C}_f$

Let $f \in \mathscr{RF}(G)$ be cyclically reduced, of length $L(f) = \alpha > 0$, and normalised, and recall the notation $\mathfrak{C}_f$ for the centraliser of f in $\mathscr{RF}(G)$. Given f, we set

$$C_f^- := \Big\{ g \in \mathfrak{C}_f : 0 < L(g) < \alpha \text{ and } \varepsilon_0(f, g) = 0 \Big\},$$

$$C_f^+ := \Big\{ g \in \mathfrak{C}_f : L(g) \geq \alpha \text{ and } \varepsilon_0(f, g) = 0 \Big\}$$

and let

$$C_f := C_f^- \cup C_f^+ = \Big\{ g \in \mathfrak{C}_f : \varepsilon_0(f, g) = 0 \Big\} - \{\mathbf{1}_G\}.$$

In our next result we concentrate on analysing the subset C_f^- of the centraliser $\mathfrak{C}_f$ of f.

Lemma 8.8 *The elements g of C_f^- are in one-to-one correspondence with the non-trivial strong periods ω of f via $g \mapsto L(g)$, the inverse map being given by $\omega \mapsto f|_{[0,\omega]}$.*

Proof Let $g \in C_f^-$ and $\omega := L(g)$. By definition of C_f^-, we have $0 < \omega < \alpha$ as well as $\varepsilon_0(f,g) = 0$ and $fg = gf$, hence also

$$\begin{aligned}\varepsilon_0(g,f) &= \tfrac{1}{2}\big(L(g)+L(f)-L(gf)\big)\\ &= \tfrac{1}{2}\big(L(f)+L(g)-L(fg)\big)\\ &= \varepsilon_0(f,g)\\ &= 0.\end{aligned}$$

Consequently, $fg = f * g$ and $gf = g * f$. Comparing values in the equation $fg = gf$, we now find the following.

(i) We have, for $0 \leq \xi < \omega$,

$$g(\xi) = (g*f)(\xi) = (f*g)(\xi) = f(\xi).$$

Moreover, since f is normalised and $\omega < \alpha$,

$$g(\omega) = g(\omega)f(0) = (g*f)(\omega) = (f*g)(\omega) = f(\omega),$$

that is,

$$g = f|_{[0,\omega]};$$

in particular, g is normalised.

(ii) We have

$$f(\xi-\omega) = (g*f)(\xi) = (f*g)(\xi) = f(\xi), \quad \omega < \xi < \alpha$$

as well as

$$f(\alpha) = f(\alpha)g(0) = (f*g)(\alpha) = (g*f)(\alpha) = f(\alpha-\omega),$$

since $g(0) = 1_G$. These last two equations together show that $\omega \in \Omega_f$.

(iii) We have

$$f(\xi-\alpha) = g(\xi-\alpha) = (f*g)(\xi) = (g*f)(\xi) = f(\xi-\omega),\ \alpha < \xi \leq \alpha+\omega,$$

since $g = f|_{[0,\omega]}$. Hence

$$f(\xi) = f(\xi+\alpha-\omega), \quad 0 < \xi \leq \omega;$$

thus $\alpha - \omega \in \Omega_f$.

Summarising, we have shown that $\omega \in \Omega_f^0 - \{0, \alpha\}$, that is, ω is a non-trivial strong period, and that the map

$$\Phi_f^- : C_f^- \to \Omega_f^0 - \{0, \alpha\}$$

given by $g \mapsto L(g)$ is injective.

Conversely, let $\omega \in \Omega_f^0$ be a non-trivial strong period of f and let $g := f|_{[0,\omega]}$. Then $g \in \mathscr{RF}(G)$ and $0 < \omega = L(g) < \alpha$. Moreover, we have

$$\varepsilon_0(f, g) = 0 = \varepsilon_0(g, f).$$

Indeed, $\varepsilon_0(f, g) = 0$ follows immediately from the definition of g and the fact that f is cyclically reduced. Furthermore, for $0 \leq \eta < \omega$,

$$g(\omega - \eta) = f(\omega - \eta) = f(\alpha - \eta),$$

by definition of g and the fact that $\alpha - \omega$ is a period of f; consequently,

$$g(\omega - \eta) f(\eta) = 1_G, \quad 0 \leq \eta \leq \varepsilon,$$

for some sufficiently small $\varepsilon > 0$ implies that

$$f(\alpha - \eta) f(\eta) = 1_G, \quad 0 \leq \eta \leq \varepsilon,$$

contradicting the fact that f is cyclically reduced. Thus, $\varepsilon_0(g, f) = 0$ also.

It follows that, for $0 \leq \xi \leq \alpha + \omega$,

$$(fg)(\xi) = (f * g)(\xi) = \begin{cases} f(\xi), & 0 \leq \xi \leq \alpha, \\ f(\xi - \alpha), & \alpha < \xi \leq \alpha + \omega, \end{cases}$$

and

$$(gf)(\xi) = (g * f)(\xi) = \begin{cases} f(\xi), & 0 \leq \xi \leq \omega, \\ f(\xi - \omega), & \omega < \xi \leq \alpha + \omega. \end{cases}$$

Moreover, since ω is a period of f we have

$$f(\xi) = f(\xi - \omega), \quad \omega < \xi \leq \alpha, \tag{8.3}$$

while the fact that $\alpha - \omega \in \Omega_f$ gives

$$f(\xi - \alpha) = f(\xi - \omega), \quad \alpha < \xi \leq \alpha + \omega. \tag{8.4}$$

Comparing the values of fg and gf for $0 \leq \xi \leq \alpha + \omega$, making use of equations (8.3) and (8.4), we find that in fact $[f, g] = \mathbf{1}_G$ as required, so that $g \in C_f^-$ and the map Φ_f^- is surjective as well. Finally, by what we have just shown, the inverse map $(\Phi_f^-)^{-1}$ is given by

$$(\Phi_f^-)^{-1}(\omega) = f|_{[0,\omega]},$$

completing the proof of the lemma. □

Corollary 8.9 *For $g \in C_f^-$ we have $\varepsilon_0(f, g^{-1}) > 0$; in particular,*

$$C_f \cap (C_f^-)^{-1} = \varnothing.$$

Proof By Lemma 8.8 we have

$$C_f^- = \Big\{ f|_{[0,\omega]} : \omega \in \Omega_f^0 - \{0, \alpha\} \Big\}.$$

Fix $g \in C_f^-$ of length ω, and choose a real number ε such that $0 < \varepsilon < \omega$. Then, for $0 \leq \eta \leq \varepsilon$,

$$\begin{aligned} f(\alpha - \eta) g^{-1}(\eta) &= f(\alpha - \eta) \big(g(\omega - \eta) \big)^{-1} \\ &= f(\alpha - \eta) \big(f(\omega - \eta) \big)^{-1} \\ &= f(\omega - \eta) \big(f(\omega - \eta) \big)^{-1} \\ &= 1_G, \end{aligned}$$

where in step 2 we have used the fact that g is a restriction of f and in step 3 the fact that $\alpha - \omega \in \Omega_f$. It follows that

$$\varepsilon_0(f, g^{-1}) = \sup \mathscr{E}(f, g^{-1}) \geq \varepsilon > 0,$$

as claimed; the particular statement is an immediate consequence. □

Remark 8.10 Since the real number ε chosen in the proof of Corollary 8.9 is arbitrary, subject only to the condition that $0 < \varepsilon < \omega$, the argument given there shows in fact that

$$\varepsilon_0(f, g^{-1}) = \omega, \quad g \in C_f^-, L(g) = \omega.$$

8.5 The subset C_f^+ of $\mathfrak{C}_f$

Our next result provides a parametrisation for the elements of the set C_f^+.

Lemma 8.11 *The elements g of C_f^+ are in one-to-one correspondence with the elements (k,ω) of the set $\mathbb{N}\times(\Omega_f^0-\{\alpha\})$ via the assignment*

$$g\mapsto\left(\left[\frac{L(g)}{\alpha}\right],L(g)-\left[\frac{L(g)}{\alpha}\right]\alpha\right). \tag{8.5}$$

The map inverse to (8.5) *is given by*

$$(k,\omega)\mapsto f^k\circ f|_{[0,\omega]}.$$

Proof Let $g\in C_f^+$ and set $\beta:=L(g)$, noting that $\beta\geq\alpha$. As before, the facts that $\varepsilon_0(f,g)=0$ and that $[f,g]=\mathbf{1}_G$ together imply that $\varepsilon_0(g,f)=0$ also, so that $fg=f*g$ and $gf=g*f$. Comparing the values of these two functions we find the following.

(i) We have

$$f(\xi)=(f*g)(\xi)=(g*f)(\xi)=g(\xi),\quad 0\leq\xi<\alpha,$$

so that $f\approx g|_{[0,\alpha]}$; in particular, g is normalised.

Using the last observation plus the fact that f is normalised, by comparing values at $\xi=\alpha$ we now find that

$$\begin{aligned}f(\alpha)=f(\alpha)g(0)&=(f*g)(\alpha)=(g*f)(\alpha)\\&=\left.\begin{cases}g(\alpha), & \beta>\alpha\\ g(\alpha)f(0), & \beta=\alpha\end{cases}\right\}=g(\alpha).\end{aligned}$$

Hence, $f=g|_{[0,\alpha]}$.

(ii) In the range $\alpha<\xi\leq\beta$ we have

$$g(\xi-\alpha)=(f*g)(\xi)=(g*f)(\xi)=\left.\begin{cases}g(\xi), & \xi<\beta\\ g(\beta)f(0), & \xi=\beta\end{cases}\right\}=g(\xi),$$

from which we conclude that $\alpha\in\Omega_g$.

(iii) In the range $\beta<\xi\leq\alpha+\beta$ we have

$$g(\xi-\alpha)=(f*g)(\xi)=(g*f)(\xi)=f(\xi-\beta)=g(\xi-\beta)$$

since $f=g|_{[0,\alpha]}$; thus $\beta-\alpha\in\Omega_g$ and therefore $\alpha\in\Omega_g^0$.

Summarising, we have shown so far that $f = g|_{[0,\alpha]}$ and that $\alpha \in \Omega_g^0$. Next, we claim that

$$g = f^k \circ f|_{[0,\beta-k\alpha]}, \tag{8.6}$$

where $k := [\beta/\alpha]$. As a first step in this direction, we will show that

$$f^\kappa = g|_{[0,\kappa\alpha]}, \quad 0 \leq \kappa \leq k, \tag{8.7}$$

using induction on κ. Equation (8.7) holds for $\kappa = 0$ since g is normalised. Assume (8.7) to hold for some value λ of κ, where $0 \leq \lambda < k$. Then we have, for $0 \leq \xi \leq (\lambda+1)\alpha$,

$$\begin{aligned} f^{\lambda+1}(\xi) &= (f^\lambda * f)(\xi) \\ &= \begin{cases} f^\lambda(\xi), & 0 \leq \xi < \lambda\alpha, \\ f^\lambda(\lambda\alpha)f(0), & \xi = \lambda\alpha, \\ f(\xi - \lambda\alpha), & \lambda\alpha < \xi \leq (\lambda+1)\alpha, \end{cases} \\ &= \begin{cases} g(\xi), & 0 \leq \xi \leq \lambda\alpha, \\ g(\xi - \lambda\alpha), & \lambda\alpha < \xi \leq (\lambda+1)\alpha, \end{cases} \\ &= g(\xi), \end{aligned}$$

that is,

$$f^{\lambda+1} = g|_{[0,(\lambda+1)\alpha]}.$$

In the computation above, we have used in step 1 the fact that f is cyclically reduced, in step 3 the facts that f is normalised and satisfies $f = g|_{[0,\alpha]}$ plus the inductive hypothesis, and in the final step the fact that $\lambda\alpha$ is a period of g, the latter assertion following from part (ii) of Lemma 8.5 and the fact that α is a (strong) period of g. This finishes the proof of (8.7). In particular, we have shown that

$$f^k = g|_{[0,k\alpha]}, \quad k = \left[\frac{\beta}{\alpha}\right]. \tag{8.8}$$

We are now in a position to establish equation (8.6). Both sides of (8.6) are of length β, and, by (8.8) plus the facts that f is a restriction of g and normalised, we have for $0 \le \xi \le \beta$

$$(f^k \circ f|_{[0,\beta-k\alpha]})(\xi) = \begin{cases} f^k(\xi), & 0 \le \xi < k\alpha, \\ f^k(k\alpha)f(0), & \xi = k\alpha, \\ f(\xi - k\alpha), & k\alpha < \xi \le \beta, \end{cases}$$

$$= \begin{cases} g(\xi), & 0 \le \xi \le k\alpha, \\ g(\xi - k\alpha), & k\alpha < \xi \le \beta, \end{cases}$$

$$= g(\xi),$$

where in the last step we have used the fact, following from part (ii) of Lemma 8.5, that $k\alpha$ is a period of g. This completes the proof of equation (8.6).

Further, since g commutes with f, so does $f|_{[0,\beta-k\alpha]}$. Indeed, since $\varepsilon_0(f,g) = 0 = \varepsilon_0(g,f)$ and f is cyclically reduced, equation (8.6) together with Corollary 2.18 (associativity of the circle product) yields

$$\begin{aligned} fg &= f \circ g \\ &= f \circ (f^k \circ f|_{[0,\beta-k\alpha]}) \\ &= (f \circ f^k) \circ f|_{[0,\beta-k\alpha]} \\ &= (f^k \circ f) \circ f|_{[0,\beta-k\alpha]} \\ &= f^k \circ (f \circ f|_{[0,\beta-k\alpha]}) \end{aligned}$$

as well as

$$\begin{aligned} gf &= g \circ f \\ &= (f^k \circ f|_{[0,\beta-k\alpha]}) \circ f \\ &= f^k \circ (f|_{[0,\beta-k\alpha]} \circ f). \end{aligned}$$

Consequently, since $[f,g] = \mathbf{1}_G$ we find that

$$f^k \circ (f \circ f|_{[0,\beta-k\alpha]}) = f^k \circ (f|_{[0,\beta-k\alpha]} \circ f),$$

implying

$$\left[f, f|_{[0,\beta-k\alpha]}\right] = \mathbf{1}_G$$

by Proposition 2.1. It follows now that

$$f|_{[0,\beta-k\alpha]} \in C_f^- \cup \{\mathbf{1}_G\};$$

thus $\beta - k\alpha \in \Omega_f^0 - \{\alpha\}$ by Lemma 8.8. All in all, we have shown so far that the map

$$\Phi_f^+ : C_f^+ \to \mathbb{N} \times \left(\Omega_f^0 - \{\alpha\}\right)$$

given by

$$\Phi_f^+(g) := \left(\left[\frac{L(g)}{\alpha}\right], L(g) - \left[\frac{L(g)}{\alpha}\right]\alpha\right)$$

is well defined and injective.

Conversely, given $(k, \omega) \in \mathbb{N} \times (\Omega_f^0 - \{\alpha\})$, we set

$$g := f^k \circ f|_{[0,\omega]}.$$

Then $g \in \mathscr{RF}(G)$ and we have $\varepsilon_0(f, g) = 0$ since f is cyclically reduced and $k \geq 1$; the fact that g commutes with f follows from Lemma 8.8. Since also

$$L(g) = k\alpha + \omega \geq \alpha,$$

we find that $g \in C_f^+$; consequently the map Φ_f^+ is surjective, hence a bijection, and the inverse map $(\Phi_f^+)^{-1}$ is given by $(k, \omega) \mapsto f^k \circ f|_{[0,\omega]}$, as claimed. The proof of Lemma 8.11 is complete. □

Corollary 8.12 *For $g \in C_f^+$ we have $\varepsilon_0(f, g^{-1}) > 0$; in particular,*

$$C_f \cap (C_f^+)^{-1} = \varnothing.$$

Proof By Lemma 8.11 we have

$$C_f^+ = \left\{f^k \circ f|_{[0,\omega]} : (k, \omega) \in \mathbb{N} \times \left(\Omega_f^0 - \{\alpha\}\right)\right\}.$$

Let $g \in C_f^+$, and let

$$\beta := L(g) = k\alpha + \omega$$

with $k \in \mathbb{N}$ and $\omega \in \Omega_f^0 - \{\alpha\}$. There are two cases. If on the one hand $\omega = 0$ then

$$g = f^k = f^{k-1} \circ f$$

since f is cyclically reduced, so $g^{-1} = f^{-1} \circ f^{-(k-1)}$ by Lemma 2.12 (inversion of star products). Choose a real number ε satisfying $0 < \varepsilon < \alpha$. Then, for $0 \leq \eta \leq \varepsilon$, we have

$$\begin{aligned} f(\alpha - \eta)g^{-1}(\eta) &= f(\alpha - \eta)f^{-1}(\eta) \\ &= f(\alpha - \eta)\big(f(\alpha - \eta)\big)^{-1} \\ &= 1_G; \end{aligned}$$

hence

$$\varepsilon_0(f, g^{-1}) = \sup \mathscr{E}(f, g^{-1}) \geq \varepsilon > 0.$$

If on the other hand $\omega > 0$, then again applying Lemma 2.12, we decompose g^{-1} as

$$g^{-1} = \left(f|_{[0,\omega]}\right)^{-1} \circ f^{-k}.$$

Choosing ε such that $0 < \varepsilon < \omega$ we find, for $0 \leq \eta \leq \varepsilon$, that

$$\begin{aligned} f(\alpha - \eta) g^{-1}(\eta) &= f(\alpha - \eta)\left(f|_{[0,\omega]}\right)^{-1}(\eta) \\ &= f(\alpha - \eta)\big(f(\omega - \eta)\big)^{-1} \\ &= f(\omega - \eta)\big(f(\omega - \eta)\big)^{-1} \\ &= 1_G, \end{aligned}$$

since $\alpha - \omega$ is a period of f. As in the case $\omega = 0$, it follows that

$$\varepsilon_0(f, g^{-1}) = \sup \mathscr{E}(f, g^{-1}) \geq \varepsilon > 0,$$

and the proof is complete. □

Remark 8.13 The proof of Corollary 8.12 actually shows that, for $g \in C_f^+$ with $\Phi_f^+(g) = (k, 0)$, we have $\varepsilon_0(f, g^{-1}) = \alpha$; with a little extra work, one sees that the same result also holds in the case where $\omega > 0$.

8.6 The subset C_f of $\mathfrak{C}_f$

We continue our analysis of the centraliser $\mathfrak{C}_f$ by investigating the subset $C_f = C_f^- \cup C_f^+$.

Lemma 8.14

(i) *We have* $C_f \cap C_f^{-1} = \varnothing$.

(ii) *If* $g_1, g_2 \in C_f$ *then* $\varepsilon_0(g_1, g_2) = 0$; *in particular, every element of* C_f *is cyclically reduced.*

Proof (i) This follows from Corollaries 8.9 and 8.12.

(ii) By Lemmas 8.8 and 8.11 we have

$$C_f = \left\{ f^k \circ f|_{[0,\omega]} : (k,\omega) \in \mathbb{N}_0 \times (\Omega_f^0 - \{\alpha\}), k+\omega > 0 \right\}.$$

Coupling this information with the fact that f is cyclically reduced it is clear that, for $g \in C_f$,

$$g(\eta) = f(\eta), \quad 0 \le \eta \le \varepsilon, \tag{8.9}$$

holds for some sufficiently small $\varepsilon > 0$. We claim that for small $\varepsilon > 0$ we also have

$$g(L(g) - \eta) = f(\alpha - \eta), \quad 0 \le \eta \le \varepsilon. \tag{8.10}$$

Suppose first that $g = f^k$ for some $k \ge 1$. Since f is cyclically reduced and normalised, Lemma 2.2 yields that

$$g(\xi) = \left\{ \begin{array}{ll} f(\xi - (j-1)\alpha), & (j-1)\alpha < \xi \le j\alpha \ (1 \le j \le k) \\ 1_G, & \xi = 0, \end{array} \right\} \ (0 \le \xi \le k\alpha),$$

implying that (8.10) holds for every ε satisfying $\varepsilon < \alpha$. If, however,

$$g = f^k \circ f|_{[0,\omega]}$$

for some $k \ge 0$ and $\omega \in \Omega_f^0 - \{0, \alpha\}$ then we have, for $0 \le \eta \le \varepsilon$ and $\varepsilon < \omega$,

$$\begin{aligned} g(L(g) - \eta) &= f|_{[0,\omega]}(\omega - \eta) \\ &= f(\omega - \eta) \\ &= f(\alpha - \eta), \end{aligned}$$

since $\alpha - \omega$ is a period of f. The assertion of part (ii) follows now from (8.9), (8.10), and the fact that f is cyclically reduced. □

Lemma 8.15 *Let $g \in \mathfrak{C}_f$, and suppose that $\varepsilon_0(f,g) > 0$. Then $g \in C_f^{-1}$.*

Proof Comparing the lengths of the functions fg and gf, we see that

$$\varepsilon_0(f,g) = \varepsilon_0(g,f) =: \varepsilon_0 > 0.$$

In particular, $\beta := L(g) > 0$ and

$$f(\alpha - \eta)g(\eta) = 1_G, \quad 0 \le \eta < \varepsilon_0;$$

hence

$$g|_{[0,\varepsilon_0]} \approx f^{-1}|_{[0,\varepsilon_0]} \tag{8.11}$$

(see (3.1) for the definition of $\approx$). We distinguish two cases.

Case 1: $0 < \beta < \alpha$. Comparing values of fg and gf in the range $0 \le \xi < \beta - \varepsilon_0$, we find that

$$f(\xi) = (fg)(\xi) = (gf)(\xi) = g(\xi),$$

that is,

$$g|_{[0,\beta-\varepsilon_0]} \approx f|_{[0,\beta-\varepsilon_0]}. \tag{8.12}$$

Suppose that $\varepsilon_0 < \beta$. Then

$$\eta_0 := \min\{\varepsilon_0, \beta - \varepsilon_0\} > 0,$$

and we infer from (8.11) and (8.12) that, for $0 \le \eta < \eta_0$,

$$f^{-1}(\eta) = g(\eta) = f(\eta);$$

thus,

$$f(\alpha - \eta) f(\eta) = 1_G, \quad 0 \le \eta < \eta_0,$$

implying

$$\varepsilon_0(f, f) \ge \eta_0 > 0,$$

and so contradicting the fact that f is cyclically reduced. Hence we must have $\varepsilon_0 = \beta$ in case 1. Consider the element $h := fg \in \mathfrak{C}_f$. We have

$$L(h) = L(f) + L(g) - 2\varepsilon_0 = \alpha - \beta \in (0, \alpha)$$

and $h \approx f|_{[0,\alpha-\beta]}$; hence $\varepsilon_0(f, h) = 0$ since f is cyclically reduced. It follows that $h \in C_f^-$ and, consequently, that

$$L(h) = \alpha - \beta \in \Omega_f^0 - \{0, \alpha\}$$

by Lemma 8.8; we conclude that β is also a (non-trivial) strong period of$f$. Now

$$\begin{aligned}(fg)(\alpha - \beta) &= (fg)(L(f) - \varepsilon_0(f, g)) \\ &= f(L(f) - \varepsilon_0(f, g)) g(\varepsilon_0(f, g)) \\ &= f(\alpha - \beta) g(\beta),\end{aligned}$$

while

$$(gf)(\alpha - \beta) = f(\alpha)$$

and therefore

$$g(\beta) = 1_G. \tag{8.13}$$

Combining (8.11) and (8.13) with the facts that f is normalised and that $\alpha-\beta$ is a period of f, we conclude that, for $0 \le \xi \le \beta$,

$$\begin{aligned} g(\xi) &= \begin{cases} f^{-1}(\xi), & 0 \le \xi < \beta, \\ g(\beta), & \xi = \beta, \end{cases} \\ &= \begin{cases} \big(f(\alpha-\xi)\big)^{-1}, & 0 \le \xi < \beta, \\ 1_G, & \xi = \beta, \end{cases} \\ &= \big(f(\beta-\xi)\big)^{-1} \\ &= \big(f|_{[0,\beta]}\big)^{-1}(\xi). \end{aligned}$$

Hence, by Lemma 8.8 we have

$$g = \big(f|_{[0,\beta]}\big)^{-1} \in (C_f^-)^{-1}.$$

Case 2: $\beta \ge \alpha$. Comparing values of fg and gf in the range $0 \le \xi < \alpha - \varepsilon_0$, we find that

$$f(\xi) = (fg)(\xi) = (gf)(\xi) = g(\xi);$$

that is,

$$f|_{[0,\alpha-\varepsilon_0]} \approx g|_{[0,\alpha-\varepsilon_0]}.$$

The last formula in conjunction with (8.11) implies that $\varepsilon_0 = \alpha$ by an argument completely analogous to that establishing $\varepsilon_0 = \beta$ in case 1. Let $k := [\beta/\alpha]$ and $\beta - k\alpha =: \omega$, so that $k \in \mathbb{N}$ and $0 \le \omega < \alpha$. We claim that

$$L(gf^\kappa) = \beta - \kappa\alpha, \quad 0 \le \kappa \le k. \tag{8.14}$$

This holds trivially if $\kappa = 0$. Assume by way of induction that equation (8.14) holds for some κ_0 satisfying $0 \le \kappa_0 < k$. Since

$$L(gf^{\kappa_0}) = \beta + \kappa_0\alpha - 2\varepsilon_0(g, f^{\kappa_0}),$$

by the fact that f is cyclically reduced, our inductive hypothesis implies that

$$\varepsilon_0(g, f^{\kappa_0}) = \kappa_0\alpha,$$

so that

$$\begin{aligned} \beta - \varepsilon_0(g, f^{\kappa_0}) &= \beta - \kappa_0\alpha \\ &= (k-\kappa_0)\alpha + \omega \\ &\ge \alpha, \end{aligned}$$

and therefore

$$(gf^{\kappa_0})|_{[0,\alpha]} \approx g|_{[0,\alpha]} \approx f^{-1}$$

by (8.11). It follows that

$$\varepsilon_0(gf^{\kappa_0}, f) = \varepsilon_0(f, gf^{\kappa_0}) = \alpha,$$

since $gf^{\kappa_0} \in \mathfrak{C}_f$, and hence that

$$\begin{aligned} L(gf^{\kappa_0+1}) &= L(gf^{\kappa_0}) + L(f) - 2\varepsilon_0(gf^{\kappa_0}, f) \\ &= \beta - \kappa_0\alpha + \alpha - 2\alpha \\ &= \beta - (\kappa_0 + 1)\alpha, \end{aligned}$$

again making use of the inductive hypothesis. Our claim (8.14) is therefore established.

In particular, we have shown that $L(gf^k) = \omega$, so $\varepsilon_0(g, f^k) = k\alpha$. If $\omega = 0$ then we have $gf^k = \mathbf{1}_G$, that is, $g = f^{-k} \in (C_f^+)^{-1}$. For, if we had $gf^k \in G_0 - \{\mathbf{1}_G\}$ then, by part (i) of Proposition 2.20,

$$f \in C_{\mathscr{RF}(G)}(gf^k) = C_{G_0}(gf^k),$$

implying that f has length 0, a contradiction. It remains to deal with the case that $\omega > 0$; then

$$gf^k \approx g|_{[0,\omega]} \approx f^{-1}|_{[0,\omega]}$$

by (8.11) and the fact that $\varepsilon_0 = \alpha$ in case 2, so that

$$\varepsilon_0(f, gf^k) \geq \omega > 0.$$

It follows now from the proof in case 1 that $gf^k \in (C_f^-)^{-1}$, hence

$$g = (f^k \circ f|_{[0,\omega]})^{-1}, \quad (k, \omega) \in \mathbb{N} \times (\Omega_f^0 - \{0, \alpha\})$$

by Lemma 8.8 and therefore $g \in (C_f^+)^{-1}$ by Lemma 8.11. □

8.7 The main result

Define a binary operation $\boxplus$ on the set $\Omega_f^0 - \{\alpha\}$ by

$$\omega_1 \boxplus \omega_2 := \left\{ \begin{array}{ll} \omega_1 + \omega_2, & \omega_1 + \omega_2 < \alpha \\ \omega_1 + \omega_2 - \alpha, & \omega_1 + \omega_2 \geq \alpha \end{array} \right\} \quad (\omega_1, \omega_2 \in \Omega_f^0 - \{\alpha\}).$$

Note that, by parts (ii) and (iv) of Lemma 8.5, $\boxplus$ is indeed defined on $\Omega_f^0 - \{\alpha\}$. In this way, the set $\Omega_f^0 - \{\alpha\}$ becomes an abelian group; in fact, sending ω to $\omega + \langle\alpha\rangle$ gives a group isomorphism

$$(\Omega_f^0 - \{\alpha\}, \boxplus) \cong \langle\Omega_f^0\rangle / \langle\alpha\rangle.$$

We now come to the main result of this chapter.

Theorem 8.16 *(The centraliser theorem) Let $f \in \mathscr{RF}(G)$ be cyclically reduced, of length $L(f) = \alpha > 0$, and normalised.*

(i) *The set*

$$C_f = \left\{ f^k \circ f|_{[0,\omega]} : (k,\omega) \in \mathbb{N}_0 \times \left(\Omega_f^0 - \{\alpha\}\right),\ k + \omega > 0 \right\}$$

forms a positive cone for the centraliser $\mathfrak{C}_f$ of f in $\mathscr{RF}(G)$, giving $\mathfrak{C}_f$ the structure of an ordered abelian group.

(ii) *Every element of $\mathfrak{C}_f$ is cyclically reduced; in particular $\mathfrak{C}_f$ is a hyperbolic subgroup of $\mathscr{RF}(G)$.*

(iii) *The mapping $\rho_f : \mathfrak{C}_f \to \langle\Omega_f^0\rangle$ given by $(f^k \circ f|_{[0,\omega]})^\sigma \mapsto \sigma(k\alpha + \omega)$ is an isomorphism of ordered abelian groups satisfying*

$$L(g) = |\rho_f(g)|, \quad g \in \mathfrak{C}_f. \tag{8.15}$$

(iv) *$\mathfrak{C}_f$ has the presentation*

$$\Big\langle x_\omega\ (\omega \in \Omega_f^0) \ \Big|\ [x_\alpha, x_\omega] = 1\ (\omega < \alpha),$$

$$x_{\omega_1} x_{\omega_2} = x_\alpha^{[(\omega_1+\omega_2)/\alpha]} x_{\omega_1 \boxplus \omega_2}\ (\omega_1, \omega_2 < \alpha) \Big\rangle. \tag{8.16}$$

Proof (i) Suppose that $g \in \mathscr{RF}(G)$ is elliptic and commutes with f. By Lemma A.54, g fixes the axis A_f of f pointwise and, by Remark 3.14, $g = \mathbf{1}_G$. Hence $\mathfrak{C}_f$ contains no elliptic element (apart from $\mathbf{1}_G$). It follows that $\mathfrak{C}_f$ is $\mathbb{R}$-free, hence commutative transitive by Proposition A.59 and thus abelian. Further, by Lemma 8.15 we have

$$\mathfrak{C}_f = C_f \cup C_f^{-1} \cup \{\mathbf{1}_G\}$$

and, in view of part (i) of Lemma 8.14, the right-hand side forms a partition of $\mathfrak{C}_f$.

Next, if $g_1, g_2 \in C_f$ then $g_1 g_2 \neq \mathbf{1}_G$ since $C_f \cap C_f^{-1} = \varnothing$, and we have

$$\varepsilon_0(f, g_1) = \varepsilon_0(f, g_2) = \varepsilon_0(g_1, g_2) = 0$$

by the definition of C_f and part (ii) of Lemma 8.14. If $L(g_1) = 0$ then g_1 is elliptic, hence $g_1 = \mathbf{1}_G$, which is impossible as $g_1 \in C_f$, so $L(g_1) > 0$; part (ii) of Lemma 2.17 then implies that $\varepsilon_0(f, g_1 g_2) = 0$. Hence $g_1 g_2 \in C_f$, so C_f is closed under taking products. Consequently, C_f forms a positive cone for the abelian group $\mathfrak{C}_f$, as claimed. Finally, the explicit formula for C_f follows immediately from Lemmas 8.8 and 8.11 (and was observed already in the proof of Lemma 8.14).

(ii) By part (ii) of Lemma 8.14, the set C_f consists entirely of cyclically reduced functions and, by Lemma 3.6(ii), the inverse of a cyclically reduced function is again cyclically reduced. The first assertion is now an immediate consequence of part (i) and the second statement follows from this, part (ii) of Proposition 3.13, and part (i) of Proposition 2.20.

(iii) With the convention that $\mathbf{1}_G$ is written as $(f^0 \circ f|_{[0,0]})^1$, not $(f^0 \circ f|_{[0,0]})^{-1}$, every element $g \in \mathfrak{C}_f$ can be written uniquely in the form

$$g = \left(f^k \circ f|_{[0,\omega]}\right)^{\sigma},$$

where $k \in \mathbb{N}_0$, $\omega \in \Omega_f^0 - \{\alpha\}$, and $\sigma \in \{1, -1\}$. Hence, the map ρ_f is well defined and, by Corollary 8.6, a bijection. Clearly ρ_f satisfies (8.15) and identifies C_f with the positive cone of $\langle \Omega_f^0 \rangle$; in particular, ρ_f is seen to be order-preserving in both directions once we know that ρ_f is a group homomorphism. For this it suffices to show the following three facts:

$$\rho_f(g^{-1}) = -\rho_f(g), \quad g \in \mathfrak{C}_f; \tag{8.17}$$

$$\rho_f(g_1 g_2) = \rho_f(g_1) + \rho_f(g_2), \quad g_1, g_2 \in C_f; \tag{8.18}$$

$$\rho_f(g_1 g_2^{-1}) = \rho_f(g_1) - \rho_f(g_2), \quad g_1, g_2 \in C_f. \tag{8.19}$$

Assertion (8.17) is clear from the definition of ρ_f and implies in particular that

$$\rho_f(\mathbf{1}_G) = 0. \tag{8.20}$$

Let $g_1, g_2 \in C_f$, $g_i = f^{k_i} \circ f|_{[0,\omega_i]}$, where $k_i \in \mathbb{N}_0$, $\omega_i \in \Omega_f^0 - \{\alpha\}$, and $k_i + \omega_i > 0$. Then we have

$$g_1 g_2 = f^{k_1+k_2} \circ (f|_{[0,\omega_1]} \circ f|_{[0,\omega_2]}) \tag{8.21}$$

since $f^{k_1}, f^{k_2}, f|_{[0,\omega_1]}, f|_{[0,\omega_2]} \in \mathfrak{C}_f$, $\mathfrak{C}_f$ is abelian, and $\alpha - \omega_1 \in \Omega_f$. Next, we claim that

$$f|_{[0,\omega_1]} \circ f|_{[0,\omega_2]} = \begin{cases} f|_{[0,\omega_1+\omega_2]}, & \omega_1+\omega_2 \le \alpha, \\ f \circ f|_{[0,\omega_1+\omega_2-\alpha]}, & \omega_1+\omega_2 > \alpha. \end{cases} \tag{8.22}$$

Indeed, by definition of the circle product we have, for $0 \le \xi \le \omega_1+\omega_2$,

$$\big(f|_{[0,\omega_1]} \circ f|_{[0,\omega_2]}\big)(\xi) = \begin{cases} f(\xi), & 0 \le \xi \le \omega_1, \\ f(\xi-\omega_1), & \omega_1 < \xi \le \omega_1+\omega_2. \end{cases}$$

If on the one hand $\omega_1+\omega_2 \le \alpha$ then

$$f(\xi-\omega_1) = f(\xi), \quad \omega_1 < \xi \le \omega_1+\omega_2,$$

since $\omega_1 \in \Omega_f$, and so

$$f|_{[0,\omega_1]} \circ f|_{[0,\omega_2]} = f|_{[0,\omega_1+\omega_2]}$$

in this case. If on the other hand $\omega_1+\omega_2 > \alpha$ then

$$f(\xi-\omega_1) = f(\xi), \quad \omega_1 < \xi \le \alpha,$$

since $\omega_1 \in \Omega_f$, while

$$f(\xi-\omega_1) = f(\xi-\alpha), \quad \alpha < \xi \le \omega_1+\omega_2,$$

since $\alpha-\omega_1 \in \Omega_f$. Thus, in this second case,

$$\begin{aligned}\big(f|_{[0,\omega_1]} \circ f|_{[0,\omega_2]}\big)(\xi) &= \begin{cases} f(\xi), & 0 \le \xi \le \alpha, \\ f(\xi-\alpha), & \alpha < \xi \le \omega_1+\omega_2, \end{cases} \\ &= \big(f \circ f|_{[0,\omega_1+\omega_2-\alpha]}\big)(\xi)\end{aligned}$$

and our claim (8.22) is proved. Combining equations (8.21) and (8.22), we now find that

$$g_1g_2 = \begin{cases} f^{k_1+k_2} \circ f|_{[0,\omega_1+\omega_2]}, & \omega_1+\omega_2 < \alpha, \\ f^{k_1+k_2+1} \circ f|_{[0,\omega_1+\omega_2-\alpha]}, & \omega_1+\omega_2 \ge \alpha. \end{cases} \tag{8.23}$$

Thus (8.23) computes a normal form for the product of two elements of C_f. Assertion (8.18) is an immediate consequence of formula (8.23).

Concerning formula (8.19), there are three cases to consider.

(a) We have $g_1g_2^{-1} = \mathbf{1}_G$. Then $g_1 = g_2$ and, by (8.20),

$$\rho_f(g_1g_2^{-1}) = \rho_f(\mathbf{1}_G) = 0 = \rho_f(g_1) - \rho_f(g_1) = \rho_f(g_1) - \rho_f(g_2);$$

thus (8.19) holds in this case.

(b) We have $g_1g_2^{-1} \in C_f$. Set

$$g_1g_2^{-1} = g_3 = f^{k_3} \circ f|_{[0,\omega_3]}$$

with $k_3 \in \mathbb{N}_0$, $\omega_3 \in \Omega_f^0 - \{\alpha\}$, and $k_3 + \omega_3 > 0$. Then

$$g_1 = g_2g_3 = \begin{cases} f^{k_2+k_3} \circ f|_{[0,\omega_2+\omega_3]}, & \omega_2+\omega_3 < \alpha, \\ f^{k_2+k_3+1} \circ f|_{[0,\omega_2+\omega_3-\alpha]}, & \omega_2+\omega_3 \geq \alpha, \end{cases}$$

by an application of (8.23). Comparing normal forms, we get

$$(k_3,\omega_3) = \begin{cases} (k_1-k_2,\omega_1-\omega_2), & \omega_2+\omega_3 < \alpha, \\ (k_1-k_2-1,\omega_1-\omega_2+\alpha), & \omega_2+\omega_3 \geq \alpha, \end{cases}$$

and thus

$$\begin{aligned} \rho_f(g_1g_2^{-1}) &= \rho_f(g_3) \\ &= \begin{cases} (k_1-k_2)\alpha+(\omega_1-\omega_2), & \omega_2+\omega_3 < \alpha, \\ (k_1-k_2-1)\alpha+(\omega_1-\omega_2+\alpha), & \omega_2+\omega_3 \geq \alpha, \end{cases} \\ &= (k_1\alpha+\omega_1)-(k_2\alpha+\omega_2) \\ &= \rho_f(g_1)-\rho_f(g_2), \end{aligned}$$

as required.

(c) We have $g_1g_2^{-1} \in C_f^{-1}$. Set $g_1g_2^{-1} = g_3^{-1}$ with $g_3 = f^{k_3} \circ f|_{[0,\omega_3]} \in C_f$. Then, again using (8.23),

$$g_2 = g_1g_3 = \begin{cases} f^{k_1+k_3} \circ f|_{[0,\omega_1+\omega_3]}, & \omega_1+\omega_3 < \alpha, \\ f^{k_1+k_3+1} \circ f|_{[0,\omega_1+\omega_3-\alpha]}, & \omega_1+\omega_3 \geq \alpha, \end{cases}$$

and so

$$(k_3,\omega_3) = \begin{cases} (k_2-k_1,\omega_2-\omega_1), & \omega_1+\omega_3 < \alpha, \\ (k_2-k_1-1,\omega_2-\omega_1+\alpha), & \omega_1+\omega_3 \geq \alpha. \end{cases}$$

Using (8.17) together with the last formula, it follows that

$$\begin{aligned}
\rho_f(g_1g_2^{-1}) &= \rho_f(g_3^{-1}) \\
&= -\rho_f(g_3) \\
&= \begin{cases} -((k_2-k_1)\alpha+(\omega_2-\omega_1)), & \omega_1+\omega_3<\alpha, \\ -((k_2-k_1-1)\alpha+(\omega_2-\omega_1+\alpha)), & \omega_1+\omega_3\geq\alpha, \end{cases} \\
&= (k_1\alpha+\omega_1)-(k_2\alpha+\omega_2) \\
&= \rho_f(g_1)-\rho_f(g_2),
\end{aligned}$$

which finishes the proof of equation (8.19) and hence of part (iii) of the theorem.

(iv) Let Γ_f be the group defined by the presentation (8.16). The mapping

$$\{x_\omega : \omega\in\Omega_f^0\}\to\langle\Omega_f^0\rangle$$

given by $x_\omega\mapsto\omega$ extends to a surjective homomorphism

$$\varphi_f:\Gamma_f\to\langle\Omega_f^0\rangle.$$

We note the following consequences of the second-type relators

$$x_{\omega_1}x_{\omega_2}=x_\alpha^{[(\omega_1+\omega_2)/\alpha]}x_{\omega_1\boxplus\omega_2},\quad \omega_1,\omega_2\in\Omega_f^0-\{\alpha\}, \tag{8.24}$$

in (8.16):

$$x_0=1, \tag{8.25}$$

$$[x_{\omega_1},x_{\omega_2}]=1,\quad (\omega_1,\omega_2\in\Omega_f^0-\{\alpha\}), \tag{8.26}$$

$$x_{\omega_1}x_{\omega_2}^{-1}=\begin{cases} x_{\omega_1-\omega_2}, & \omega_1\geq\omega_2 \\ x_{\omega_2-\omega_1}^{-1}, & \omega_1<\omega_2 \end{cases}\quad (\omega_1,\omega_2\in\Omega_f^0-\{\alpha\}), \tag{8.27}$$

$$x_\alpha x_\omega^{-1}=\begin{cases} x_{\alpha-\omega}, & \omega>0 \\ x_\alpha, & \omega=0 \end{cases}\quad (\omega\in\Omega_f^0-\{\alpha\}). \tag{8.28}$$

Now we claim that every element w of the group Γ_f can be written in the form

$$w=(x_\alpha^k x_\omega)^\sigma,\quad k\in\mathbb{N}_0,\ \omega\in\Omega_f^0-\{\alpha\},\ \sigma\in\{1,-1\},$$

which immediately implies that $\ker(\varphi_f)=1$. Indeed, let w be a word in the generators x_ω with $\omega \in \Omega_f^0$. First, using the relations $[x_\alpha, x_\omega] = 1$, collect all occurrences of $x_\alpha^{\pm 1}$ in a single left-most factor, so that we have

$$w = x_\alpha^\rho x_{\omega_1}^{\pm 1} \cdots x_{\omega_r}^{\pm 1}, \quad (\rho \in \mathbb{Z},\ \omega_1, \ldots, \omega_r \in \Omega_f^0 - \{\alpha\}).$$

Next, making use of relations (8.24) and (8.27) and the fact that the generators commute, the product $x_{\omega_1}^{\pm 1} \cdots x_{\omega_r}^{\pm 1}$ can be rewritten in finitely many steps as $x_\alpha^m x_\omega^\sigma$, with $m \in \mathbb{Z}$, $\omega \in \Omega_f^0 - \{\alpha\}$, and $\sigma \in \{1, -1\}$, so that now

$$w = x_\alpha^{\rho'} x_\omega^\sigma, \quad (\rho' = \rho + m \in \mathbb{Z},\ \omega \in \Omega_f^0 - \{\alpha\},\ \sigma \in \{1, -1\}).$$

If $\rho' = 0$ then $w = (x_\alpha^0 x_\omega)^\sigma$ is of the required form. Also, if $\rho' > 0$ and $\sigma = +1$ then $w = (x_\alpha^{\rho'} x_\omega)^1$ is of the required form, and if $\rho' < 0$ and $\sigma = -1$ then we can write $w = (x^{-\rho'} x_\omega)^{-1}$, which is again of the required form. Thus it remains only to consider the cases where ρ' and σ are of opposite sign. If $\rho' > 0$ and $\sigma = -1$ then we rewrite w as follows:

$$w = x_\alpha^{\rho'} x_\omega^{-1} = x_\alpha^{\rho'-1} x_\alpha x_\omega^{-1} = \begin{cases} x_\alpha^{\rho'-1} x_{\alpha-\omega}, & \omega > 0, \\ x_\alpha^{\rho'} x_0, & \omega = 0, \end{cases} \tag{8.29}$$

using relations (8.25) and (8.28). Finally, if $\rho' < 0$ and $\sigma = +1$ then

$$w = \left(x_\alpha^{-\rho'} x_\omega^{-1}\right)^{-1} = \begin{cases} \left(x_\alpha^{-\rho'-1} x_{\alpha-\omega}\right)^{-1}, & \omega > 0, \\ \left(x_\alpha^{-\rho'} x_0\right)^{-1}, & \omega = 0, \end{cases}$$

where we have applied equation (8.29) to the expression $x_\alpha^{-\rho'} x_\omega^{-1}$.

Having gone through all cases, it is established that our arbitrary word w in the generators x_ω, $\omega \in \Omega_f^0$, can always be written in the desired form $(x_\alpha^k x_\omega)^\sigma$ with $k \in \mathbb{N}_0$, $\omega \in \Omega_f^0 - \{\alpha\}$, and $\sigma \in \{1, -1\}$. It follows that $\varphi_f : \Gamma_f \to \langle \Omega_f^0 \rangle$ is an isomorphism; thus, (8.16) is a presentation for $\langle \Omega_f^0 \rangle$ and hence, by part (iii), for the centraliser $\mathfrak{C}_f$. □

8.8 The case when $\mathfrak{C}_f$ is cyclic

Theorem 8.16 allows us in particular to characterise those hyperbolic functions f whose centraliser $\mathfrak{C}_f$ in $\mathscr{RF}(G)$ is cyclic.

Corollary 8.17 *Let f be as in Theorem 8.16, and set*

$$\omega_0 := \inf\left(\Omega_f^0 - \{0\}\right).$$

Then the following assertions are equivalent.

(i) *The set Ω_f^0 is finite.*

(ii) *We have $\omega_0 \in \Omega_f^0 - \{0\}$.*

(iii) *The centraliser $\mathfrak{C}_f$ is cyclic.*

Moreover, if (i)–(iii) hold then $\alpha = k_0\omega_0$ for some positive integer k_0, we have $f = f_0^{k_0}$ with $f_0 := f|_{[0,\omega_0]}$, the positive cone C_f of f consists of the positive powers of f_0, that is,

$$C_f = \left\{f_0^k : k \in \mathbb{N}\right\},$$

and the centraliser $\mathfrak{C}_f$ is given by

$$\mathfrak{C}_f = \langle f_0 \rangle.$$

Proof Clearly, (i) implies (ii). Next, assume (ii), that is, $\omega_0 \in \Omega_f^0 - \{0\}$, set $k_0 := [\alpha/\omega_0]$, and consider $\omega' := \alpha - k_0\omega_0$. Then $k_0 \in \mathbb{N}$, $0 \leq \omega' < \omega_0$, and $\omega' \in \Omega_f^0$ by parts (ii) and (iii) of Lemma 8.5. This forces $\omega' = 0$ by the definition of ω_0; that is, $\alpha = k_0\omega_0$. A similar argument shows that Ω_f^0 cannot contain any point which is not an integral multiple of ω_0. Indeed, let $\omega_1 \in \Omega_f^0$ be such that $\omega_1 > \omega_0$, and set $k_1 := [\omega_1/\omega_0]$. Then $k_1 \in \mathbb{N}$ and

$$0 \leq \omega_1' := \omega_1 - k_1\omega_0 < \omega_0.$$

Again applying parts (ii) and (iii) of Lemma 8.5, we see that ω_1' is a strong period of f, forcing $\omega_1' = 0$ by the definition of ω_0. Since, however, again using part (ii) of Lemma 8.5, the number $k\omega_0$ must be a strong period of f for $k \in \{0,1,2,\ldots,k_0\}$, we conclude that

$$\Omega_f^0 = \left\{0,\, \omega_0,\, 2\omega_0,\, \ldots,\, (k_0-1)\omega_0,\, \alpha\right\}$$

and hence that $\langle \Omega_f^0 \rangle = \langle \omega_0 \rangle$. In particular, we see that (ii) $\Rightarrow$ (i).

Translating our observations back by means of the isomorphism ρ_f of part (iii) of Theorem 8.16, we find that $f = f_0^{k_0}$ where $f_0 = f|_{[0,\omega_0]}$, that $C_f =$

$\{f_0^k : k \in \mathbb{N}\}$, and that $\mathfrak{C}_f = \langle f_0 \rangle$. This shows that (ii) $\Rightarrow$ (iii) and establishes the claims concerning α, f, C_f, and $\mathfrak{C}_f$ under the assumption that assertion (ii) holds.

(iii) $\Rightarrow$ (ii). Suppose that $\omega_0 \notin \Omega_f^0 - \{0\}$. Then Ω_f^0 contains a strictly decreasing sequence $\{\omega_\kappa\}_{\kappa \geq 1}$ of points converging to ω_0; in particular, $\{\omega_\kappa\}$ is a Cauchy sequence. It follows that $\langle \Omega_f^0 \rangle$, and hence $\mathfrak{C}_f$, is not cyclic. □

8.9 An application: the non-existence of soluble normal subgroups

As an application of Theorem 8.16, we will show the following.

Proposition 8.18 *The only soluble normal subgroup of $\mathscr{RF}(G)$ is the trivial group $\{\mathbf{1}_G\}$.*

We shall require two simple lemmas.

Lemma 8.19 *Let ξ_1, ξ_2 be real numbers, and suppose that there exists $\varepsilon > 0$ such that*

$$(\xi_1 + \eta)^2 \in \mathbb{Q} \iff (\xi_2 + \eta)^2 \in \mathbb{Q}, \quad |\eta| < \varepsilon. \tag{8.30}$$

Then we have $\xi_1 = \xi_2$.

Proof Moving ξ_1, ξ_2 slightly (in the same direction and by the same amount), and replacing ε by a smaller positive number if necessary, we can take ξ_1 in (8.30) to be rational. If q is a rational number satisfying $0 < q < \varepsilon$ then $(\xi_1 \pm q)^2$ is rational. Invoking condition (8.30) with $\eta = q$ and $\eta = -q$, we find that

$$(\xi_2 + q)^2, (\xi_2 - q)^2 \in \mathbb{Q}.$$

Subtracting yields that $4q\xi_2$ is rational, hence $\xi_2 \in \mathbb{Q}$ since q is rational and non-zero.

Now let r be a rational number such that $r \neq 0$ and such that $r\sqrt{2} \in (\xi_1, \xi_1 + \varepsilon)$. Applying condition (8.30) with $\eta = r\sqrt{2} - \xi_1$, we obtain

$$(\xi_2 + r\sqrt{2} - \xi_1)^2 = (\xi_2 - \xi_1)^2 + 2r^2 + 2r(\xi_2 - \xi_1)\sqrt{2} \in \mathbb{Q}.$$

Since $r \neq 0$, assuming that $\xi_1 \neq \xi_2$ would imply that $\sqrt{2}$ is rational, a contradiction. Thus we must have $\xi_1 = \xi_2$ as claimed. □

Lemma 8.20 *Let G be non-trivial, and let $f \in \mathscr{RF}(G)$ be a function of length $L(f) = \alpha > 0$. Then there exists $t \in \mathscr{RF}(G)$ such that $L(t) > 0$ and such that*

$$\varepsilon_0(t, f) = 0 = \varepsilon_0(tf, t^{-1}). \tag{8.31}$$

Proof This is easy if $|G| \geq 3$. In this case, fix a non-trivial element $x \in G$, choose $c \in G$ such that $c \neq (f(0))^{-1}, f(\alpha)$, and let t be the function of length 1 given by

$$t(\xi) = \begin{Bmatrix} x, & 0 \leq \xi < 1 \\ c, & \xi = 1 \end{Bmatrix} \qquad (\xi \in [0,1]).$$

Then t is reduced since it does not assume the value 1_G at any interior point, and, by construction, we have

$$t(1)f(0) = cf(0) \neq 1_G \neq f(\alpha)c^{-1} = f(\alpha)t^{-1}(0).$$

It follows that

$$\varepsilon_0(t, f) = 0 = \varepsilon_0(f, t^{-1}),$$

which in turn implies that $\varepsilon_0(tf, t^{-1}) = 0$ by part (i) of Lemma 2.17, applied to the functions t, f, t^{-1}, and using the fact that $\alpha > 0$.

Thus, for the rest of the argument we may focus on the case where

$$G = C_2 = \{1_G, x\}.$$

If $f(0) = f(\alpha)$ then the previous construction still works, as we need to avoid only one value for $t(1)$; so we may assume that $g(0) = x$ and $g(\alpha) = 1_G$, where g is one of f and f^{-1}. If g takes the value 1_G arbitrarily close to zero, that is, if for every $\varepsilon > 0$ there exists ξ_ε such that $0 < \xi_\varepsilon < \varepsilon$ and $g(\xi_\varepsilon) = 1_G$, then the function t of length 1 given by

$$t(\xi) = x, \quad 0 \leq x \leq 1,$$

satisfies

$$\varepsilon_0(t, g) = 0 = \varepsilon_0(g, t^{-1})$$

and hence also (8.31) with f replaced by g, by another application of Lemma 2.17(i). If $g = f$, we are finished in this case; otherwise, inverting the equation

$$tf^{-1}t^{-1} = t \circ f^{-1} \circ t^{-1}$$

by means of Lemma 2.12 (inversion of star products) shows that t also satisfies condition (8.31) itself. Hence, from now on we may further assume that

$$g(\xi) = x, \quad 0 \le \xi \le \varepsilon_1,$$

for some small $\varepsilon_1 > 0$. Define a function $t \in \mathscr{F}(G)$ of length 1 via

$$t(\xi) = \left\{ \begin{matrix} x, & \xi^2 \in \mathbb{Q} \\ & \\ 1_G, & \xi^2 \notin \mathbb{Q} \end{matrix} \right\} \quad (\xi \in [0,1]). \tag{8.32}$$

Then, in particular,

$$t^{-1}(0) = t(1) = x;$$

thus $\varepsilon_0(g, t^{-1}) = 0$, as required. Further, denoting by $\bar{\mathbb{Q}}$ the algebraic closure of $\mathbb{Q}$ in $\mathbb{R}$, the set $(\mathbb{R} - \bar{\mathbb{Q}}) \cap [0,1]$ is dense in the unit interval, so we also have

$$\varepsilon_0(t,g) = \sup \mathscr{E}(t,g) = 0.$$

Another application of part (i) of Lemma 2.17, and if necessary of Lemma 2.12, finishes the proof, provided that we can show that t as defined in (8.32) is in fact reduced. Suppose that there exist $\xi_0 \in (0,1)$ and a real number $\varepsilon > 0$ such that $t(\xi_0) = 1_G$ and such that

$$t(\xi_0 + \eta)t(\xi_0 - \eta) = 1_G, \quad 0 \le \eta \le \varepsilon,$$

that is, such that t has a cancelling ε-neighbourhood around an interior point ξ_0 of its domain, where the function t takes the value 1_G. Since $G = C_2$ this means that

$$t(\xi_0 + \eta) = t(\xi_0 - \eta), \quad 0 \le \eta \le \varepsilon,$$

which in turn, by the definition of the function t, is equivalent to the assertion that

$$(\xi_0 + \eta)^2 \in \mathbb{Q} \iff (\xi_0 - \eta)^2 \in \mathbb{Q}, \quad |\eta| \le \varepsilon.$$

However, by Lemma 8.19 the last displayed formula implies $\xi_0 = -\xi_0$, that is, $\xi_0 = 0$, contradicting the fact that ξ_0 is an interior point of the unit interval. Hence t is reduced and the proof is complete. □

Armed with Lemma 8.20, we are now ready for the proof of Proposition 8.18.

Proof of Proposition 8.18 We may assume that G is non-trivial, since the assertion clearly holds for $G = \{1_G\}$. Next, we note that it is enough to show that $\mathscr{RF}(G)$ does not contain a non-trivial abelian normal subgroup. Indeed, suppose that

$$\mathscr{N} \trianglelefteq \mathscr{RF}(G)$$

is a non-trivial soluble normal subgroup and that $\mathscr{RF}(G)$ does not contain a non-trivial abelian normal subgroup. Let

$$\mathscr{N} > \mathscr{N}' > \mathscr{N}'' > \cdots > \mathscr{N}^{(r)} > \mathscr{N}^{(r+1)} = \{\mathbf{1}_G\}$$

be the derived series of $\mathscr{N}$, with $\mathscr{N}^{(r)} \neq \{\mathbf{1}_G\}$ and $r \geq 0$. Then $\mathscr{N}^{(r)}$ is abelian and, like all the other terms of the derived series, is characteristic in $\mathscr{N}$ and hence normal in $\mathscr{RF}(G)$; thus $\mathscr{N}^{(r)} = \{\mathbf{1}_G\}$, since by assumption $\mathscr{RF}(G)$ does not contain a non-trivial abelian normal subgroup, a contradiction. Consequently, if $\mathscr{RF}(G)$ does not contain a non-trivial abelian normal subgroup then there is no non-trivial soluble normal subgroup in $\mathscr{RF}(G)$.

Now let $\mathscr{N}$ be a non-trivial abelian normal subgroup of $\mathscr{RF}(G)$.

1. Suppose that $\mathscr{N}$ contains a non-trivial elliptic element as well as a hyperbolic element. Then, by normality, $\mathscr{N}$ contains a non-trivial element g of length 0 and an element f of positive length. Moreover, since $L(g) = 0$ we have

$$\varepsilon_0(f,g) = 0 = \varepsilon_0(g,f),$$

and so $fg = f * g$ and $gf = g * f$ by Lemma 2.8. It follows that

$$(gf)(0) = (g * f)(0) = g(0)f(0) \neq f(0) = (f * g)(0) = (fg)(0),$$

contradicting the fact that $\mathscr{N}$ is abelian. Hence, $\mathscr{N}$ either consists entirely of elliptic elements, or all non-trivial elements of $\mathscr{N}$ are hyperbolic.

2. Next, suppose that all elements of $\mathscr{N}$ are elliptic. Then, by Proposition 3.25, $\mathscr{N}$ is a bounded normal subgroup of $\mathscr{RF}(G)$ and thus $\mathscr{N} = \{\mathbf{1}_G\}$ by Corollary 3.27, contradicting the fact that $\mathscr{N}$ is non-trivial.

3. Hence, we may assume that $\mathscr{N}$ is hyperbolic. Pick a non-trivial element f of our abelian normal subgroup $\mathscr{N}$; then, by Lemma 8.2, we can conjugate f into a function $f_1 \in \mathscr{N}$ which is cyclically reduced, of positive length, and normalised. Moreover, since $\mathscr{N}$ is abelian we have

$$\mathscr{N} \leq C_{\mathscr{RF}(G)}(f_1).$$

Now f_1 satisfies all the hypotheses of Theorem 8.16; thus part (ii) of that theorem asserts that $C_{\mathscr{RF}(G)}(f_1)$, and hence also $\mathscr{N}$, consists entirely of cyclically reduced elements. However, by Lemma 8.20 there exists $t \in \mathscr{RF}(G)$ such that $L(t) > 0$ and $tf_1t^{-1} = t \circ f_1 \circ t^{-1}$, and we have that $tf_1t^{-1} \in \mathscr{N}$ by the normality of $\mathscr{N}$. Furthermore, by Lemma 2.16 (visible cancellation) we have

$$(tf_1t^{-1})^2 = ((t\circ f_1)\circ t^{-1})(t\circ(f_1\circ t^{-1})) = (t\circ f_1)(f_1\circ t^{-1})$$

and therefore

$$\begin{aligned} L((tf_1t^{-1})^2) &= L(tf_1^2t^{-1}) \\ &\leq 2L(t)+2L(f_1) \\ &< 4L(t)+2L(f_1) \\ &= 2L(tf_1t^{-1}), \end{aligned}$$

since $L(t) > 0$. It follows that tf_1t^{-1} is not cyclically reduced, a final contradiction finishing the proof of Proposition 8.18. □

Remarks 8.21

(i) The theme of Lemma 8.19 will be taken up again in Chapter 9 in connection with the theory of test functions; see Lemma 9.5, which provides a non-trivial refinement of Lemma 8.19.

(ii) In Chapter 10 we shall see a different proof of Proposition 8.18, this time spelling out an explicit obstruction to the solubility of a non-trivial normal subgroup in $\mathscr{RF}(G)$ (a free subgroup of large rank); see part (v) of Theorem 10.10.

(iii) Proposition 8.18 may also be refined in another way as follows.[2]

Proposition 8.22 *Suppose that $\mathscr{H} \leq \mathscr{RF}(G)$ does not contain a non-abelian free subgroup. Then $\mathscr{H}$ consists entirely of elliptic elements (thus it is contained in tG_0t^{-1} for some $t \in \mathscr{RF}(G)$, by Proposition* 3.25).

The proof of Proposition 8.22 needs more of the theory of Λ-trees than we have developed in this book and thus is omitted.

8.10 More on centralisers

Let F be a free group. Then, as is well known, F is commutative transitive, as defined in the paragraph before Proposition A.59. [3] Equivalently, the binary

[2] The authors are indebted to an anonymous referee for this observation.
[3] See Proposition 2.17 in Chapter 1 of Lyndon and Schupp [30].

relation on the set $F - \{1\}$ given by

$$a \leftrightarrow b :\Longleftrightarrow a \text{ and } b \text{ commute}$$

is an equivalence relation. This is also equivalent to: the family of sets

$$C_F(a) - \{1\}, \quad a \in F - \{1\}$$

partitions the set $F - \{1\}$ of non-trivial elements of F. We call this the *centraliser partition property* of F. The last result of this chapter establishes an analogue of this fact for the hyperbolic elements of $\mathscr{RF}(G)$, thus providing further evidence that the philosophy enunciated in the introduction,

the hyperbolic elements of a non-trivial $\mathscr{RF}$-group behave analogously to the non-trivial elements of a (large) free group

is valid, and a helpful guiding principle. We are going to show the following.

Proposition 8.23 *Let G be a non-trivial group, and let $f, g \in \mathscr{RF}(G)$ be hyperbolic elements. Then the following assertions are equivalent.*

(i) *The centralisers of f and g in $\mathscr{RF}(G)$ coincide.*

(ii) *The elements f and g commute.*

(iii) *We have $C_{\mathscr{RF}(G)}(f) \cap C_{\mathscr{RF}(G)}(g) \neq \{\mathbf{1}_G\}$.*

(iv) *Choose $t \in \mathscr{RF}(G)$ such that $f_1 := tft^{-1}$ is cyclically reduced and normalised, and set $g_1 := tgt^{-1}$. Then one of g_1, g_1^{-1} (denoted as g_1') is cyclically reduced and normalised and (at least) one of the following holds:*

(a) $g_1' = f_1|_{[0,L(g_1)]}$ *and* $L(g_1) \in \Omega^0_{f_1}$*;*

(b) $f_1 = (g_1')|_{[0,L(f_1)]}$ *and* $L(f_1) \in \Omega^0_{g_1'}$*;*

(c) $(g_1')^{-1} = f_1|_{[0,L(g_1)]}$ *and* $L(g_1) \in \Omega^0_{f_1}$*;*

(d) $f_1^{-1} = (g_1')|_{[0,L(f_1)]}$ *and* $L(f_1) \in \Omega^0_{g_1'}$*.*

As an immediate consequence of Proposition 8.23 we have the following analogue of the centraliser partition property of free groups, which was alluded to above.

Corollary 8.24 ***(The centraliser partition property for $\mathscr{RF}$-groups)*** *Suppose that G is non-trivial. Then the sets*

$$C_{\mathscr{RF}(G)}(f) - \{\mathbf{1}_G\}$$

for hyperbolic functions f form a partition of the set $\mathscr{RF}(G) - \bigcup_{t \in \mathscr{RF}(G)} tG_0t^{-1}$ *of hyperbolic elements; equivalently, the binary relation*

$$f \leftrightarrow g \;:\Longleftrightarrow\; f \text{ and } g \text{ commute}, \quad f, g \in \mathscr{RF}(G) - \bigcup_{t \in \mathscr{RF}(G)} tG_0t^{-1},$$

is an equivalence relation on the set $\mathscr{RF}(G) - \bigcup_{t \in \mathscr{RF}(G)} tG_0t^{-1}$.

Furthermore, as an application of Corollary 8.24, we have the following result.

Corollary 8.25 *Suppose that G is abelian and non-trivial. Then* $\mathscr{RF}(G) - \{\mathbf{1}_G\}$ *has a partition given by self-centralising abelian subgroups.*

Proof We have already seen (for instance, in the proof of Proposition 8.18) that a non-trivial element of length 0 cannot commute with a reduced function of positive length; thus, if $U \leq G_0$ is a non-trivial subgroup then

$$C_{\mathscr{RF}(G)}(U) = C_{G_0}(U).$$

Consequently, if G is assumed to be abelian then we have

$$C_{\mathscr{RF}(G)}(G_0) = \zeta_1(G_0) = G_0.$$

Hence the sets

$$tG_0t^{-1} - \{\mathbf{1}_G\}, \quad t \in \mathscr{RF}(G),$$

together with the sets

$$C_{\mathscr{RF}(G)}(f) - \{\mathbf{1}_G\}, \quad f \in \mathscr{RF}(G) - E(G),$$

form a partition of the set $\mathscr{RF}(G) - \{\mathbf{1}_G\}$ and the corresponding subgroups tG_0t^{-1} and $C_{\mathscr{RF}(G)}(f)$ are abelian and self-centralising, as claimed. □

Before setting out on the proof of Proposition 8.23, we shall need to establish a crucial lemma.

Lemma 8.26 *Let* $f, g \in \mathscr{RF}(G)$ *be cyclically reduced and normalised elements of positive length* α *and* β*, respectively, and suppose that the centralisers of f and g in* $\mathscr{RF}(G)$ *coincide. Then (at least) one of the following assertions holds:*

(i) $g = f|_{[0,\beta]}$ *and* $\beta \in \Omega_f^0$;

(ii) $f = g|_{[0,\alpha]}$ *and* $\alpha \in \Omega_g^0$;

(iii) $g^{-1} = f|_{[0,\beta]}$ *and* $\beta \in \Omega_f^0$;

(iv) $f^{-1} = g|_{[0,\alpha]}$ *and* $\alpha \in \Omega_g^0$.

Proof Since f and g are cyclically reduced, of positive length, and normalised, Theorem 8.16 applies to both $C_{\mathscr{RF}(G)}(f)$ and $C_{\mathscr{RF}(G)}(g)$. Further, since

$$C_{\mathscr{RF}(G)}(f) = C_{\mathscr{RF}(G)}(g)$$

by assumption, we obtain from part (i) of Theorem 8.16 two equations of the form

$$g^{\delta} = f^k \circ f|_{[0,\omega]}, \tag{8.33}$$

$$f^{\varepsilon} = g^{\ell} \circ g|_{[0,\omega']}, \tag{8.34}$$

where $k,\ell \in \mathbb{N}_0$, $\omega \in \Omega_f^0 - \{\alpha\}$, $\omega' \in \Omega_g^0 - \{\beta\}$, $k+\omega, \ell+\omega' > 0$, and $\delta,\varepsilon \in \{1,-1\}$.

Suppose first that $\varepsilon = -\delta$; to fix our ideas, say $\delta = 1$ and $\varepsilon = -1$. Inserting equation (8.33) into (8.34), multiplying from the left by f, and using the facts that f is cyclically reduced, that $[f, f|_{[0,\omega]}] = \mathbf{1}_G$, and that $\alpha - \omega$ is a period of f, we obtain the equation

$$\underbrace{f\circ\cdots\circ f}_{(k\ell+1)\text{ times}} \circ \underbrace{f|_{[0,\omega]}\circ\cdots\circ f|_{[0,\omega]}}_{\ell\text{ times}} \circ \big(f^k\circ f|_{[0,\omega]}\big)\big|_{[0,\omega']} = \mathbf{1}_G,$$

which is impossible, since the left-hand side is of length at least

$$(k\ell+1)\alpha > 0.$$

The case where $\delta = -1$ and $\varepsilon = 1$ is disposed of in a similar way, by interchanging f and g. Hence, we must have $\delta = \varepsilon$; to fix our ideas, say $\delta = 1 = \varepsilon$. If $k\ell = 0$ then at least one of (i) and (ii) holds, with β or α respectively a non-trivial period in this case. Now suppose that $k\ell \geq 1$. Then, inserting (8.33) into (8.34), multiplying from the left by f^{-1}, and using again the facts that f is cyclically reduced, that $[f, f|_{[0,\omega]}] = \mathbf{1}_G$, and that $\alpha - \omega \in \Omega_f$, we obtain an equation

$$\underbrace{f\circ\cdots\circ f}_{(k\ell-1)\text{ times}} \circ \underbrace{f|_{[0,\omega]}\circ\cdots\circ f|_{[0,\omega]}}_{\ell\text{ times}} \circ \big(f^k\circ f|_{[0,\omega]}\big)\big|_{[0,\omega']} = \mathbf{1}_G. \tag{8.35}$$

Comparing lengths in (8.35), we are led to the equation

$$(k\ell-1)\alpha + \ell\omega + \omega' = 0, \tag{8.36}$$

whose only solution (k,ℓ,ω,ω'), with k,ℓ satisfying $k\ell \geq 1$, is given by $k = \ell = 1$ and $\omega = \omega' = 0$ so that $f = g$ in this case, which fits (i) with $\beta = \alpha$ and (ii) with $\alpha = \beta$.

Finally, suppose that $\delta = \varepsilon = -1$. Repeating the discussion of the situation where $k\ell = 0$, we find that in this case (at least) one of (ii) and (iv) holds, with β or α respectively a non-trivial strong period. If $k\ell \geq 1$ then, using the fact that f^{-1} is cyclically reduced (since f is so), an argument similar to that leading to equation (8.35) now yields the equation

$$\underbrace{f^{-1}\circ\cdots\circ f^{-1}}_{(k\ell-1)\text{ times}}\circ\underbrace{(f|_{[0,\omega]})^{-1}\circ\cdots\circ(f|_{[0,\omega]})^{-1}}_{\ell\text{ times}}\circ\big(f^{-k}\circ(f|_{[0,\omega]})^{-1}\big)\big|_{[0,\omega']} = \mathbf{1}_G.$$

Comparing lengths, we again derive equation (8.36), which now leads to $g^{-1} = f$; this fits both cases (iii) and (iv). □

Proof of Proposition 8.23 The implications (i) $\Rightarrow$ (ii) $\Rightarrow$ (iii) are obvious; thus it suffices to show the implications

$$\text{(iii)} \Rightarrow \text{(ii)} \Rightarrow \text{(i)} \Leftrightarrow \text{(iv)}.$$

(ii) $\Rightarrow$ (i). This follows from the fact that the centralisers of hyperbolic elements in $\mathscr{RF}(G)$ are abelian. Indeed suppose that $[f,g] = \mathbf{1}_G$, so that $g \in C_{\mathscr{RF}(G)}(f)$, and let $h \in C_{\mathscr{RF}(G)}(f)$ be arbitrary. Then h and g both lie in the centraliser of f in $\mathscr{RF}(G)$ and, since f is hyperbolic, $C_{\mathscr{RF}(G)}(f)$ is abelian by Theorem 8.16; in particular, $[h,g] = \mathbf{1}_G$ so $h \in C_{\mathscr{RF}(G)}(g)$. This shows that

$$C_{\mathscr{RF}(G)}(f) \leq C_{\mathscr{RF}(G)}(g),$$

and the reverse inclusion is established in a similar way.

(iii) $\Rightarrow$ (ii). This follows from the fact that the centralisers of hyperbolic elements are both abelian and hyperbolic, by Theorem 8.16. To be more explicit, let

$$h \in C_{\mathscr{RF}(G)}(f) \cap C_{\mathscr{RF}(G)}(g)$$

be a non-trivial element. As f is assumed to be hyperbolic, the set $C_{\mathscr{RF}(G)}(f) - \{\mathbf{1}_G\}$ consists entirely of hyperbolic elements; in particular, h itself is hyperbolic and, consequently, $C_{\mathscr{RF}(G)}(h)$ is abelian. Now, f and g are both contained in $C_{\mathscr{RF}(G)}(h)$, whence $[f,g] = \mathbf{1}_G$, as claimed.

(i) $\Leftrightarrow$ (iv). Let $f,g \in \mathscr{RF}(G)$ be hyperbolic elements, and suppose that

$$C_{\mathscr{RF}(G)}(f) = C_{\mathscr{RF}(G)}(g).$$

Choose $t \in \mathscr{RF}(G)$ according to Lemma 8.2, that is, such that $f_1 := tft^{-1}$ is cyclically reduced, of positive length, and normalised. Thus Theorem 8.16 applies to the centraliser of f_1 in $\mathscr{RF}(G)$. Set $g_1 := tgt^{-1}$. Then $L(g_1) > 0$, since g is hyperbolic, and one of g_1, g_1^{-1} (denoted as g_1') is contained in the positive cone C_{f_1} of f_1; consequently, by parts (i) and (ii) of Theorem 8.16, g'_1 is cyclically reduced and normalised as well. Also, we have

$$C_{\mathscr{RF}(G)}(f_1) = C_{\mathscr{RF}(G)}(g'_1),$$

since

$$C_{\mathscr{RF}(G)}(f) = t^{-1}C_{\mathscr{RF}(G)}(f_1)t$$

and

$$C_{\mathscr{RF}(G)}(g) = t^{-1}C_{\mathscr{RF}(G)}(g_1)t = t^{-1}C_{\mathscr{RF}(G)}(g'_1)t.$$

Applying Lemma 8.26 to the elements f_1 and g'_1, we find that one of the conditions

$$(g'_1)^{\pm 1} = f_1|_{[0,L(g_1)]}, \quad L(g_1) \in \Omega^0_{f_1},$$

$$f_1^{\pm 1} = (g'_1)|_{[0,L(f_1)]}, \quad L(f_1) \in \Omega^0_{g'_1},$$

must hold, whence (iv). Conversely, suppose that assertion (iv) holds. Then, by Theorem 8.16(i), we have

$$\left[(g'_1)^{\delta}, f_1^{\varepsilon}\right] = \mathbf{1}_G,$$

where $\delta, \varepsilon \in \{1, -1\}$, which implies that

$$[g_1, f_1] = \mathbf{1}_G.$$

Since f_1, g_1 are hyperbolic elements of $\mathscr{RF}(G)$, the implication (ii) $\Rightarrow$ (i), which has already been proved, when applied to these elements allows us to conclude that

$$C_{\mathscr{RF}(G)}(f_1) = C_{\mathscr{RF}(G)}(g_1),$$

implying

$$C_{\mathscr{RF}(G)}(f) = C_{\mathscr{RF}(G)}(g)$$

by the definitions of f_1 and g_1. Hence, (iv) $\Rightarrow$ (i) holds as well and the proof of Proposition 8.23 is complete. □

8.11 Exercises

As is to be expected, the set of (strong) periods of a function does not behave well when one is forming products or inverting. The following three exercises illustrate this point but also demonstrate that some limited results may still be obtained in this direction. For Exercise 8.2. note that the definition of a period, which in Definition 8.3 was given only for reduced functions f, makes perfect sense for $f \in \mathscr{F}(G)$.

8.1. Let $f \in \mathscr{RF}(G)$ be a function of length $\alpha > 0$. Show that

$$\Omega_f \cap \{\xi \in [0,\alpha] : f(\xi) = f(0)\} = \Omega_{f^{-1}} \cap \{\xi \in [0,\alpha] : f^{-1}(\xi) = f^{-1}(0)\} \tag{8.37}$$

and

$$\Omega_f^0 \cap \{\xi \in [0,\alpha] : f(\xi) = f(0) = f(\alpha - \xi)\}$$

$$= \Omega_{f^{-1}}^0 \cap \{\xi \in [0,\alpha] : f^{-1}(\xi) = f^{-1}(0) = f^{-1}(\alpha - \xi)\}. \tag{8.38}$$

8.2. Let $f_1, f_2 \in \mathscr{RF}(G)$ be functions of positive lengths α_1, α_2, respectively, and suppose that $f_2(0) = 1_G$. Compute $\Omega_{f_1 * f_2}$.

8.3. Let $f \in \mathscr{RF}(G)$ be a function of length $\alpha > 0$, and let $g \in G_0 - \{\mathbf{1}_G\}$. Show that

$$\Omega_{gf} = \Omega_f \quad \text{and} \quad \Omega_{gf}^0 = \Omega_f^0,$$

while

$$\Omega_{fg} \cap \Omega_f = \{0, \alpha\}.$$

Remark 8.27 The asymmetry observed in Exercise 8.3 with regard to multiplication by a non-trivial G_0-element as well as the fact that in general we have $\Omega_f \neq \Omega_{f^{-1}}$ (cf. Exercise 8.1), are due to a certain lack of symmetry in our definition of periods; they are consequences of the fact that 0 is not allowed as a comparison point in part (i) of Definition 8.3, while $L(f) = \alpha$ is. These pathologies can be avoided by altering the definition of a period slightly, as follows:

a point $\omega \in [0,\alpha]$ is a period of f if and only if

$$\forall \xi_1, \xi_2 \in (0,\alpha) : |\xi_1 - \xi_2| = \omega \longrightarrow f(\xi_1) = f(\xi_2). \tag{8.39}$$

Redefining periods via (8.39), while keeping intact the definition of strong periods, would indeed resolve the pathologies just mentioned (see the next exercise). However, the main significance of periods lies in their role in connection with the centralisers of hyperbolic elements of $\mathscr{RF}$-groups, and one would have to check carefully what happens to the contents of this chapter with this new definition of a period; that is, one would have to check whether Theorem 8.16 remains intact.

8.4. Adopting condition (8.39) as the definition of a period, while keeping Definition 8.3(ii) intact, establish the following results.

(i) If $f \in \mathscr{RF}(G)$ is a function of positive length then we have $\Omega_f = \Omega_{f^{-1}}$ and $\Omega_f^0 = \Omega_{f^{-1}}^0$.

(ii) If $f \in \mathscr{RF}(G)$ has positive length and $g \in G_0 - \{\mathbf{1}_G\}$ then $\Omega_{gf} = \Omega_f = \Omega_{fg}$ and $\Omega_{gf}^0 = \Omega_f^0 = \Omega_{fg}^0$.

8.5. Let $f = t \circ g \circ t^{-1}$, where $L(t) > 0$ and $g \in G_0 - \{\mathbf{1}_G\}$. Adopting condition (8.39) as the definition of a period, compute the periods of f.

8.6. Let $f \in \mathscr{RF}(G)$ be a hyperbolic element, and let $c_0(f)$ be the normalised core of f as introduced in part (ii) of Definition 3.10. Demonstrate the following criterion for the divisibility of the centraliser of a hyperbolic element.

Divisibility criterion *The centraliser $C_{\mathscr{RF}(G)}(f)$ of a hyperbolic function $f \in \mathscr{RF}(G)$ is divisible if and only if the set $\Omega_{c_0(f)}^0$ of strong periods of the normalised core $c_0(f)$ of f is closed under division by positive integers.*

9

Test functions: basic theory and first applications

9.1 Introduction

In the course of the previous chapters, we have repeatedly come upon questions concerning the structure of $\mathscr{R}\mathscr{F}$-groups that could not be satisfactorily resolved at that stage. For instance, it was shown in Chapter 5 that $\mathscr{R}\mathscr{F}(G)$ is not generated by its elliptic elements provided that G contains an element g satisfying $g \neq g^{-1}$ (that is, a non-involution); see Corollary 5.9. However, for G an elementary abelian 2-group this question had to remain open, since in that case all exponent sums are trivial.

In Chapter 6 it was shown among other things that two groups G and H having the same number of involutions, as well as the same number of non-involutions, satisfy $|\mathscr{R}\mathscr{F}(G)| = |\mathscr{R}\mathscr{F}(H)|$ (see Corollary 6.2), but the actual cardinality of $\mathscr{R}\mathscr{F}(G)$ could not be determined there. In the same chapter it was shown that, under this hypothesis concerning the involutions and non-involutions of two groups G and H, we also have

$$\mathscr{R}\mathscr{F}(G)/E(G) \cong \mathscr{R}\mathscr{F}(H)/E(H).$$

However, the isomorphism type of the quotient $\mathscr{R}\mathscr{F}_0(G) = \mathscr{R}\mathscr{F}(G)/E(G)$ itself remained a complete mystery, so much so that, up to this stage, it is not even clear whether $\mathscr{R}\mathscr{F}_0(G)$ might perhaps be a free group, in which case $\mathscr{R}\mathscr{F}(G)$ would turn out to be a split extension of $E(G)$ by $\mathscr{R}\mathscr{F}_0(G)$.

Similarly, up to now the structure of the abelianised groups

$$\overline{\mathscr{R}\mathscr{F}(G)} = \mathscr{R}\mathscr{F}(G)/[\mathscr{R}\mathscr{F}(G),\mathscr{R}\mathscr{F}(G)]$$

and

$$\overline{\mathscr{R}\mathscr{F}_0(G)} = \mathscr{R}\mathscr{F}(G)/\big(E(G)[\mathscr{R}\mathscr{F}(G),\mathscr{R}\mathscr{F}(G)]\big).$$

is completely unknown. Could these groups perhaps turn out to be free abelian?

Further, in Chapter 8 it remained open whether all non-trivial subgroups of the additive reals are realised (up to isomorphism) as centralisers of the hyperbolic elements in $\mathscr{R}\mathscr{F}(G)$.

Our aim in this and the following chapter is to introduce and discuss two basic new concepts, those of a *test function* and of a family of pairwise *locally incompatible* functions. These new concepts and the techniques developed around them make it possible to answer all the questions above, as well as some related questions. For instance, we shall be able to prove that, for any group G,

$$|\mathscr{R}\mathscr{F}(G)| = |G|^{2^{\aleph_0}}$$

and that

$$[\mathscr{R}\mathscr{F}(G) : E(G)] = |\mathscr{R}\mathscr{F}(G)|;$$

in particular, for G non-trivial, $\mathscr{R}\mathscr{F}(G)$ is never generated by its elliptic elements. Moreover, it is now possible to show that:

- both $\mathscr{R}\mathscr{F}(G)$ and $\mathscr{R}\mathscr{F}_0(G)$ contain a free subgroup of rank $|G|^{2^{\aleph_0}}$, but are not free;
- every non-trivial torsion-free abelian group of rank at most $2^{\aleph_0}$ is realised (up to isomorphism) as the centraliser of a hyperbolic element in $\mathscr{R}\mathscr{F}(G)$;
- $\overline{\mathscr{R}\mathscr{F}(G)}$ and $\overline{\mathscr{R}\mathscr{F}_0(G)}$ both contain (the additive group of) a $\mathbb{Q}$-vector space of dimension $|G|^{2^{\aleph_0}}$, so that these abelianised groups are not free abelian while still containing large free abelian subgroups.

Further, we can now demonstrate that every non-trivial normal subgroup $\mathscr{N} \trianglelefteq \mathscr{R}\mathscr{F}(G)$ contains a free subgroup of rank $|G|^{2^{\aleph_0}}$; in particular, $|\mathscr{N}| = |\mathscr{R}\mathscr{F}(G)|$. The existence of such a large free subgroup also provides an explicit obstruction to the solubility of $\mathscr{N}$; compare Section 8.9, where the non-existence in $\mathscr{R}\mathscr{F}(G)$ of non-trivial soluble normal subgroups was established by a different method. Moreover, a similar and almost equally strong result can be shown to hold for the normal subgroups of the quotient group $\mathscr{R}\mathscr{F}_0(G)$.

The purpose of Chapters 9 and 10 is to explain these new concepts, introduce their machinery and to sketch some of the less technical arguments leading to the conclusions mentioned above; full details may be found in the paper [38] by Müller and Schlage-Puchta.

We now describe the contents of the present chapter. Section 9.2 introduces the concept of a test function; test functions are certain special elements of $\mathscr{RF}$-groups which, roughly speaking, do not look locally like their own inverses. We show that test functions are cyclically reduced and that non-trivial powers of a test function are again test functions.

In Section 9.3 we establish a purely arithmetic result (Lemma 9.5) that allows us to settle the existence problem for test functions by exhibiting, in a uniform way, a concrete example of a test function in every non-trivial $\mathscr{RF}$-group. Lemma 9.5 will also play a role in the proof of Theorem 10.1, one of the main results of the next chapter; this theorem ensures the existence of large families of pairwise locally incompatible test functions with prescribed centraliser. The local incompatibility of two functions $f, g \in \mathscr{F}(G)$ means, roughly speaking, that locally f does not look like g or g^{-1}; it is this notion of independence which turns out to be the most suitable when we are dealing with families of test functions and products involving them.

In Section 9.4, which is long and technically rather involved, we explain how a given test function $f \in \mathscr{RF}(G)$ gives rise to a surjective homomorphism

$$\lambda_f : \mathscr{RF}(G) \to \mathbb{R}$$

satisfying $\lambda_f(E(G)) = 0$. As in the case of exponent sums (see Chapter 5), our construction employs Lebesgue measure theory but is based on a different idea, namely the *local comparison* of a function $g \in \mathscr{RF}(G)$ with a fixed test function and its inverse (whence the name 'test function'). Coupled with the existence of test functions (as demonstrated in Section 9.3), we thus find in particular that non-trivial $\mathscr{RF}$-groups are never generated by their elliptic elements only; see Corollary 9.9. We also deduce that the test functions of $\mathscr{RF}(G)$ are not contained in the normal subgroup $E(G)[\mathscr{RF}(G), \mathscr{RF}(G)]$ and that the centraliser $C_{\mathscr{RF}(G)}(f)$ of a test function $f \in \mathscr{RF}(G)$, apart from the identity element, consists only of test functions, from which we conclude in particular the remarkable fact that

$$C_{\mathscr{RF}(G)}(f) \cap E(G)[\mathscr{RF}(G), \mathscr{RF}(G)] = \{\mathbf{1}_G\}.$$

The second fundamental notion, that of the local incompatibility of a pair of functions, is introduced at the beginning of Section 9.5. Among other things, it is shown that two locally incompatible functions $f, g \in \mathscr{F}(G)$ satisfy $\varepsilon_0(f, g) = 0$, that is, they do not exhibit any cancellation; see Lemma 9.13(i). The main result of Section 9.5 shows that, if $f_1, f_2, \ldots, f_k \in \mathscr{RF}(G)$ are pairwise locally incompatible test functions, and if $\gamma_1, \gamma_2, \ldots, \gamma_k$ are non-zero integers, then $f_1^{\gamma_1} f_2^{\gamma_2} \cdots f_k^{\gamma_k}$ is also a test function; see Proposition 9.18. This last result will play an important role in the proof of Proposition 10.9 and hence that of

Theorem 10.10, which establishes our main structural conclusions concerning the groups $\mathscr{RF}(G)$ and their quotients $\mathscr{RF}_0(G)$ (see the results mentioned in the first half of this introduction).

Sections 9.6 and 9.7 centre around families $\mathscr{F} = \{f_\sigma\}_{\sigma \in S}$ of pairwise locally incompatible test functions. In Section 9.6 we show that, among other things, the system of centralisers $\{C_{\mathscr{RF}(G)}(f_\sigma)\}_{\sigma \in S}$ associated with $\mathscr{F}$ generates a hyperbolic subgroup of $\mathscr{RF}(G)$ isomorphic to the free product of these centralisers; see Corollary 9.24. In Section 9.7, we investigate the homomorphism $\lambda_S : \mathscr{RF}(G) \to \mathbb{R}^S$ obtained by stringing together the maps λ_{f_σ} introduced in Section 9.4 for all $\sigma \in S$; our main result (Theorem 9.26) characterises the image of the map λ_S as the subspace $\ell^1(S)$ of $\mathbb{R}^S$. This result should be compared with a corresponding, much less successful, attempt in Chapter 5; see Proposition 5.10.

9.2 Test functions: definition and first properties

We begin with the following basic definition.

Definition 9.1 A function $f \in \mathscr{F}(G)$ is called a *test function* if it has positive length and there do not exist a real number $\varepsilon > 0$ and points $\xi_1, \xi_2 \in (0, L(f))$ such that

$$f(\xi_1 + \eta) = f^{-1}(\xi_2 + \eta), \quad |\eta| < \varepsilon.$$

Since f and f^{-1} occur symmetrically in Definition 9.1, it follows that the inverse of a test function is also a test function. A first observation is that test functions are automatically reduced. In fact, more than that is true.

Lemma 9.2 *Test functions are cyclically reduced.*

Proof Let $f \in \mathscr{F}(G)$ be a test function of length α. We shall first show that f is reduced. Suppose for a contradiction that there exists an interior point ξ_0 of the domain $[0, \alpha]$ such that $f(\xi_0) = 1_G$ and a cancelling ε-neighbourhood for f around ξ_0. This means that, for $|\eta| < \varepsilon$,

$$f(\xi_0 + \eta) = \big(f(\xi_0 - \eta)\big)^{-1} = f^{-1}(\alpha - \xi_0 + \eta) = f^{-1}(\xi_0' + \eta),$$

where $\xi_0' := \alpha - \xi_0$ is also an interior point of the domain of f. Since the resulting equation,

$$f(\xi_0 + \eta) = f^{-1}(\xi_0' + \eta), \quad |\eta| < \varepsilon,$$

contradicts the definition of a test function, we deduce that f is reduced, as claimed.

Now suppose that f is not cyclically reduced, that is, there exists $\varepsilon > 0$ such that

$$f(\alpha - \eta)f(\eta) = 1_G, \quad 0 \leq \eta \leq \varepsilon.$$

Rewriting the last equation as $f(\eta) = f^{-1}(\eta)$ and letting $\eta = \varepsilon/2 + \eta'$, we find that

$$f\left(\frac{\varepsilon}{2} + \eta'\right) = f^{-1}\left(\frac{\varepsilon}{2} + \eta'\right), \quad |\eta'| < \frac{\varepsilon}{2},$$

again contradicting the definition of a test function. □

Our next lemma is the first in a series of results constructing new test functions from old ones.

Lemma 9.3 *Let f be a test function of length α, let k be a non-negative integer, and let α' be a real number such that $0 \leq \alpha' < \alpha$. Then the function $g = f^k f|_{[0,\alpha']}$ is also a test function provided that $k + \alpha' > 0$.*

Corollary 9.4 *If f is a test function and k is a non-zero integer then f^k is also a test function.*

Since the inverse of a test function is also a test function, Corollary 9.4 follows immediately from Lemma 9.3 in the case when $\alpha' = 0$.

Proof of Lemma 9.3 *(Sketch)* By Lemma 9.2 f is cyclically reduced, so that

$$g = \underbrace{f \circ \cdots \circ f}_{k \text{ times}} \circ f|_{[0,\alpha']};$$

in particular, $L(g) = k\alpha + \alpha'$. Assume for a contradiction that there exist $\varepsilon > 0$ and points $\xi_1, \xi_2 \in (0, L(g))$ such that

$$g(\xi_1 + \eta) = g^{-1}(\xi_2 + \eta), \quad |\eta| < \varepsilon, \tag{9.1}$$

and distinguish the cases (i) $k = 0$, (ii) $\alpha' = 0$, and (iii) $k > 0$ and $0 < \alpha' < \alpha$. We shall give the details for case (ii), in which $g = f^k$ for some $k \geq 1$; see Lemma 13 in Müller and Schlage-Puchta [38] for the remaining cases.

By Lemma 2.2 with k replaced by $k - 1$ and $f_j = f$ for $j = 1, 2, \ldots, k$, we find that the values of g may be computed using the formula

$$g(\xi) = \begin{cases} f(\xi \bmod \alpha), & \xi \notin \{\alpha, 2\alpha, \ldots, k\alpha\} \\ f(\alpha)f(0), & \xi \in \{\alpha, 2\alpha, \ldots, (k-1)\alpha\} \\ f(\alpha), & \xi = k\alpha \end{cases} \quad (\xi \in [0, k\alpha]). \tag{9.2}$$

Clearly, by moving ξ_1 and ξ_2 slightly if necessary and decreasing ε accordingly, we may assume that

$$\xi_1, \xi_2 \notin \{\alpha, 2\alpha, \ldots, (k-1)\alpha\}.$$

Let

$$(\mu - 1)\alpha < \xi_1 < \mu\alpha$$

and

$$(\nu - 1)\alpha < \xi_2 < \nu\alpha$$

for integers μ, ν satisfying $1 \leq \mu, \nu \leq k$. Then, according to formula (9.2), for $|\eta| < \varepsilon$ and sufficiently small ε we have

$$g(\xi_1 + \eta) = f(\xi_1 - (\mu - 1)\alpha + \eta),$$

while

$$\begin{aligned} g^{-1}(\xi_2 + \eta) &= \big(g(k\alpha - \xi_2 - \eta)\big)^{-1} \\ &= \big(f(\nu\alpha - \xi_2 - \eta)\big)^{-1} \\ &= f^{-1}(\xi_2 - (\nu - 1)\alpha + \eta). \end{aligned}$$

We thus find from (9.1) that

$$f(\xi_1 - (\mu - 1)\alpha + \eta) = f^{-1}(\xi_2 - (\nu - 1)\alpha + \eta), \quad |\eta| < \varepsilon,$$

which contradicts the fact that f is a test function, since

$$0 < \xi_1 - (\mu - 1)\alpha,\ \xi_2 - (\nu - 1)\alpha < \alpha.$$

□

9.3 Existence of test functions

The purpose of this short section is to show that test functions do in fact exist; this is all that is needed to ensure the availability of the maps λ_f to be discussed in Section 9.4 below. In fact much more is true: given any proper subgroup Λ of $(\mathbb{R},+)$, there exists a family $\mathfrak{F}$ of pairwise locally incompatible test functions in $\mathscr{R}\mathscr{F}(G)$ such that $|\mathfrak{F}| = |G|^{(\mathbb{R}:\Lambda)}$ and such that $C_{\mathscr{R}\mathscr{F}(G)}(f) \cong \Lambda$ for each $f \in \mathfrak{F}$; see Theorem 10.1.

We shall need the following purely arithmetic result, which will also play a role in the proof of Theorem 10.1.

Lemma 9.5 *Let ξ_1, ξ_2 be real numbers, at least one of which is rational. Suppose that there exists $\varepsilon > 0$ such that*

$$(\xi_1 + \eta)^2 \in \mathbb{Q} \iff (\xi_2 + \eta)^2 \in \mathbb{Q}, \quad \eta \in (0, \varepsilon) \cap \bar{\mathbb{Q}}, \tag{9.3}$$

where $\bar{\mathbb{Q}}$ denotes the algebraic closure of $\mathbb{Q}$ in $\mathbb{R}$. Then $\xi_1 = \xi_2$.

Proof It is enough to establish the desired conclusion (that $\xi_1 = \xi_2$) under the (formally stronger) hypotheses that at least one of ξ_1, ξ_2 is rational and that

$$(\xi_1 + \eta)^2 \in \mathbb{Q} \iff (\xi_2 + \eta)^2 \in \mathbb{Q}, \quad \eta \in (-\varepsilon, \varepsilon) \cap \bar{\mathbb{Q}}. \tag{9.4}$$

Indeed, suppose that ξ_1, ξ_2 meet the hypotheses of Lemma 9.5. Then, choosing $\eta_1 \in (0, \varepsilon/2] \cap \mathbb{Q}$ and setting $\xi_i' := \xi_i + \eta_1$, at least one of ξ_1', ξ_2' is still rational and we have

$$(\xi_1' + \eta)^2 \in \mathbb{Q} \iff (\xi_2' + \eta)^2 \in \mathbb{Q}, \quad \eta \in (-\eta_1, \eta_1) \cap \bar{\mathbb{Q}}.$$

The (formally) weaker version of Lemma 9.5 now yields $\xi_1' = \xi_2'$, hence also $\xi_1 = \xi_2$.

We shall now prove this formally weaker version of Lemma 9.5. Suppose without loss of generality that ξ_1 is rational. If q is a rational number satisfying $0 < q < \varepsilon$ then $(\xi_1 \pm q)^2 \in \mathbb{Q}$. Invoking condition (9.4) with $\eta = q$ and $\eta = -q$, we find that $(\xi_2 + q)^2, (\xi_2 - q)^2 \in \mathbb{Q}$. Subtracting yields that $4q\xi_2$ is rational, thus $\xi_2 \in \mathbb{Q}$, since q is rational and non-zero.

Now let r be a rational number such that $r \neq 0$ and $r\sqrt{2} \in (\xi_1, \xi_1 + \varepsilon)$. Applying condition (9.4) with

$$\eta = r\sqrt{2} - \xi_1 \in \bar{\mathbb{Q}} \cap (0, \varepsilon),$$

we obtain that

$$(\xi_2 + r\sqrt{2} - \xi_1)^2 = (\xi_2 - \xi_1)^2 + 2r^2 + 2r(\xi_2 - \xi_1)\sqrt{2} \in \mathbb{Q}.$$

Since $r \neq 0$, the assumption that $\xi_1 \neq \xi_2$ implies that $\sqrt{2}$ is rational, a contradiction. Hence $\xi_1 = \xi_2$, as claimed. □

Corollary 9.6 *Let ξ_1, ξ_2 be real numbers, and suppose that there exists $\varepsilon > 0$ such that*

$$(\xi_1+\eta)^2 \in \mathbb{Q} \iff (\xi_2+\eta)^2 \in \mathbb{Q}, \quad \eta \in (0,\varepsilon).$$

Then we have $\xi_1 = \xi_2$.

Proof Choose $\eta_1 \in (0,\varepsilon)$ such that $\xi_1' := \xi_1 + \eta_1$ is rational, and set $\xi_2' := \xi_2 + \eta_1$. Then ξ_1' and ξ_2' satisfy

$$(\xi_1'+\eta)^2 \in \mathbb{Q} \iff (\xi_2'+\eta)^2 \in \mathbb{Q}, \quad \eta \in (0,\varepsilon-\eta_1).$$

By Lemma 9.5 we have $\xi_1' = \xi_2'$, hence also $\xi_1 = \xi_2$. □

We can now give a uniform construction exhibiting a concrete test function in each non-trivial $\mathscr{RF}$-group. Let G be a non-trivial group, and let $x \in G - \{1_G\}$ be any non-trivial element. We claim that the function f_0 of length 1 given by

$$f_0(\xi) = \left\{\begin{matrix} x, & \xi^2 \in \mathbb{Q} \\ 1_G & \text{otherwise} \end{matrix}\right\} \quad (\xi \in [0,1]),$$

is a test function. To see this, suppose for a contradiction that there exist $\varepsilon > 0$ and points $\xi_1, \xi_2 \in (0,1)$ such that

$$f_0(\xi_1+\eta) = f_0^{-1}(\xi_2+\eta), \quad |\eta| < \varepsilon. \tag{9.5}$$

Choosing η in equation (9.5) in such a way that $\xi_1 + \eta$ is rational, we observe that (9.5) is impossible unless $x^2 = 1_G$. Thus we may suppose that $x = x^{-1}$ is a non-trivial involution, in which case (9.5) simplifies to

$$f_0(\xi_1+\eta) = f_0(\xi_2'-\eta), \quad |\eta| < \varepsilon, \tag{9.6}$$

where $\xi_2' := 1 - \xi_2$. Assertion (9.6) in turn is equivalent to the statement that

$$(\xi_1+\eta)^2 \in \mathbb{Q} \iff (-\xi_2'+\eta)^2 \in \mathbb{Q}, \quad |\eta| < \varepsilon.$$

However, in view of Corollary 9.6 the last assertion implies that $\xi_1 = -\xi_2'$, which is impossible since $\xi_1, \xi_2' > 0$. Hence f_0 is a test function as claimed.

9.4 The maps λ_f

Here we are going to explain how a given test function $f \in \mathscr{RF}(G)$ gives rise to a surjective homomorphism $\lambda_f : \mathscr{RF}(G) \to \mathbb{R}$ with $\lambda_f(E(G)) = 0$. Our construction is based on the idea of *locally comparing* a function $g \in \mathscr{RF}(G)$ with a fixed test function and its inverse.

9.4.1 Definition of λ_f

Given a test function $f \in \mathscr{RF}(G)$ of length α and an arbitrary element $g \in \mathscr{F}(G)$ of length β, say, define sets $\mathscr{M}_f^+(g)$ and $\mathscr{M}_f^-(g)$ by

$$\mathscr{M}_f^+(g) := \Big\{\xi \in (0,\beta) : \exists \varepsilon > 0, \exists \xi' \in (0,\alpha) \text{ such that } g(\xi+\eta) = f(\xi'+\eta) \text{ for all } |\eta| < \varepsilon\Big\}$$

and

$$\mathscr{M}_f^-(g) := \Big\{\xi \in (0,\beta) : \exists \varepsilon > 0, \exists \xi' \in (0,\alpha) \text{ such that } g(\xi+\eta) = f^{-1}(\xi'+\eta) \text{ for all } |\eta| < \varepsilon\Big\}$$

(of course, our notation is intended to imply that all the function values written down are actually defined). The following is more or less immediate from the definition of the sets $\mathscr{M}_f^+(g)$ and $\mathscr{M}_f^-(g)$.

Lemma 9.7 *Let $f \in \mathscr{RF}(G)$ be any fixed test function, and let $L(f) = \alpha$. Then:*

(i) $\mathscr{M}_f^+(g) \cap \mathscr{M}_f^-(g) = \varnothing, \quad g \in \mathscr{F}(G)$;

(ii) $\mathscr{M}_f^+(f|_{[0,\beta]}) = (0,\beta)$ *and* $M_f^-(f|_{[0,\beta]}) = \varnothing, \quad 0 \le \beta \le \alpha$;

(iii) $\mathscr{M}_f^+(g) = \varnothing = \mathscr{M}_f^-(g), \quad g \in G_0$.

Since the sets $\mathscr{M}_f^+(g), \mathscr{M}_f^-(g)$ are defined by open conditions (that is, conditions invariant under slight perturbations of the point considered), $\mathscr{M}_f^+(g)$ and $\mathscr{M}_f^-(g)$ are open sets and hence Lebesgue measurable. Given a fixed test function $f \in \mathscr{RF}(G)$, we define a function $\lambda_f : \mathscr{RF}(G) \to \mathbb{R}$ by

$$\lambda_f(g) := \mu\big(\mathscr{M}_f^+(g)\big) - \mu\big(\mathscr{M}_f^-(g)\big), \quad g \in \mathscr{RF}(G),$$

where μ denotes Lebesgue measure. We note that, by part (iii) of Lemma 9.7, we have

$$\lambda_f(G_0) = 0. \tag{9.7}$$

Also, by part (ii) of Lemma 9.7,

$$\lambda_f(f|_{[0,\beta]}) = \beta, \quad 0 \le \beta \le L(f). \tag{9.8}$$

9.4.2 The main result for the maps λ_f

The main result for the maps λ_f is the following.

Theorem 9.8 *For each fixed test function $f \in \mathscr{RF}(G)$, the map*

$$\lambda_f : \mathscr{RF}(G) \to \mathbb{R}$$

defined in subsection 9.4.1 *is a surjective group homomorphism whose kernel contains $E(G)$.*

We shall establish Theorem 9.8 in three steps: first, we will show that λ_f respects inverses; then, we shall prove that the equation

$$\lambda_f(gh) = \lambda_f(g) + \lambda_f(h) \tag{9.9}$$

holds for $g, h \in \mathscr{RF}(G)$ with $\varepsilon_0(g,h) = 0$; finally, making use of both these results we shall show that equation (9.9) holds for arbitrary elements $g, h \in \mathscr{RF}(G)$, that is, that λ_f is a group homomorphism. Once this is accomplished, the surjectivity of λ_f will follow from the fact (see equation (9.8)) that

$$[0, L(f)] \subseteq \lambda_f(\mathscr{RF}(G)),$$

while our claim that $\lambda_f(E(G)) = 0$ is an obvious consequence of (9.9) and (9.7), plus the fact that $E(G)$ is the normal closure of G_0.

Step 1 *For each test function $f \in \mathscr{RF}(G)$, we have*

$$\lambda_f(g^{-1}) = -\lambda_f(g), \quad g \in \mathscr{RF}(G). \tag{9.10}$$

Suppose that $L(g) = \beta$. Then, by definition,

$$\mathscr{M}_f^+(g^{-1}) = \Big\{\xi \in (0,\beta) : \exists \varepsilon > 0, \exists \xi_1 \in (0,\alpha) \text{ such that } g^{-1}(\xi+\eta) = f(\xi_1+\eta) \text{ for all } |\eta| < \varepsilon\Big\}.$$

Since for $\xi \in \mathscr{M}_f^+(g^{-1})$ with corresponding ξ_1, ε, and $|\eta| < \varepsilon$ we have

$$\big(g(\beta-\xi-\eta)\big)^{-1} = g^{-1}(\xi+\eta) = f(\xi_1+\eta) = \big(f^{-1}(\alpha-\xi_1-\eta)\big)^{-1},$$

we can see that, for $\xi \in (0,\beta)$,

$$\begin{aligned}\xi \in \mathscr{M}_f^+(g^{-1}) &\iff \exists \varepsilon > 0, \exists \xi_1' \in (0,\alpha) \text{ such that} \\ &\qquad g(\xi'+\eta') = f^{-1}(\xi_1'+\eta') \text{ for all } |\eta'| < \varepsilon \\ &\iff \xi' \in \mathscr{M}_f^-(g),\end{aligned}$$

where $\xi' := \beta - \xi$, $\xi_1' := \alpha - \xi_1$, and $\eta' := -\eta$. We thus find that

$$\mathscr{M}_f^+(g^{-1}) = -\big(\mathscr{M}_f^-(g)\big) + \beta, \tag{9.11}$$

and replacing g by g^{-1} in (9.11) yields the corresponding formula

$$\mathscr{M}_f^-(g^{-1}) = -\big(\mathscr{M}_f^+(g)\big) + \beta. \tag{9.12}$$

Formulae (9.11) and (9.12) together with well-known invariance properties of Lebesgue measure now give

$$\begin{aligned}\lambda_f(g^{-1}) &= \mu\big(\mathscr{M}_f^+(g^{-1})\big) - \mu\big(\mathscr{M}_f^-(g^{-1})\big) \\ &= \mu\big(\mathscr{M}_f^-(g)\big) - \mu\big(\mathscr{M}_f^+(g)\big) \\ &= -\lambda_f(g),\end{aligned}$$

as claimed.

Step 2 Let $g,h \in \mathscr{F}(G)$ with $L(g) = \beta$ and $L(h) = \gamma$. We claim that

$$\mathscr{M}_f^+(g*h) - \{\beta\} = \mathscr{M}_f^+(g) \cup \big(\mathscr{M}_f^+(h) + \beta\big) \tag{9.13}$$

and

$$\mathscr{M}_f^-(g*h) - \{\beta\} = \mathscr{M}_f^-(g) \cup \big(\mathscr{M}_f^-(h) + \beta\big). \tag{9.14}$$

Indeed, for $\xi \in (0,\beta+\gamma)$, we have

$$\begin{aligned}
&\xi \in \mathscr{M}_f^+(g*h) - \{\beta\} \\
&\iff \Big(\exists \varepsilon > 0, \exists \xi' \in (0,\alpha) \text{ such that } (\xi-\varepsilon,\xi+\varepsilon) \subseteq (0,\beta) \text{ and} \\
&\qquad (g*h)(\xi+\eta) = f(\xi'+\eta) \text{ for all } |\eta| < \varepsilon\Big) \text{ or} \\
&\qquad \Big(\exists \varepsilon > 0, \exists \xi' \in (0,\alpha) \text{ such that } (\xi-\varepsilon,\xi+\varepsilon) \subseteq (\beta,\beta+\gamma) \text{ and} \\
&\qquad (g*h)(\xi+\eta) = f(\xi'+\eta) \text{ for all } |\eta| < \varepsilon\Big) \\
&\iff \Big(\exists \varepsilon > 0, \exists \xi' \in (0,\alpha) \text{ such that} \\
&\qquad g(\xi+\eta) = f(\xi'+\eta) \text{ for all } |\eta| < \varepsilon\Big) \text{ or} \\
&\qquad \Big(\exists \varepsilon > 0, \exists \xi' \in (0,\alpha) \text{ such that} \\
&\qquad h(\xi-\beta+\eta) = f(\xi'+\eta) \text{ for all } |\eta| < \varepsilon\Big) \\
&\iff \xi \in \mathscr{M}_f^+(g) \text{ or } \xi-\beta \in \mathscr{M}_f^+(h) \\
&\iff \xi \in \mathscr{M}_f^+(g) \cup \big(\mathscr{M}_f^+(h)+\beta\big),
\end{aligned}$$

whence (9.13). A similar argument establishes equation (9.14). Since a singleton set has measure 0, and since Lebesgue measure is invariant under translations as well as (finitely) additive, we infer from (9.13) and (9.14) that

$$\mu\big(\mathscr{M}_f^+(g*h)\big) = \mu\big(\mathscr{M}_f^+(g)\big) + \mu\big(\mathscr{M}_f^+(h)\big) \tag{9.15}$$

and

$$\mu\big(\mathscr{M}_f^-(g*h)\big) = \mu\big(\mathscr{M}_f^-(g)\big) + \mu\big(\mathscr{M}_f^-(h)\big). \tag{9.16}$$

Combining formulae (9.15) and (9.16) with Remark 2.9, we find that, for $g,h \in \mathscr{RF}(G)$ such that $\varepsilon_0(g,h) = 0$,

$$\begin{aligned}
\lambda_f(gh) &= \mu\big(\mathscr{M}_f^+(gh)\big) - \mu\big(\mathscr{M}_f^-(gh)\big) \\
&= \mu\big(\mathscr{M}_f^+(g*h)\big) - \mu\big(\mathscr{M}_f^-(g*h)\big) \\
&= \Big(\mu\big(\mathscr{M}_f^+(g)\big) + \mu\big(\mathscr{M}_f^+(h)\big)\Big) - \Big(\mu\big(\mathscr{M}_f^-(g)\big) + \mu\big(\mathscr{M}_f^-(h)\big)\Big) \\
&= \lambda_f(g) + \lambda_f(h).
\end{aligned}$$

Step 3 Now let $g,h \in \mathscr{RF}(G)$ be arbitrary, and apply Lemma 2.15 (visibility of cancellation) to write $g = g_1 \circ c$, $h = c^{-1} \circ h_1$, so that $gh = g_1 \circ h_1$. Then, using the results of steps 1 and 2, we obtain

$$\begin{aligned}\lambda_f(gh) &= \lambda_f(g_1 \circ h_1)\\ &= \lambda_f(g_1) + \lambda_f(h_1)\\ &= \lambda_f(g_1) + \lambda_f(c) + \lambda_f(c^{-1}) + \lambda_f(h_1)\\ &= \lambda_f(g_1 \circ c) + \lambda_f(c^{-1} \circ h_1)\\ &= \lambda_f(g) + \lambda_f(h),\end{aligned}$$

so that equation (9.9) holds in general, that is, λ_f is a group homomorphism as claimed.

9.4.3 Some consequences of Theorem 9.8

As an immediate consequence of Theorem 9.8 and the existence of test functions (see Section 9.3), we find that non-trivial $\mathscr{RF}$-groups are never generated by their elliptic elements.

Corollary 9.9 *Let G be a non-trivial group. Then the quotient group $\mathscr{RF}(G)/E(G)$ maps homomorphically onto $\mathbb{R}$; in particular, $\mathscr{RF}(G)$ is not generated by its elliptic elements.*

We also obtain the following.

Corollary 9.10 *If $f \in \mathscr{RF}(G)$ is a test function then f is not contained in the normal subgroup $E(G)[\mathscr{RF}(G), \mathscr{RF}(G)]$.*

Proof By equation (9.8) with $\beta = L(f)$ we have $\lambda_f(f) = L(f) > 0$, while Theorem 9.8 tells us that $\lambda_f(E(G)[\mathscr{RF}(G), \mathscr{RF}(G)]) = 0$. □

Corollary 9.11 *Let f be a test function. Then every non-trivial element of the centraliser $C_{\mathscr{RF}(G)}(f)$ of f in $\mathscr{RF}(G)$ is itself a test function; in particular, we have*

$$C_{\mathscr{RF}(G)}(f) \cap E(G)[\mathscr{RF}(G), \mathscr{RF}(G)] = \{\mathbf{1}_G\}. \qquad (9.17)$$

Proof If $f \in \mathscr{RF}(G)$ is cyclically reduced, of positive length, and normalised so that $f(0) = 1_G$ then, according to Theorem 8.16, the elements of $C_{\mathscr{RF}(G)}(f)$ are of the form

$$\left(f^k \circ f|_{[0,\alpha']}\right)^{\pm 1}, \quad (k \in \mathbb{N}_0,\ 0 \leq \alpha' < L(f)),$$

with α' subject to certain further restrictions that do not matter for the present purpose.

Now let f be a test function such that $f(0) = 1_G$. Then f is cyclically reduced by Lemma 9.2 and we have $L(f) > 0$ by definition, so that the above description of centraliser elements applies. Since the inverse of a test function is also a test function, Lemma 9.3 ensures that, for f a normalised test function, $C_{\mathscr{RF}(G)}(f) - \{\mathbf{1}_G\}$ consists entirely of test functions. For an arbitrary test function f, we conjugate by a G_0-element x to make $\tilde{f} = xfx^{-1}$ normalised, obtain that $C_{\mathscr{RF}(G)}(\tilde{f}) - \{\mathbf{1}_G\}$ consists entirely of test functions, and then conjugate back to obtain the same conclusion for f itself. The particular statement follows from the main assertion just proved plus Corollary 9.10. □

9.5 Locally incompatible test functions

We shall need a notion of independence for test functions. The natural concept in this context turns out to be that of local incompatibility, which is introduced in the following definition in somewhat greater generality.

Definition 9.12 Two functions $f_1, f_2 \in \mathscr{F}(G)$ of lengths α_1 and α_2, respectively, are called *locally compatible* (loc. comp.) if there exist $\varepsilon > 0$ and points $\xi_i \in (0, \alpha_i)$ such that we have either

$$f_1(\xi_1 + \eta) = f_2(\xi_2 + \eta), \quad |\eta| < \varepsilon,$$

or

$$f_1(\xi_1 + \eta) = f_2^{-1}(\xi_2 + \eta), \quad |\eta| < \varepsilon.$$

If f_1 and f_2 both have positive length but are not locally compatible, they are called *locally incompatible* (loc. incomp.).

If two functions $f_1, f_2 \in \mathscr{RF}(G)$ are locally compatible, we also say, slightly varying the above terminology, that f_1 *is locally compatible to* f_2. We observe, however, that local compatibility is a *symmetric* relation on $\mathscr{F}(G)$; that is, we have

$$f_1 \text{ loc. comp. } f_2 \Longrightarrow f_2 \text{ loc. comp. } f_1, \quad (f_1, f_2 \in \mathscr{F}(G)). \tag{9.18}$$

A first observation concerning locally incompatible functions is as follows.

Lemma 9.13

(i) *If* $f_1, f_2 \in \mathscr{F}(G)$ *are locally incompatible then* $\varepsilon_0(f_1, f_2) = 0$.

(ii) *If* $f_1, f_2 \in \mathscr{F}(G)$ *are locally incompatible then so are the functions* f_1^{-1} *and* f_2, *as are the functions* f_1^{-1} *and* f_2^{-1}.

Proof We shall show (i), leaving the verification of (ii) as an exercise for the reader. Let $L(f_i) = \alpha_i > 0$, and suppose that $\varepsilon_0(f_1, f_2) > 0$. Then there exists $\varepsilon > 0$ such that

$$f_1(\alpha_1 - \eta) f_2(\eta) = 1_G, \quad 0 \leq \eta \leq \varepsilon,$$

that is,

$$f_2(\eta) = f_1^{-1}(\eta), \quad 0 \leq \eta \leq \varepsilon.$$

It follows that

$$f_2\Big(\frac{\varepsilon}{2} + \eta'\Big) = f_1^{-1}\Big(\frac{\varepsilon}{2} + \eta'\Big), \quad |\eta'| < \frac{\varepsilon}{2};$$

hence f_2 is locally compatible to f_1, so f_1 is locally compatible to f_2 by (9.18), contradicting our assumption. □

In the remainder of this section we are going to establish a somewhat technical result to the effect that every finite product of the form $\prod_j f_j^{\gamma_j}$ with nonzero exponents γ_j in pairwise locally incompatible test functions f_j is also a test function; see Proposition 9.18. This result will be put to good use in Section 10.7, where we shall derive certain structural properties of the groups $\mathscr{RF}(G)$ and their quotients $\mathscr{RF}(G)/E(G)$.

Lemma 9.14 *If* $f, g \in \mathscr{RF}(G)$ *are locally incompatible test functions then* fg *is also a test function.*

Proof Let $L(f) = \alpha$, $L(g) = \beta$, and set $h = fg$. By Lemma 9.13(i) we have $\gamma := L(h) = \alpha + \beta$ and

$$h(\xi) = \left\{ \begin{array}{ll} f(\xi), & 0 \leq \xi < \alpha \\ f(\alpha)g(0), & \xi = \alpha \\ g(\xi - \alpha), & \alpha < \xi \leq \gamma \end{array} \right\} \quad (\xi \in [0, \gamma]). \tag{9.19}$$

Suppose for a contradiction that there exist $\varepsilon > 0$ and points $\xi_1, \xi_2 \in (0, \gamma)$ such that

$$h(\xi_1 + \eta) = h^{-1}(\xi_2 + \eta), \quad |\eta| < \varepsilon. \tag{9.20}$$

We may assume without loss of generality that $\xi_1 \neq \alpha$ and $\xi_2 \neq \beta$. Suppose first that $\xi_1 \in (0,\alpha)$. Then, for ε sufficiently small, equation (9.19) yields

$$h(\xi_1+\eta) = f(\xi_1+\eta), \quad |\eta| < \varepsilon,$$

while

$$h^{-1}(\xi_2+\eta) = \big(h(\gamma-\xi_2-\eta)\big)^{-1} = \begin{Bmatrix} f^{-1}(\xi_2-\beta+\eta), & \xi_2 > \beta \\ g^{-1}(\xi_2+\eta), & \xi_2 < \beta \end{Bmatrix} \quad (|\eta| < \varepsilon).$$

We deduce from (9.20) that

$$f(\xi_1+\eta) = \begin{Bmatrix} f^{-1}(\xi_2-\beta+\eta), & \xi_2 > \beta \\ g^{-1}(\xi_2+\eta), & \xi_2 < \beta \end{Bmatrix} \quad (|\eta| < \varepsilon).$$

In the first case the corresponding assertion contradicts the fact that f is a test function, while the assertion corresponding to the second case contradicts our hypothesis that f and g are locally incompatible. The case where $\xi_1 \in (\alpha,\gamma)$ is similar and so is omitted. □

Lemma 9.15 *Let $f_1, f_2 \in \mathscr{RF}(G)$ be locally incompatible cyclically reduced functions, and let γ_1, γ_2 be non-zero integers. Then $f_1^{\gamma_1}$ and $f_2^{\gamma_2}$ are locally incompatible.*

Proof In view of part (ii) of Lemma 9.13, it is enough to consider the case when $\gamma_1, \gamma_2 \in \mathbb{N}$. Let $L(f_i) = \alpha_i$, and set $g := f_1^{\gamma_1}$ and $h := f_2^{\gamma_2}$. Since f_1, f_2 are of positive length and cyclically reduced we have $L(g) = \gamma_1\alpha_1 > 0$ and $L(h) = \gamma_2\alpha_2 > 0$, and g is given by the formula (cf. (9.2))

$$g(\xi) = \begin{Bmatrix} f_1(\xi \bmod \alpha_1), & \xi \notin \{\alpha_1, 2\alpha_1, \ldots, \gamma_1\alpha_1\} \\ f_1(\alpha_1)f_1(0), & \xi \in \{\alpha_1, 2\alpha_1, \ldots, (\gamma_1-1)\alpha_1\} \\ f_1(\alpha_1), & \xi = \gamma_1\alpha_1 \end{Bmatrix} \quad (\xi \in [0,\gamma_1\alpha_1]),$$

with a corresponding formula for h. Suppose for a contradiction that there exist $\varepsilon > 0$ and points $\xi_i \in (0,\gamma_i\alpha_i)$ such that either

$$g(\xi_1+\eta) = h(\xi_2+\eta), \quad |\eta| < \varepsilon, \tag{9.21}$$

or

$$g(\xi_1+\eta) = h^{-1}(\xi_2+\eta), \quad |\eta| < \varepsilon. \tag{9.22}$$

We may clearly assume without loss of generality that

$$\xi_i \notin \{\alpha_i, 2\alpha_i, \ldots, (k_i-1)\alpha_i\}, \quad i = 1,2.$$

Let

$$(j_1-1)\alpha_1 < \xi_1 < j_1\alpha_1$$

and

$$(j_2-1)\alpha_2 < \xi_2 < j_2\alpha_2$$

hold for integers j_1, j_2 satisfying respectively $1 \le j_1 \le \gamma_1$ and $1 \le j_2 \le \gamma_2$. Then we have, for sufficiently small ε,

$$\begin{aligned} g(\xi_1+\eta) &= f_1(\xi_1-(j_1-1)\alpha_1+\eta), \quad |\eta|<\varepsilon, \\ h(\xi_2+\eta) &= f_2(\xi_2-(j_2-1)\alpha_2+\eta), \quad |\eta|<\varepsilon, \\ h^{-1}(\xi_2+\eta) &= f_2^{-1}(\xi_2-(j_2-1)\alpha_2+\eta), \quad |\eta|<\varepsilon. \end{aligned}$$

Hence, we find from (9.21) that

$$f_1(\xi_1-(j_1-1)\alpha_1+\eta) = f_2(\xi_2-(j_2-1)\alpha_2+\eta), \quad |\eta|<\varepsilon,$$

while (9.22) implies that

$$f_1(\xi_1-(j_1-1)\alpha_1+\eta) = f_2^{-1}(\xi_2-(j_2-1)\alpha_2+\eta), \quad |\eta|<\varepsilon,$$

both assertions contradicting our hypothesis that f_1 and f_2 are locally incompatible. □

Lemma 9.16 *Let $f_1, f_2, f_3 \in \mathscr{F}(G)$ be pairwise locally incompatible functions. Then $f_1 f_2$ and f_3 are locally incompatible.*

Proof For $i = 1,2,3$, set $L(f_i) = \alpha_i > 0$ and let $g := f_1 f_2$. By the first part of Lemma 9.13 we have

$$\beta := L(g) = \alpha_1 + \alpha_2 > 0$$

and

$$g(\xi) = \left\{ \begin{array}{ll} f_1(\xi), & 0 \le \xi < \alpha_1 \\ f_1(\alpha_1) f_2(0), & \xi = \alpha_1 \\ f_2(\xi-\alpha_1), & \alpha_1 < \xi \le \beta \end{array} \right\} \quad (\xi \in [0,\beta]).$$

Suppose for a contradiction that there exist $\varepsilon > 0$ and points $\xi_1 \in (0,\beta)$, $\xi_2 \in (0,\alpha_3)$ such that either

$$g(\xi_1+\eta) = f_3(\xi_2+\eta), \quad |\eta|<\varepsilon, \tag{9.23}$$

or

$$g(\xi_1+\eta) = f_3^{-1}(\xi_2+\eta), \quad |\eta|<\varepsilon. \tag{9.24}$$

We may assume without loss of generality that $\xi_1 \neq \alpha_1$, so that there are only two cases, $\xi_1 < \alpha_1$ and $\xi_1 > \alpha_1$. Suppose that $\xi_1 \in (0, \alpha_1)$. Then, for sufficiently small ε, we have

$$g(\xi_1 + \eta) = f_1(\xi_1 + \eta), \quad |\eta| < \varepsilon,$$

and (9.23), (9.24) imply that either

$$f_1(\xi_1 + \eta) = f_3(\xi_2 + \eta), \quad |\eta| < \varepsilon,$$

or

$$f_1(\xi_1 + \eta) = f_3^{-1}(\xi_2 + \eta), \quad |\eta| < \varepsilon.$$

Both assertions contradict the fact that f_1 and f_3 are locally incompatible. The case where $\xi_1 \in (\alpha_1, \beta)$ is similar and so is omitted. □

At this stage, immediate induction arguments allow us to generalise first Lemma 9.16 and then Lemma 9.14 to finitely many factors, thereby establishing the following.

Lemma 9.17

(i) *Let $k \geq 2$ be an integer, and let $f_1, f_2, \ldots, f_k \in \mathscr{F}(G)$ be pairwise locally incompatible functions. Then $f_1 f_2 \cdots f_{k-1}$ and f_k are locally incompatible.*

(ii) *Let $f_1, f_2, \ldots, f_k \in \mathscr{R}\mathscr{F}(G)$ be pairwise locally incompatible test functions, where $k \geq 1$. Then $f_1 f_2 \cdots f_k$ is also a test function.*

We now come to the main result of this section.

Proposition 9.18 *For $k \geq 1$, let $f_1, f_2, \ldots, f_k \in \mathscr{R}\mathscr{F}(G)$ be pairwise locally incompatible test functions, and let $\gamma_1, \gamma_2, \ldots, \gamma_k$ be non-zero integers. Then the element $f_1^{\gamma_1} f_2^{\gamma_2} \cdots f_k^{\gamma_k}$ is also a test function.*

Proof By Corollary 9.4 plus Lemmas 9.2 and 9.15, $f_1^{\gamma_1}, f_2^{\gamma_2}, \ldots, f_k^{\gamma_k}$ form a set of pairwise locally incompatible test functions. The result follows now from part (ii) of Lemma 9.17. □

9.6 A subgroup theorem

Definition 9.19 A subgroup $\mathscr{H} \leq \mathscr{R}\mathscr{F}(G)$ is called *hyperbolic* if the set $\mathscr{H} - \{\mathbf{1}_G\}$ consists entirely of hyperbolic elements, that is, of functions satisfying inequality (4.1).

It seems difficult to decide in general, when a family $\{\mathscr{H}_\sigma\}_{\sigma\in S}$ of hyperbolic subgroups $\mathscr{H}_\sigma \leq \mathscr{R}\mathscr{F}(G)$ has the property that its group-theoretic join

$$\mathscr{H} = \langle \mathscr{H}_\sigma : \sigma \in S \rangle$$

is also hyperbolic. In this section we shall consider a particular situation, where this can be shown to be the case. Our scenario concerns a family of hyperbolic subgroups, whose non-trivial elements are test functions, with the property that any two of these subgroups are independent in a sense made precise in part (ii) of the following definition.

Definition 9.20 Let $\{\mathscr{H}_\sigma\}_{\sigma\in S}$ be a family of hyperbolic subgroups $\mathscr{H}_\sigma \leq \mathscr{R}\mathscr{F}(G)$ with bijective indexing, and let $\mathscr{H} \leq \mathscr{R}\mathscr{F}(G)$ be a hyperbolic subgroup.

(i) The subgroup $\mathscr{H}$ is said to satisfy condition (T) if the set $\mathscr{H} - \{\mathbf{1}_G\}$ consists entirely of test functions.

(ii) The family $\{\mathscr{H}_\sigma\}_{\sigma\in S}$ is said to satisfy condition (LI) if

$$f_1 \in \mathscr{H}_{\sigma_1} - \{\mathbf{1}_G\},\ f_2 \in \mathscr{H}_{\sigma_2} - \{\mathbf{1}_G\},\ \text{and } \sigma_1 \neq \sigma_2 \Longrightarrow f_1 \text{ loc. incomp. } f_2.$$

We are going to show the following.

Theorem 9.21 *Let $\{\mathscr{H}_\sigma\}_{\sigma\in S}$ be a family of hyperbolic subgroups, $\mathscr{H}_\sigma \leq \mathscr{R}\mathscr{F}(G)$, with bijective indexing, and let $\mathscr{H} = \langle \mathscr{H}_\sigma : \sigma \in S \rangle$ be the subgroup of $\mathscr{R}\mathscr{F}(G)$ generated by the subgroups $\mathscr{H}_\sigma$. If the family $\{\mathscr{H}_\sigma\}_{\sigma\in S}$ satisfies condition* (LI), *and if every subgroup $\mathscr{H}_\sigma$ meets condition* (T), *then $\mathscr{H}$ is hyperbolic and $\mathscr{H} \cong \ast_{\sigma\in S}\mathscr{H}_\sigma$.*

For convenience, we now introduce a second (weaker) notion of independence for families of subgroups in $\mathscr{R}\mathscr{F}(G)$.

Definition 9.22 A family $\{\mathscr{H}_\sigma\}_{\sigma\in S}$ of subgroups $\mathscr{H}_\sigma \leq \mathscr{R}\mathscr{F}(G)$ with bijective indexing is called *independent* if

$$f_1 \in \mathscr{H}_{\sigma_1} - \{\mathbf{1}_G\}, f_2 \in \mathscr{H}_{\sigma_2} - \{\mathbf{1}_G\} \text{ and } \sigma_1 \neq \sigma_2 \Longrightarrow \varepsilon_0(f_1,f_2) = 0.$$

We note that a family $\{\mathscr{H}_\sigma\}_{\sigma\in S}$ satisfying property (LI) is independent in view of part (i) of Lemma 9.13.

The main purpose of the notion of independence of families of subgroups as introduced in Definition 9.22 is to ensure that the subgroups in question generate their free product.

Lemma 9.23 *Let $\{\mathscr{H}_\sigma\}_{\sigma\in S}$ be an independent family of subgroups $\mathscr{H}_\sigma \leq \mathscr{RF}(G)$, in the sense of Definition 9.22, with bijective indexing; let $\mathscr{H} = \langle \mathscr{H}_\sigma : \sigma \in S\rangle$ be their join, and suppose that $\mathscr{H}_\sigma \cap G_0 = \{\mathbf{1}_G\}$ for all $\sigma \in S$. Then $\mathscr{H} \cap G_0 = \{\mathbf{1}_G\}$ and $\mathscr{H} \cong \ast_{\sigma\in S}\mathscr{H}_\sigma$.*

Proof Let $h = h_1h_2\cdots h_r \in \mathscr{H}$ be an arbitrary element, where $r \in \mathbb{N}_0$, $h_\rho \in \mathscr{H}_{\sigma_\rho} - \{\mathbf{1}_G\}$ for $\rho = 1,2,\ldots,r$, and $\sigma_\rho \neq \sigma_{\rho+1}$ for $1 \leq \rho < r$. Since the family $\{\mathscr{H}_\sigma\}_{\sigma\in S}$ is independent, we have

$$\varepsilon_0(h_\rho, h_{\rho+1}) = 0, \quad 1 \leq \rho < r.$$

Furthermore, since each h_ρ, being non-trivial, has positive length, it follows by induction that $h = h_1 \circ h_2 \circ \cdots \circ h_r$ and thus

$$L(h) = L(h_1 \circ h_2 \circ \cdots \circ h_r) = \sum_{\rho=1}^{r} L(h_\rho).$$

Consequently, the assumption that $L(h) = 0$ forces $r = 0$ and thus $h = \mathbf{1}_G$; in particular, the system of subgroups $\{\mathscr{H}_\sigma\}_{\sigma\in S}$ generates its free product, that is, we have $\mathscr{H} \cong \ast_{\sigma\in S}\mathscr{H}_\sigma$ as claimed. We also deduce that $\mathscr{H} \cap G_0 = \{\mathbf{1}_G\}$. □

Proof of Theorem 9.21 Suppose that $\{\mathscr{H}_\sigma\}_{\sigma\in S}$ satisfies condition (LI) and that each subgroup $\mathscr{H}_\sigma$ meets condition (T). Since test functions have positive length by definition and condition (LI) ensures the independence of the system of subgroups $\{\mathscr{H}_\sigma\}_{\sigma\in S}$, Lemma 9.23 applies, to give $\mathscr{H} \cap G_0 = \{\mathbf{1}_G\}$ and $\mathscr{H} \cong \ast_{\sigma\in S}\mathscr{H}_\sigma$. In order to see that $\mathscr{H}$ is hyperbolic, consider the equation

$$h_1h_2\cdots h_r = h = txt^{-1}, \tag{9.25}$$

where $h_1, h_2, \ldots, h_r$ are test functions such that $h_\rho, h_{\rho+1}$ are locally incompatible for all $1 \leq \rho < r$ and $x \in G_0 - \{\mathbf{1}_G\}$. We must have $\tau := L(t) > 0$ since $\mathscr{H} \cap G_0 = \{\mathbf{1}_G\}$; and we may assume that $r \geq 3$ since test functions, being being cyclically reduced and of positive length, are hyperbolic (see Proposition 3.13) and since, for $r = 2$, the product h_1h_2 is also a test function by Lemma 9.14. Set $\gamma := L(h) = 2\tau$, and let $\gamma_\rho := L(h_\rho) > 0$ for $1 \leq \rho \leq r$. Evaluating h via the right-hand side of (9.25), we find from the definition of the star product that, on the one hand

$$h(\xi) = ((t \circ x) \circ t^{-1})(\xi) = \left\{\begin{array}{ll} t(\xi), & 0 \le \xi < \tau \\ t(\tau)x(0)t(\tau)^{-1}, & \xi = \tau \\ t(2\tau - \xi)^{-1}, & \tau < \xi \le \gamma \end{array}\right\} \quad (0 \le \xi \le \gamma). \tag{9.26}$$

On the other hand, evaluating h via the left-hand side of (9.25) by means of Lemma 2.2, we find that

$$h_\rho(\xi) = h(\xi + \xi_{\rho-1}), \quad \rho \in [r], 0 < \xi < \gamma_\rho, \tag{9.27}$$

where, for $0 \le \rho \le r$, we set $\xi_\rho = \sum_{i=1}^{\rho} \gamma_i$ in accordance with the notation of that lemma. We now claim that

$$\tau \in \{\xi_1, \xi_2, \ldots, \xi_{r-1}\}. \tag{9.28}$$

In order to see this, suppose for a contradiction that $\xi_{\rho_0 - 1} < \tau < \xi_{\rho_0}$ for some ρ_0 with $1 \le \rho_0 \le r$. Combining formulae (9.26) and (9.27) we find that

$$h_{\rho_0}(\xi) = \left\{\begin{array}{ll} t(\xi + \xi_{\rho_0-1}), & 0 < \xi < \tau - \xi_{\rho_0-1} \\ t(\tau)x(0)t(\tau)^{-1}, & \xi = \tau - \xi_{\rho_0-1} \\ t(2\tau - \xi - \xi_{\rho_0-1})^{-1}, & \tau - \xi_{\rho_0-1} < \xi < \gamma_{\rho_0} \end{array}\right\} \quad (0 < \xi < \gamma_{\rho_0}). \tag{9.29}$$

Suppose that

$$|\xi_{\rho_0-1} - \tau| \le |\tau - \xi_{\rho_0}|,$$

pick a point $\zeta_0 \in (0, \tau - \xi_{\rho_0-1})$, and set

$$\zeta_1 := \zeta_0 + 2\xi_{\rho_0-1} - 2\tau + \gamma_{\rho_0}.$$

Then

$$\zeta_1 > \gamma_{\rho_0} - 2|\xi_{\rho_0-1} - \tau| \ge 0,$$

so that $\gamma_{\rho_0} - \zeta_1 < \gamma_{\rho_0}$ and

$$\begin{aligned} \gamma_{\rho_0} - \zeta_1 &= 2\tau - 2\xi_{\rho_0-1} - \zeta_0 \\ &> 2(\tau - \xi_{\rho_0-1}) - (\tau - \xi_{\rho_0-1}) \\ &= \tau - \xi_{\rho_0-1}. \end{aligned}$$

Using (9.29), we thus find that, for $\varepsilon > 0$ sufficiently small,

$$h_{\rho_0}(\zeta_0 + \eta) = t(\zeta_0 + \xi_{\rho_0-1} + \eta), \quad |\eta| < \varepsilon,$$

and, again for $|\eta| < \varepsilon$, that

$$\begin{aligned} h_{\rho_0}^{-1}(\zeta_1+\eta) &= h_{\rho_0}(\gamma_{\rho_0}-\zeta_1-\eta)^{-1} \\ &= t(2\tau-\gamma_{\rho_0}+\zeta_1+\eta-\xi_{\rho_0-1}) \\ &= t(\zeta_0+\xi_{\rho_0-1}+\eta) \\ &= h_{\rho_0}(\zeta_0+\eta). \end{aligned}$$

The resulting equation,

$$h_{\rho_0}(\zeta_0+\eta) = h_{\rho_0}^{-1}(\zeta_1+\eta), \quad |\eta| < \varepsilon,$$

contradicts the fact that h_{ρ_0} is a test function, thus proving (9.28) in the case where $|\xi_{\rho_0-1}-\tau| \leq |\tau-\xi_{\rho_0}|$. If $|\xi_{\rho_0-1}-\tau| > |\tau-\xi_{\rho_0}|$ then we choose $\zeta_0 \in (\tau-\xi_{\rho_0-1}, \gamma_{\rho_0})$ and obtain the same contradiction, with ζ_1 defined as above. This establishes our claim (9.28).

Now let $\tau = \xi_{\rho_0}$ with $1 \leq \rho_0 \leq r-1$, according to (9.28), and consider the functions h_{ρ_0} and h_{ρ_0+1}. Suppose first that $\gamma_{\rho_0} \leq \gamma_{\rho_0+1}$, choose a point $\zeta \in (0, \gamma_{\rho_0})$, and set $\zeta' := \zeta + \gamma_{\rho_0+1} - \gamma_{\rho_0}$. Then

$$0 < \zeta \leq \zeta' < \gamma_{\rho_0+1},$$

so that ζ' lies in the interior of the domain of the function h_{ρ_0+1}; now, using (9.26) and (9.27), we can compute that, for small $\varepsilon > 0$ and $|\eta| < \varepsilon$,

$$h_{\rho_0}(\zeta+\eta) = h(\zeta+\xi_{\rho_0-1}+\eta) = t(\zeta+\xi_{\rho_0-1}+\eta)$$

while

$$\begin{aligned} h_{\rho_0+1}^{-1}(\zeta'+\eta) &= h_{\rho_0+1}(\gamma_{\rho_0+1}-\zeta'-\eta)^{-1} \\ &= h(\gamma_{\rho_0+1}+\xi_{\rho_0}-\zeta'-\eta)^{-1} \\ &= t(2\tau-\gamma_{\rho_0+1}-\xi_{\rho_0}+\zeta'+\eta) \\ &= t(\zeta+\xi_{\rho_0-1}+\eta) \\ &= h_{\rho_0}(\zeta+\eta). \end{aligned}$$

It follows that the functions h_{ρ_0} and h_{ρ_0+1} are locally compatible, contradicting the fact that h_{ρ_0}, h_{ρ_0+1}, being adjacent functions on the left-hand side of (9.25), are locally incompatible. Finally, it remains to consider the case when $\gamma_{\rho_0} > \gamma_{\rho_0+1}$. In this case, we choose $\zeta' \in (0, \gamma_{\rho_0+1})$, set $\zeta := \zeta'$, so that $\zeta \in (0, \gamma_{\rho_0})$, and then obtain, for $|\eta| < \varepsilon$ and small $\varepsilon > 0$,

$$h_{\rho_0}^{-1}(\zeta+\eta) = t(\tau-\zeta-\eta)^{-1} = h_{\rho_0+1}(\zeta'+\eta),$$

which again contradicts the fact that the functions h_{ρ_0}, h_{ρ_0+1} are locally incompatible. This final contradiction now shows that equation (9.25) with $x \in G_0 - \{\mathbf{1}_G\}$ is indeed impossible, implying that $\mathscr{H}$ is hyperbolic as claimed. $\square$

As a first illustration, consider an arbitrary family $\{f_\sigma\}_{\sigma\in S}$ of pairwise locally incompatible test functions with bijective indexing, and let $\mathscr{H}_\sigma := \langle f_\sigma \rangle$. By Corollary 9.4, each $\mathscr{H}_\sigma$ meets condition (T), and by Lemma 9.15 the family of subgroups $\{\mathscr{H}_\sigma\}_{\sigma\in S}$ also satisfies condition (LI). It follows from Theorem 9.21 that the subgroup $\mathscr{H} = \langle f_\sigma : \sigma \in S\rangle \leq \mathscr{RF}(G)$ is hyperbolic and free, with basis $\{f_\sigma\}_{\sigma\in S}$; see part (i) of Proposition 10.9, where a stronger result is established by a different argument.

Our main application of Theorem 9.21, however, is the following.

Corollary 9.24 *Let $\{f_\sigma\}_{\sigma\in S}$ be a family of pairwise locally incompatible test functions, with bijective indexing. Then the corresponding family $\{C_{\mathscr{RF}(G)}(f_\sigma)\}_{\sigma\in S}$ of centralisers generates a hyperbolic subgroup of $\mathscr{RF}(G)$ isomorphic to the free product $\ast_{\sigma\in S} C_{\mathscr{RF}(G)}(f_\sigma)$.*

According to Corollary 9.11 the centraliser of a test function satisfies condition (T). Moreover, it is possible to generalise Lemma 9.15 to the following.

Lemma 9.25 *Let g_1, g_2 be locally incompatible cyclically reduced functions of length β_1 and β_2 respectively, let γ_1, γ_2 be non-negative integers, and let β_1', β_2' be real numbers such that $0 \leq \beta_i' < \beta_i$. Then the functions $h_1 := g_1^{\gamma_1} g_1|_{[0,\beta_1']}$ and $h_2 := g_2^{\gamma_2} g_2|_{[0,\beta_2']}$ are also locally incompatible, provided that $\gamma_1 + \beta_1' > 0$ and $\gamma_2 + \beta_2' > 0$.*

See Lemma 6.1 in Müller [36] for the (rather technical) proof of Lemma 9.25.

Combining Lemma 9.25 with part (i) of Theorem 8.16, and making use of Lemmas 9.2 and 9.13(ii), it is now straightforward to show that the family of centralisers $\{C_{\mathscr{RF}(G)}(f_\sigma)\}_{\sigma\in S}$ also satisfies property (LI) in Definition 9.20, so that Corollary 9.24 follows from Theorem 9.21; see Corollary 2 in [36] for more details.

9.7 The maps λ_S

Given a set S, we denote by $\ell^1(S)$ the subspace of the real vector space $\mathbb{R}^S$ consisting of those functions $f : S \to \mathbb{R}$ which are absolutely summable, that is, which satisfy

$$\|f\| := \sum_{\sigma \in S} |f(\sigma)| < \infty.$$

Here, as usual, the sum $\sum_{\sigma \in S} x_\sigma$ of an arbitrary family $\{x_\sigma\}_{\sigma \in S}$ of non-negative real numbers x_σ is defined as

$$\sum_{\sigma \in S} x_\sigma := \sup_{\substack{T \subseteq S \\ |T| < \infty}} \sum_{\sigma \in T} x_\sigma. \tag{9.30}$$

Clearly, $\ell^1(S)$ is completely determined by the cardinality of S. More formally, if S_1, S_2 are sets such that $|S_1| = |S_2|$ and if $\varphi : S_1 \to S_2$ is a bijection then the map $\Phi : \ell^1(S_1) \to \ell^1(S_2)$ given by $\Phi(f) := f \circ \varphi^{-1}$ is a norm-preserving $\mathbb{R}$-linear isomorphism; in particular, $|S_1| = |S_2|$ implies that $\ell^1(S_1)$ and $\ell^1(S_2)$ are isomorphic as abelian groups. Also, concerning the cardinality of $\ell^1(S)$, we have

$$|\ell^1(S)| = \begin{cases} 2^{\aleph_0}, & 0 < |S| < 2^{\aleph_0}, \\ |S|, & |S| \geq 2^{\aleph_0}; \end{cases} \tag{9.31}$$

see Exercise 9.11. For the present purposes, $\mathbb{R}^S$ and $\ell^1(S)$ will be viewed simply as additive abelian groups.

The aim of this section is to exhibit (large) abelian quotients of the group

$$\mathscr{RF}_0(G) = \mathscr{RF}(G)/E(G)$$

that are of the form $\ell^1(S)$ for certain sets S. Our principal tool will again be sets of pairwise locally incompatible test functions. The precise connection is given by the following result.[1]

Theorem 9.26 *Let $\{f_\sigma\}_{\sigma \in S}$ be a family of pairwise locally incompatible test functions such that $f_\sigma(0) = 1_G = f_\sigma(L(f_\sigma))$ for every $\sigma \in S$. Define a homomorphism $\lambda_S : \mathscr{RF}(G) \to \mathbb{R}^S$ via*

$$\lambda_S(g)(\sigma) = \lambda_{f_\sigma}(g), \quad g \in \mathscr{RF}(G), \sigma \in S,$$

where λ_{f_σ} is as in subsection 9.4.1. *Then the image of λ_S equals $\ell^1(S)$.*

Corollary 9.27 *The map λ_S defined in Theorem* 9.26 *induces a homomorphism $\bar{\lambda}_S : \mathscr{RF}_0(G) \to \mathbb{R}^S$ with image $\ell^1(S)$.*

Proof Since $\lambda_{f_\sigma}(g) = 0$ for all $g \in E(G)$ and $\sigma \in S$, by Theorem 9.8, we have $\lambda_S(E(G)) = 0$, so λ_S factors through $\mathscr{RF}_0(G)$. □

[1] The material of this section is taken from previously unpublished joint work of J.-C. Schlage-Puchta and the second-named author of the present text.

Proof of Theorem 9.26 Let $L(f_\sigma) = \alpha_\sigma > 0$, and let $g \in \mathscr{RF}(G)$ be a function of length $L(g) = \beta$. We want to show that $\lambda_S(g) \in \ell^1(S)$. For $\sigma \in S$, in the notation of Section 9.4 we set

$$I_\sigma = \mathscr{M}^+_{f_\sigma}(g) \cup \mathscr{M}^-_{f_\sigma}(g).$$

As a union of open sets, I_σ is open and thus Lebesgue measurable, and we have

$$\begin{aligned}|\lambda_{f_\sigma}(g)| &= \left|\mu\left(\mathscr{M}^+_{f_\sigma}(g)\right) - \mu\left(\mathscr{M}^-_{f_\sigma}(g)\right)\right| \\ &\leq \mu\left(\mathscr{M}^+_{f_\sigma}(g)\right) + \mu\left(\mathscr{M}^-_{f_\sigma}(g)\right) = \mu(I_\sigma),\end{aligned}$$

where we have used part (i) of Lemma 9.7 in the last step. Moreover, since the test functions f_σ are locally incompatible, given $\sigma_1, \sigma_2 \in S$ with $\sigma_1 \neq \sigma_2$ and $g \in \mathscr{RF}(G)$ the four sets $\mathscr{M}^+_{f_{\sigma_1}}(g) \cap \mathscr{M}^+_{f_{\sigma_2}}(g)$, $\mathscr{M}^+_{f_{\sigma_1}}(g) \cap \mathscr{M}^-_{f_{\sigma_2}}(g)$, $\mathscr{M}^-_{f_{\sigma_1}}(g) \cap \mathscr{M}^+_{f_{\sigma_2}}(g)$, and $\mathscr{M}^-_{f_{\sigma_1}}(g) \cap \mathscr{M}^-_{f_{\sigma_2}}(g)$ are all empty, so that we have

$$I_{\sigma_1} \cap I_{\sigma_2} = \varnothing, \quad \sigma_1, \sigma_2 \in S, \sigma_1 \neq \sigma_2.$$

Consequently,

$$\begin{aligned}\|\lambda_S(g)\| &= \sum_{\sigma \in S} |\lambda_S(g)(\sigma)| = \sum_{\sigma \in S} |\lambda_{f_\sigma}(g)| \leq \sum_{\sigma \in S} \mu(I_\sigma) = \sup_{\substack{T \subseteq S \\ |T| < \infty}} \sum_{\sigma \in T} \mu(I_\sigma) \\ &= \sup_{\substack{T \subseteq S \\ |T| < \infty}} \mu\left(\bigcup_{\sigma \in T} I_\sigma\right) \leq \mu\left(\bigcup_{\sigma \in S} I_\sigma\right) \leq \mu([0, \beta]) = \beta < \infty,\end{aligned}$$

from which we see that $\lambda_S(g) \in \ell^1(S)$; hence, since $g \in \mathscr{RF}(G)$ was chosen arbitrarily,

$$\lambda_S(\mathscr{RF}(G)) \subseteq \ell^1(S). \tag{9.32}$$

Conversely, let $f \in \ell^1(S)$ be a given real and absolutely summable function on S. Then, for all but countably many $\sigma \in S$, we have $f(\sigma) = 0$.[2] Let

$$\operatorname{supp} f = \{\sigma_1, \sigma_2, , \ldots\}$$

(a finite or countably infinite list) be the collection of indices $\sigma \in S$ for which $f(\sigma) \neq 0$. Let

$$\beta_n := \sum_{j=1}^{n} |f(\sigma_j)|, \quad (n \geq 0);$$

[2] See Exercise 9.10 and also Exercise 19 in Section 4, Chapter 2 of Royden [43] for a closely related result.

in particular, $\beta_0 = 0$. Set

$$\beta := \|f\| = \lim_{n\to\infty} \beta_n$$

and let

$$s_j := \operatorname{sgn} f(\sigma_j) \in \{1, -1\}$$

be the sign of $f(\sigma_j)$, the understanding being that $n \le |\operatorname{supp} f|$ and $\beta = \beta_{|\operatorname{supp} f|}$ if the set $\operatorname{supp} f$ is finite. Define a function $g_f : [0, \beta] \to G$ via

$$g_f(\xi) := \begin{cases} f_{\sigma_n}^{s_n}(\xi - \beta_{n-1} \bmod \alpha_{\sigma_n}), & \xi \in [\beta_{n-1}, \beta_n), n \ge 1 \\ 1_G, & \xi = \beta \end{cases}, \qquad (0 \le \xi \le \beta).$$

We claim that g_f is reduced. Suppose for a contradiction that there exists $\varepsilon > 0$ and a point $\xi_0 \in (0, \beta)$, such that $g_f(\xi_0) = 1_G$ and

$$g_f(\xi_0 + \eta)\, g_f(\xi_0 - \eta) = 1_G, \quad 0 < \eta < \varepsilon.$$

We now distinguish three cases.

(i) *For some integer n with $2 \le n < |\operatorname{supp} f|$ we have $\xi_0 = \beta_{n-1}$* (that is, ξ_0 is one of the separation points β_i in the interior of the interval $[0, \beta]$). Choosing $\varepsilon \le \min\{\alpha_{\sigma_n}, |f(\sigma_n)|\}$ we have

$$g_f(\xi_0 + \eta) = f_{\sigma_n}^{s_n}(\eta), \quad 0 < \eta < \varepsilon,$$

whereas, for $r \in \mathbb{N}_0$ such that

$$r\alpha_{\sigma_{n-1}} < |f(\sigma_{n-1})| \le (r+1)\alpha_{\sigma_{n-1}}$$

and $\varepsilon \le |f(\sigma_{n-1})| - r\alpha_{\sigma_{n-1}}$, we have

$$g_f(\xi_0 - \eta) = f_{\sigma_{n-1}}^{s_{n-1}}(|f(\sigma_{n-1})| - r\alpha_{\sigma_{n-1}} - \eta), \quad 0 < \eta < \varepsilon.$$

From the last two displayed equations we deduce that, for sufficiently small ε,

$$f_{\sigma_{n-1}}^{s_{n-1}}\left(|f(\sigma_{n-1})| - r\alpha_{\sigma_{n-1}} - \frac{\varepsilon}{2} + \eta'\right) = f_{\sigma_n}^{-s_n}\left(\alpha_{\sigma_n} - \frac{\varepsilon}{2} + \eta'\right), \quad |\eta'| < \frac{\varepsilon}{2},$$

which implies that the functions $f_{\sigma_{n-1}}$ and f_{σ_n} are locally compatible, contradicting our choice of the family $\{f_\sigma\}_{\sigma\in S}$.

(ii) *We have $\xi_0 \in (\beta_{n-1}, \beta_n)$ for some positive integer n, and $\xi_0 - \beta_{n-1}$ is not an integral multiple of α_{σ_n}.* Let

$$\beta_{n-1} + r\alpha_{\sigma_n} < \xi_0 < \beta_{n-1} + (r+1)\alpha_{\sigma_n}, \quad r \in \mathbb{N}_0.$$

Choosing ε sufficiently small, we now find that

$$f_{\sigma_n}^{s_n}(\xi_0 - \beta_{n-1} - r\alpha_{\sigma_n}) = 1_G$$

and that

$$f_{\sigma_n}^{s_n}(\xi_0 - \beta_{n-1} - r\alpha_{\sigma_n} + \eta) f_{\sigma_n}^{s_n}(\xi_0 - \beta_{n-1} - r\alpha_{\sigma_n} - \eta) = 1_G, \quad 0 < \eta < \varepsilon,$$

which taken together imply that f_{σ_n} is not reduced; this contradicts the fact that f_{σ_n} is a test function, in view of Lemma 9.2.

(iii) *We have* $\xi_0 \in (\beta_{n-1}, \beta_n)$ *for some* $n \in \mathbb{N}$, *and* $\xi_0 - \beta_{n-1} = r\alpha_{\sigma_n}$ *for some positive integer* r. Choosing $\varepsilon \leq \min\{\alpha_{\sigma_n}, \beta_n - \xi_0\}$, we have

$$g_f(\xi_0 + \eta) = f_{\sigma_n}^{s_n}(\eta), \quad 0 < \eta < \varepsilon,$$

as well as

$$g_f(\xi_0 - \eta) = f_{\sigma_n}^{s_n}(\alpha_{\sigma_n} - \eta), \quad 0 < \eta < \varepsilon,$$

from which we conclude that

$$f_{\sigma_n}^{s_n}(\alpha_{\sigma_n} - \eta) f_{\sigma_n}^{s_n}(\eta) = 1_G, \quad 0 < \eta < \varepsilon. \tag{9.33}$$

Moreover, we have

$$f_{\sigma_n}^{s_n}(0) = g_f(\xi_0) = 1_G$$

while the equation

$$f_{\sigma_n}^{s_n}(\alpha_{\sigma_n}) = 1_G$$

is valid by our assumption concerning the initial and terminal values of the functions f_σ. Hence, equation (9.33) also holds for $\eta = 0$, implying that

$$\varepsilon_0(f_{\sigma_n}^{s_n}, f_{\sigma_n}^{s_n}) = \sup \mathscr{E}(f_{\sigma_n}^{s_n}, f_{\sigma_n}^{s_n}) \geq \varepsilon > 0,$$

which contradicts the fact that f_{σ_n} (and hence $f_{\sigma_n}^{s_n}$) is a test function and is thus cyclically reduced by Lemma 9.2. This final contradiction finishes the proof of the fact that the function g_f is reduced.

Finally, to compute $\lambda_S(g_f)$, we note that, by definition of the function g_f,

$$\mathscr{M}_{f_\sigma}^+(g_f) \doteq \begin{cases} (\beta_{n-1}, \beta_n), & \sigma = \sigma_n \text{ and } s_n = 1 \text{ for some } n \in \mathbb{N}, \\ \varnothing & \text{otherwise} \end{cases} \tag{9.34}$$

and

$$\mathscr{M}_{f_\sigma}^-(g_f) \doteq \begin{cases} (\beta_{n-1}, \beta_n), & \sigma = \sigma_n \text{ and } s_n = -1 \text{ for some } n \in \mathbb{N}, \\ \varnothing & \text{otherwise,} \end{cases} \tag{9.35}$$

where the symbol $\doteq$ indicates the equality of sets up to finite sets, that is,

$$S_1 \doteq S_2 \;:\Longleftrightarrow\; |S_1 \Delta S_2| < \infty.$$

It follows that, for $\sigma \in S$,

$$\begin{aligned}
\lambda_S(g_f)(\sigma) &= \lambda_{f_\sigma}(g_f) \\
&= \mu\big(\mathscr{M}^+_{f_\sigma}(g_f)\big) - \mu\big(\mathscr{M}^-_{f_\sigma}(g_f)\big) \\
&= \begin{cases} 0, & \sigma \notin \operatorname{supp} f, \\ |f(\sigma_n)|, & \sigma = \sigma_n \text{ and } s_n = 1 \text{ for some } n \in \mathbb{N}, \\ -|f(\sigma_n)|, & \sigma = \sigma_n \text{ and } s_n = -1 \text{ for some } n \in \mathbb{N}, \end{cases} \\
&= f(\sigma),
\end{aligned}$$

so that indeed $\lambda_S(g_f) = f$, as required. This shows that

$$\lambda_S(\mathscr{RF}(G)) \supseteq \ell^1(S), \tag{9.36}$$

which together with (9.32) proves the theorem. □

Remark 9.28 Our hypothesis that $f_\sigma(0) = 1_G = f_\sigma(L(f(\sigma))$, like any other hypothesis concerning the initial and terminal values of test functions, is not a serious restriction in the context of Theorem 9.26. Indeed, if $\{f_\sigma\}_{\sigma \in S}$ is a family of pairwise locally incompatible test functions and if we let

$$f'_\sigma := a_\sigma f_\sigma b_\sigma, \quad \sigma \in S,$$

where $a_\sigma, b_\sigma \in G_0$, then $\{f'_\sigma\}_{\sigma \in S}$ is also a family of pairwise locally incompatible test functions. Moreover, the isomomorphism type of a group of the form $\ell^1(S)$ depends only on the cardinality of the set S; see the opening remarks of this section.

9.8 Exercises

9.1. Fill in the missing details, concerning cases (i) and (iii), in the proof of Lemma 9.3.

9.2. Prove Lemma 9.7.

9.3. Prove the implication (9.18); that is, show that local compatibility is a symmetric relation.

9.4. Establish part (ii) of Lemma 9.13.

9.5. Prove Lemma 9.14 in the case where $\alpha < \xi_1 < \gamma$.

9.6. Establish Lemma 9.16 in the case where $\alpha_1 < \xi_1 < \beta$.

9.7. Prove Lemma 9.17.

9.8. Establish Lemma 9.25.

9.9. Making use of Lemma 9.25, show that the collection of centralisers associated with a family of pairwise locally incompatible test functions enjoys Property (LI), introduced in Definition 9.20.

9.10. Let $\{x_\sigma\}_{\sigma\in S}$ be a family of non-negative real numbers such that

$$\sum_{\sigma\in S} x_\sigma < \infty.$$

Show that we must have $x_\sigma = 0$ for all but at most countably many indices σ. Here, the sum $\sum_{\sigma\in S} x_\sigma$ is defined as in (9.30).

9.11. Prove formula (9.31) concerning the cardinality of the space $\ell^1(S)$.

9.12. Give the details to establish formulae (9.34) and (9.35).

9.13. Show that $\mathscr{RF}(G)$ is generated (as a semigroup) by its hyperbolic elements. Conclude that a homomorphism $\varphi : \mathscr{RF}(G) \to \Gamma$, which is trivial on normalized cyclically reduced functions of positive length, is trivial on $\mathscr{RF}(G)$. [Hint: since the span $\mathscr{H}_0(G)$ of all hyperbolic elements of $\mathscr{RF}(G)$ is normal in $\mathscr{RF}(G)$, it suffices to show that $G_0 \subseteq \mathscr{H}_0(G)$.]

9.14. Let $f, g \in \mathscr{RF}(G)$ be locally incompatible test functions. Show that the centraliser of fg in $\mathscr{RF}(G)$ is cyclic.

9.15. Let G be a group, and let $\mathscr{S}(G)$ denote the subgroup of $\mathscr{RF}(G)$ generated by the constant functions. Show that, with an appropriate definition of the notion of step function, $\mathscr{S}(G)$ consists precisely of the step functions over G. Also, assuming the axiom of choice, we have $|\mathscr{S}(G)| = \max\{|G|, 2^{\aleph_0}\}$. Show that $\mathscr{S}(G) \leq E(G)$ if and only if G is an elementary abelian 2-group.

10
Test functions: existence theorem and further applications

10.1 Introduction

The present chapter continues our introduction to the theory of test functions, as conceived in Müller and Schlage-Puchta [38]; we also discuss a number of important applications. The main result concerning test functions as such is Theorem 10.1, which exhibits large families of pairwise locally incompatible test functions with given centraliser. In Section 10.2 we state the precise result and explain two immediate applications:

- every non-trivial torsion-free abelian group of rank at most $2^{\aleph_0}$ (that is, every non-trivial subgroup of the additive reals) is realised up to isomorphism as the centraliser of a test function (Corollary 10.3);
- we have $|\mathscr{RF}(G)| = |G|^{2^{\aleph_0}}$; see Corollary 10.4.

The proof of Theorem 10.1 is sketched in Section 10.3, while an alternative approach to the cardinality of $\mathscr{RF}(G)$ is explained in Section 10.4, this time relying on some rather paradoxical properties of Cantor's discontinuum and the existence of a *countable* family of pairwise locally incompatible test functions.

As we know from Chapter 4, $\mathscr{RF}$-groups and their associated $\mathbb{R}$-trees are universal with respect to inclusion for free $\mathbb{R}$-tree actions; see Theorems 4.10 and 6.4. Since $(\mathbb{R},+)$ acts on itself (viewed as an $\mathbb{R}$-tree) freely by translation and since the class of $\mathbb{R}$-free groups is closed under taking free products (see Proposition A.58), it follows in particular that every free product $\Lambda = \ast_{i\in I}\Lambda_i$ of (non-trivial) real groups embeds as a hyperbolic subgroup in $\mathscr{RF}(G)$ for some suitable group G. In Section 10.5 we consider a modified embedding problem in which some non-trivial group G is fixed in advance and the following question is asked: under what condition does a free product $\Lambda = \ast_{i\in I}\Lambda_i$ of non-trivial real groups embed as a hyperbolic subgroup in $\mathscr{RF}(G)$ *for this given group* G? We show that this is the case if and only if $|I| \leq |G|^{2^{\aleph_0}}$, that is, if and only if Λ can be embedded in $\mathscr{RF}(G)$ as a mere subset (see Theorem 10.5). The proof of Theorem 10.5 is based on Corollaries 10.2–10.4 and exploits the gap between $|\mathscr{RF}(G)|$ and the cardinality of the continuum.

The main result of this chapter, as far as the group-theoretic structure of $\mathscr{RF}(G)$ and its equally mysterious quotient $\mathscr{RF}_0(G) = \mathscr{RF}(G)/E(G)$ is concerned, is Theorem 10.10; this establishes a variety of assertions, in particular concerning the normal subgroups and quotients of $\mathscr{RF}(G)$, most of which were mentioned in the introduction to Chapter 9. Again, the proof makes essential use of Theorem 10.1; other ingredients include the structure theorem for divisible (abelian) groups, a result of Greenberg concerning normal subgroups in finitely generated free groups, as well as a number of results from this and earlier chapters.

10.2 Incompatible test functions with prescribed centraliser

In this section we shall discuss an important and powerful existence theorem for families of pairwise locally incompatible test functions with prescribed centraliser, that is, Theorem 10.1 below. In what follows the axiom of choice (AC) will be tacitly assumed, in particular as a hypothesis for the results of this section. Also, the indexing of families of functions will automatically be assumed to be injective.

By a *real* group we mean any subgroup of the additive reals $(\mathbb{R},+)$. A real group Λ is called *proper* if $0 < \Lambda < \mathbb{R}$.

Theorem 10.1 *Let* G *be a non-trivial group, and let* Λ *be a proper real group. Then there exists a family* $\mathfrak{F}$ *of pairwise locally incompatible normalised test functions in* $\mathscr{RF}(G)$ *such that* $|\mathfrak{F}| = |G|^{(\mathbb{R}:\Lambda)}$ *and such that* $\langle\Omega_f^0\rangle = \Lambda$ *for all* $f \in \mathfrak{F}$, *where* Ω_f^0 *is as in Definition* 8.3.

We observe that, since test functions are cyclically reduced and of positive length, the fact that in Theorem 10.1 we have $\langle\Omega_f^0\rangle = \Lambda$ for all $f \in \mathfrak{F}$ implies via part (iii) of Theorem 8.16 that the length function L induces an isomorphism $C_{\mathscr{RF}(G)}(f) \to \Lambda$ for each $f \in \mathfrak{F}$.

Before turning to the proof of Theorem 10.1 (which is sketched in Section 10.3), we list a number of important consequences.

Corollary 10.2 *Suppose that G is non-trivial. Then there exists a family $\{f_\sigma\}_{\sigma\in S}$ of pairwise locally incompatible test functions in $\mathscr{RF}(G)$ with $|S| = |G|^{2^{\aleph_0}}$ and $L(f_\sigma) = \alpha_\sigma$ for each $\sigma \in S$, where $\{\alpha_\sigma\}_{\sigma\in S}$ is any given family of positive real numbers indexed (not necessarily injectively) by the elements of S.*

Proof Choose any non-trivial real group Λ in Theorem 10.1 with $|\Lambda|$ countable, to obtain a family $\{\bar{f}_\sigma\}_{\sigma\in S}$ of pairwise locally incompatible test functions where $|S| = |G|^{2^{\aleph_0}}$. Scaling the functions $\bar{f}_\sigma$ appropriately, by setting

$$f_\sigma(\xi) := \bar{f}_\sigma(L(\bar{f}_\sigma)\xi\alpha_\sigma^{-1}), \quad \sigma \in S, \xi \in [0, \alpha_\sigma],$$

the new family $\{f_\sigma\}_{\sigma\in S}$ meets the requirements of the corollary. □

Corollary 10.3 *Let G be a non-trivial group, and let A be a non-trivial torsion-free abelian group of rank at most $2^{\aleph_0}$. Then there exists a test function $f \in \mathscr{RF}(G)$ such that $C_{\mathscr{RF}(G)}(f) \cong A$.*

Proof Such a group A can be embedded into $\mathbb{R}$ as a proper subgroup Λ; see Exercise 10.1. Let $\mathfrak{F}$ be a family of test functions as described in Theorem 10.1 with respect to Λ. Then $\mathfrak{F} \neq \varnothing$, and every function $f \in \mathfrak{F}$ satisfies

$$C_{\mathscr{RF}(G)}(f) \cong \Lambda \cong A,$$

as required. □

Corollary 10.4 *We have $|\mathscr{RF}(G)| = |G|^{2^{\aleph_0}}$.*

Proof The assertion clearly holds if $G = \{1_G\}$; thus we may suppose that G is non-trivial so that, by Cantor's theorem concerning the cardinality of power sets and the comparison theorem for cardinal powers,[1]

$$|G|^{2^{\aleph_0}} \geq 2^{2^{\aleph_0}} > 2^{\aleph_0}.$$

Moreover, by the definition of $\mathscr{F}(G)$, we have

$$|\mathscr{F}(G)| = |G|^{2^{\aleph_0}} \tag{10.1}$$

(see Section 4, Chapter X in Sierpiński [47] and Exercise 10.2). Hence, since

[1] See, for instance, Theorem 1 in Section 3, Chapter VIII of Sierpiński [47].

$\mathscr{RF}(G)$ is a subset of $\mathscr{F}(G)$, we deduce that

$$|\mathscr{RF}(G)| \leq |G|^{2^{\aleph_0}}. \tag{10.2}$$

The reverse inequality,

$$|\mathscr{RF}(G)| \geq |G|^{2^{\aleph_0}}, \tag{10.3}$$

follows immediately from Corollary 10.2 together with the fact, implied by Lemma 9.2, that test functions are reduced as G is non-trivial. Inequalities (10.2) and (10.3) together now yield our claim. □

10.3 Proof of Theorem 10.1

In this section we will sketch the proof of Theorem 10.1; for full details see Section 6 in Müller and Schlage-Puchta [38]. It is well known that a subgroup of the additive reals is either cyclic or dense in $\mathbb{R}$ (see, for instance, Lemma 1.3 in Chapter 1 of Chiswell [10]), and we shall distinguish cases accordingly.

(i) The case where Λ is cyclic. Let α be any positive real number, and set $\Lambda := \langle \alpha \rangle$. Our task is to construct a family $\mathfrak{F}$ of pairwise locally incompatible normalized test functions in $\mathscr{RF}(G)$ of size $|\mathfrak{F}| = |G|^{2^{\aleph_0}}$, such that $\langle \Omega_f^0 \rangle = \Lambda$ for each $f \in \mathfrak{F}$. Let x be any non-trivial element of G, and define a function

$$g : ([0,\alpha) \cap \bar{\mathbb{Q}}) \cup \{\alpha\} \to G$$

via

$$g(\xi) := \left\{ \begin{array}{ll} x, & \xi^2 \in \mathbb{Q} \text{ and } \xi \neq 0, \alpha \\ 1_G, & \text{otherwise} \end{array} \right\} \quad (\xi \in ([0,\alpha) \cap \bar{\mathbb{Q}}) \cup \{\alpha\}),$$

where, as in Lemma 9.5, $\bar{\mathbb{Q}}$ is the set of real numbers that are algebraic over $\mathbb{Q}$. We note in particular that $g(0) = 1_G$. Introduce an equivalence relation $\sim$ on $(0,\alpha)$ by setting

$$\xi_1 \sim \xi_2 \;:\Longleftrightarrow\; \xi_1 - \xi_2 \in \bar{\mathbb{Q}}, \quad 0 < \xi_1, \xi_2 < \alpha,$$

and, using the axiom of choice, choose a complete set of representatives $\mathfrak{T}$ for the quotient

$$((0,\alpha) - \bar{\mathbb{Q}})/\sim.$$

Since $\bar{\mathbb{Q}}$ is countable,[2] we have $|\mathfrak{T}| = 2^{\aleph_0}$. Given an arbitrary (set-theoretic) function $h : \mathfrak{T} \to G$, we define a function $f_h : [0,\alpha] \to G$ via

$$f_h(\xi) := \left\{ \begin{matrix} g(\xi), & \xi \in ([0,\alpha) \cap \bar{\mathbb{Q}}) \cup \{\alpha\} \\ h(\tau), & \xi \sim \tau \ (\tau \in \mathfrak{T}) \end{matrix} \right\} \qquad (0 \leq \xi \leq \alpha);$$

in particular,

$$f_h(0) = g(0) = 1_G.$$

We claim that every function f_h obtained in this way is in fact a test function. To see this, suppose for a contradiction that there exist $\varepsilon > 0$ and points $\xi_1, \xi_2 \in (0,\alpha)$ such that

$$f_h(\xi_1 + \eta) = f_h^{-1}(\xi_2 + \eta), \quad |\eta| < \varepsilon, \tag{10.4}$$

holds for some $h \in G^{\mathfrak{T}}$. We may clearly suppose without loss of generality that ξ_1 is rational. Choose a positive integer n such that $\sqrt[3]{2}/n < \varepsilon$, and note that $(\sqrt[3]{2}/n + \xi_1)^2 \notin \mathbb{Q}$, as the set $\{1, 2^{1/3}, 2^{2/3}\}$ is linearly independent over $\mathbb{Q}$. Since we have $\sqrt[3]{2}/n + \xi_1 \in \bar{\mathbb{Q}} \cap (0,\alpha)$ but $(\sqrt[3]{2}/n + \xi_1)^2 \notin \mathbb{Q}$, the definitions of the functions f_h and g together imply that

$$f_h(\sqrt[3]{2}/n + \xi_1) = g(\sqrt[3]{2}/n + \xi_1) = 1_G \neq x = g(\xi_1) = f_h(\xi_1).$$

Using equation (10.4) we find that

$$f_h(\alpha - \xi_2 - \sqrt[3]{2}/n) = 1_G \neq x^{-1} = f_h(\alpha - \xi_2);$$

however, since on $(0,\alpha) - \bar{\mathbb{Q}}$ the function f_h is constant on cosets modulo $\bar{\mathbb{Q}}$, this shows that $\alpha - \xi_2 \in \bar{\mathbb{Q}}$. If $x^2 \neq 1_G$, setting $\eta = 0$ in (10.4) immediately gives rise to a contradiction:

$$\begin{aligned} x = g(\xi_1) = f_h(\xi_1) &= f_h^{-1}(\xi_2) \\ &= \big(f_h(\alpha - \xi_2)\big)^{-1} = \big(g(\alpha - \xi_2)\big)^{-1} \in \{1_G, x^{-1}\}, \end{aligned}$$

while, for $x^2 = 1_G$, we find from (10.4) that

$$(\xi_1 + \eta)^2 \in \mathbb{Q} \iff (\xi_2 - \alpha + \eta)^2 \in \mathbb{Q}, \quad \eta \in (-\varepsilon, \varepsilon) \cap \bar{\mathbb{Q}}.$$

By Lemma 9.5 we conclude that $\xi_1 = \xi_2 - \alpha$, which is impossible since $\xi_1, \xi_2 \in (0,\alpha)$. Hence,

$$\mathfrak{F} = \{f_h : h \in G^{\mathfrak{T}}\}$$

[2] See Cantor [7] for the original proof of this fact. This was the first of Cantor's publications on set theory; it also contains his first proof (by means of nested intervals) of the fact that the real numbers are uncountable, implying the existence of infinitely many transcendental numbers in each non-empty interval.

is a family of normalised test functions, as claimed. Also, since $|\mathfrak{T}| = 2^{\aleph_0}$, we have $|\mathfrak{F}| = |G|^{2^{\aleph_0}}$, as required; and, by arguments similar to the above, one can also show that

$$f_{h_1} \text{ loc. comp. } f_{h_2} \implies h_1 = h_2, \quad (h_1, h_2 \in G^{\mathfrak{T}}). \tag{10.5}$$

It remains to check the assertion concerning the strong periods of the functions f_h; that is, we have to show that

$$\langle \Omega^0_{f_h} \rangle = \langle \alpha \rangle = \Lambda, \quad h \in G^{\mathfrak{T}}. \tag{10.6}$$

By definition,

$$\{0, \alpha\} \subseteq \Omega^0_{f_h}, \quad h \in G^{\mathfrak{T}}.$$

Suppose that, for some $h \in G^{\mathfrak{T}}$, there exists a period $\omega \in \Omega_{f_h}$ with $0 < \omega < \alpha$. Then, if ω is algebraic, we have

$$g(\xi) = f_h(\xi) = f_h(\xi + \omega) = g(\xi + \omega), \quad \xi \in (0, \alpha - \omega) \cap \bar{\mathbb{Q}},$$

which is equivalent to

$$\xi^2 \in \mathbb{Q} \iff (\omega + \xi)^2 \in \mathbb{Q}, \quad \xi \in (0, \alpha - \omega) \cap \bar{\mathbb{Q}}.$$

By Lemma 9.5, we deduce that $\omega = 0$, a contradiction.

Now suppose that ω is transcendental, and let $\tau_\omega \in \mathfrak{T}$ be such that $\omega \sim \tau_\omega$. Then we have

$$g(\xi) = f_h(\xi) = f_h(\xi + \omega) = h(\tau_\omega), \quad \xi \in (0, \alpha - \omega) \cap \bar{\mathbb{Q}};$$

that is, g would have to be constant on the algebraic points of the interval $(0, \alpha - \omega)$, which is clearly impossible. It follows that, for every $h \in G^{\mathfrak{T}}$,

$$\Omega^0_{f_h} \subseteq \Omega_{f_h} \subseteq \{0, \alpha\} \subseteq \Omega^0_{f_h},$$

which implies (10.6).

(ii) The case where Λ is dense in $\mathbb{R}$. Since $(\mathbb{R}, +)$, being divisible, has no maximal subgroups we can find a subgroup Λ' such that $\Lambda < \Lambda' < \mathbb{R}$ and $(\Lambda' : \Lambda) \leq \aleph_0$; see Szele [49]. Let x be any fixed non-trivial element of G, let α be a fixed positive element of Λ, and let ξ_0 be a fixed real number in $(\mathbb{R} - \Lambda') \cap [0, \alpha]$. We shall consider functions $f : [0, \alpha] \to G$ satisfying the following three conditions:

(a) f is constant on each non-trivial coset of Λ' in $\mathbb{R}$;

(b) we have

$$f(\xi) = \begin{Bmatrix} 1_G, & \xi \in \Lambda \\ x, & \xi \in \Lambda' - \Lambda \end{Bmatrix}, \quad (\xi \in \Lambda' \cap [0,\alpha]),$$

in particular, $f(0) = 1_G$;

(c) we have $f(\xi_0/2) = 1_G$ and $f(\alpha - \xi_0/2) = x$.

Since $\xi_0 \notin \Lambda'$, we have $\xi_0/2, \alpha - \xi_0/2 \notin \Lambda'$; moreover, these two points are in distinct cosets modulo Λ'. Hence the third condition does not conflict with conditions (a) and (b). Denote by $\mathfrak{F}$ the family of all functions $f : [0,\alpha] \to G$ meeting conditions (a)–(c).

Since Λ is dense in $\mathbb{R}$, so is each Λ'-coset; hence the number $|\mathfrak{F}|$ of such functions f equals the number of functions

$$\mathbb{R}/\Lambda' - \{[0], [\xi_0/2], [-\xi_0/2]\} \longrightarrow G$$

and, as $(\mathbb{R} : \Lambda') = \infty$, the three missing cosets do not change the cardinality of the domain; thus there are $|G|^{(\mathbb{R}:\Lambda')}$ such functions. Moreover, as Λ' was chosen such that $(\Lambda' : \Lambda) \leq \aleph_0$ we have $(\mathbb{R} : \Lambda') = (\mathbb{R} : \Lambda)$, so that

$$|\mathfrak{F}| = |G|^{(\mathbb{R}:\Lambda')} = |G|^{(\mathbb{R}:\Lambda)},$$

as required.

We claim that each function $f \in \mathfrak{F}$ is a test function. Suppose for a contradiction that there exist $f \in \mathfrak{F}$, $\varepsilon > 0$, and points $\xi_1, \xi_2 \in (0,\alpha)$ such that

$$f(\xi_1 + \eta) = f^{-1}(\xi_2 + \eta), \quad |\eta| < \varepsilon. \tag{10.7}$$

Since Λ is dense in $\mathbb{R}$, we may assume without loss of generality that $\xi_1 \in \Lambda$. Furthermore, since $\Lambda' - \Lambda$ is dense in $\mathbb{R}$, [3] we can choose $\eta_1 \in (-\varepsilon, \varepsilon)$ such that $\xi_1 + \eta_1 \in \Lambda' - \Lambda$; in particular, $\eta_1 \in \Lambda'$. Thus, applying (10.7) together with condition (b) and setting $\xi_2' := \alpha - \xi_2$, we find that

$$f(\xi_2' - \eta_1) = f(\alpha - \xi_2 - \eta_1) = \big(f^{-1}(\xi_2 + \eta_1)\big)^{-1} = \big(f(\xi_1 + \eta_1)\big)^{-1} = x^{-1}.$$

[3] Otherwise we would have $\Lambda' \cap (a,b) = \Lambda \cap (a,b)$ for some non-empty open interval (a,b), implying $\Lambda = \Lambda'$, since Λ and hence Λ' are dense in $\mathbb{R}$, and a dense subgroup is generated by its intersection with any non-empty open interval.

Again making use of equation (10.7) and condition (b), it follows that

$$f(\xi_2') = \big(f^{-1}(\xi_2)\big)^{-1} = \big(f(\xi_1)\big)^{-1} = 1_G \neq x^{-1} = f(\xi_2' - \eta_1),$$

showing that f is not constant on the coset $\xi_2' + \Lambda'$. Since by conditions (a) and (b) the only coset modulo Λ' on which f is not constant is Λ' itself, we deduce that $\xi_2' \in \Lambda'$, therefore also $\xi_2 \in \Lambda'$, as $\alpha \in \Lambda$. Using the fact that Λ' (and hence each of its cosets) is dense in $\mathbb{R}$, we can now choose $\eta_2 \in (-\varepsilon, \varepsilon)$ such that $\xi_1 + \eta_2 \in \xi_0/2 + \Lambda'$; in particular, $\eta_2 \in \xi_0/2 + \Lambda'$. Then, by (10.7) together with conditions (a) and (c), we obtain

$$\begin{aligned} 1_G &= f(\xi_0/2) = f(\xi_1 + \eta_2) = f^{-1}(\xi_2 + \eta_2) \\ &= \big(f(\xi_2' - \eta_2)\big)^{-1} = \big(f(\alpha - \xi_0/2)\big)^{-1} = x^{-1}, \end{aligned}$$

a contradiction. Hence $\mathfrak{F}$ is a family of test functions, as claimed. Moreover, by arguments similar to those above, one can also show that

$$f_1 \text{ loc. comp. } f_2 \Longrightarrow f_1 = f_2, \quad f_1, f_2 \in \mathfrak{F}. \tag{10.8}$$

It remains to see that

$$\langle \Omega_f^0 \rangle = \Lambda, \quad f \in \mathfrak{F}. \tag{10.9}$$

Since by construction f is constant on cosets modulo Λ, every element of the set $\Lambda \cap [0, \alpha]$ is a period of f and, since $\alpha \in \Lambda$, each $\omega \in \Lambda \cap [0, \alpha]$ is in fact a strong period. As Λ is dense in $\mathbb{R}$, it follows that the strong periods of f generate a subgroup of $\mathbb{R}$ containing Λ. Now let ω be a period of f with $\omega < \alpha$, and choose points $\xi_1, \xi_2 \in (0, \alpha - \omega)$ such that $\xi_1 \in \Lambda$ and $\xi_2 \in \Lambda' - \Lambda$. Then

$$f(\xi_1 + \omega) = f(\xi_1) = 1_G$$

while

$$f(\xi_2 + \omega) = f(\xi_2) = x.$$

Since $\xi_1 - \xi_2 \in \Lambda'$, the points $\xi_1 + \omega$ and $\xi_2 + \omega$ lie in the same coset modulo Λ', and, as f is constant on non-trivial Λ'-cosets, we must have $\omega \in \Lambda'$. Moreover, since $f(\xi_1 + \omega) = 1_G$ and $\xi_1 \in \Lambda$, we have $\omega \in \Lambda$. We conclude that every period of f is in fact contained in Λ; thus,

$$\Lambda \leq \langle \Omega_f^0 \rangle \leq \langle \Omega_f \rangle \leq \Lambda,$$

which implies (10.9).

10.4 The cardinality of $\mathscr{RF}(G)$ revisited

Here, we give a second proof of Corollary 10.4 concerning the cardinality of the group $\mathscr{RF}(G)$, this time based on properties of the Cantor discontinuum.[4] We note that, in contrast with the proof given in Section 10.2, this second proof uses only the existence of a *countably infinite* family of pairwise incompatible test functions.

We may assume that G is non-trivial and concentrate on the proof of inequality (10.3). In order to establish the latter inequality we shall produce a set of reduced functions of cardinality $|G|^{2^{\aleph_0}}$, as follows. Let $\mathscr{C} \subset [0,1]$ be the Cantor discontinuum, that is, the set of real numbers in $[0,1]$ whose triadic expansion avoids the digit 1. It is well known that $\mathscr{C}$ is compact and of cardinality $|\mathscr{C}| = 2^{\aleph_0}$ and that $[0,1] - \mathscr{C}$ is dense in $[0,1]$. We shall construct a map $f : [0,1] - \mathscr{C} \to G$ such that, for every (set-theoretic) function $g : \mathscr{C} \to G$, the map $f \oplus g : [0,1] \to G$ given by

$$(f \oplus g)(\xi) := \begin{Bmatrix} f(\xi), & \xi \notin \mathscr{C} \\ g(\xi), & \xi \in \mathscr{C} \end{Bmatrix} \qquad (0 \leq \xi \leq 1)$$

is reduced. In this way, inequality (10.3) again follows and, together with (10.2), establishes Corollary 10.4.

Let $\{C_\mu\}_{\mu \in \mathbb{N}_0}$ be an enumeration of the connected components of $[0,1] - \mathscr{C}$, say $C_\mu = (a_\mu, b_\mu)$, let $\{\alpha_\mu\}_{\mu \in \mathbb{N}_0}$ be the corresponding sequence of interval lengths $\alpha_\mu = b_\mu - a_\mu$, and let $\{f_\mu\}_{\mu \in \mathbb{N}_0}$ be a sequence of pairwise locally incompatible test functions such that $L(f_\mu) = \alpha_\mu$ for all $\mu \in \mathbb{N}_0$ (such a sequence exists by Corollary 10.2, since G is non-trivial). Then we define a function f on $[0,1] - \mathscr{C}$ via

$$f(\xi) := f_\mu(\xi - a_\mu), \quad \xi \in C_\mu.$$

Let $g : \mathscr{C} \to G$ be an arbitrary map. We claim that $f \oplus g : [0,1] \to G$ is reduced. Suppose otherwise. Then there exists $\xi_0 \in (0,1)$ and $\varepsilon > 0$ such that $(f \oplus g)(\xi_0) = 1_G$ and

$$(f \oplus g)(\xi_0 + \eta)(f \oplus g)(\xi_0 - \eta) = 1_G, \quad 0 < \eta < \varepsilon. \tag{10.10}$$

Suppose first that $\xi_0 \notin \mathscr{C}$. Then $\xi_0 \in C_\mu$ for some $\mu \in \mathbb{N}_0$, and we find that

$$f_\mu(\xi_0 - a_\mu) = 1_G \tag{10.11}$$

[4] See p. 590 in Cantor [8]. In fact, this construction was introduced by H. J. S. Smith [48], some years before Cantor did so.

as well as

$$f_\mu(\xi_0 - a_\mu + \eta) f_\mu(\xi_0 - a_\mu - \eta) = 1_G, \quad 0 < \eta < \varepsilon', \tag{10.12}$$

where $\varepsilon' := \min\{\varepsilon, \xi_0 - a_\mu, b_\mu - \xi_0\}$. Assertions (10.11) and (10.12) imply, however, that f_μ is not reduced, contradicting the fact that f_μ is a test function in view of Lemma 9.2.

Now suppose that $\xi_0 \in \mathscr{C}$. Then we claim that there exist points $\xi_1, \xi_2 \in (0,1)$ satisfying

(i) $0 < \xi_0 - \varepsilon < \xi_1 < \xi_0 < \xi_2 < \xi_0 + \varepsilon < 1$,

(ii) $|\xi_1 - \xi_0| = |\xi_2 - \xi_0|$,

(iii) $\xi_1, \xi_2 \notin \mathscr{C}$.

Indeed, by the density of $[0,1] - \mathscr{C}$ in $[0,1]$ there exists ξ_1' such that $\xi_0 - \varepsilon < \xi_1' < \xi_0$ and $\xi_1' \notin \mathscr{C}$. Let $\xi_1' \in C_{\mu_1}$, choose $\varepsilon'' > 0$ such that

$$(\xi_1' - \varepsilon'', \xi_1' + \varepsilon'') \subseteq C_{\mu_1} \cap (\xi_0 - \varepsilon, \xi_0),$$

and let $\xi_2' := \xi_0 + |\xi_1' - \xi_0|$. Again making use of the density property of $[0,1] - \mathscr{C}$, there exists a point $\xi_2 \notin \mathscr{C}$ such that $\xi_2' - \varepsilon'' < \xi_2 < \xi_2'$. Let $\xi_1 := \xi_1' + |\xi_2 - \xi_2'|$. Then $\xi_1 \in C_{\mu_1}$; in particular, $\xi_1 \notin \mathscr{C}$ and the points ξ_1, ξ_2 satisfy properties (i)–(iii) by construction.

We have $\xi_i \in C_{\mu_i}$ $(i = 1,2)$ for certain indices $\mu_1, \mu_2 \in \mathbb{N}_0$ such that $\mu_1 \neq \mu_2$, and (10.10) now implies that

$$f_{\mu_1}(\xi_1 - a_{\mu_1} + \eta) = f_{\mu_2}^{-1}(\alpha_{\mu_2} - \xi_2 + a_{\mu_2} + \eta), \quad |\eta| < \tilde{\varepsilon}, \tag{10.13}$$

where $\tilde{\varepsilon} > 0$ is chosen such that

$$(\xi_1 - \tilde{\varepsilon}, \xi_1 + \tilde{\varepsilon}) \subseteq (\xi_0 - \varepsilon, \xi_0) \cap C_{\mu_1}$$

and

$$(\xi_2 - \tilde{\varepsilon}, \xi_2 + \tilde{\varepsilon}) \subseteq (\xi_0, \xi_0 + \varepsilon) \cap C_{\mu_2}.$$

However, equation (10.13) implies that f_{μ_1} and f_{μ_2} are locally compatible, contradicting the choice of the sequence $\{f_\mu\}_{\mu \in \mathbb{N}_0}$ since $\mu_1 \neq \mu_2$. Hence $f \oplus g$ is reduced for every choice of map $g : \mathscr{C} \to G$, as claimed.

10.5 An embedding theorem

As we saw in Chapter 4, $\mathscr{RF}$-groups and their associated $\mathbb{R}$-trees are universal with respect to inclusion for free $\mathbb{R}$-tree actions; see Theorems 4.10 and 6.4. Since $(\mathbb{R},+)$ acts on itself (viewed as an $\mathbb{R}$-tree) freely by translation, and as the class of $\mathbb{R}$-free groups is closed under taking free products (see Proposition A.58), it follows in particular that every free product $\Lambda = \ast_{i\in I}\Lambda_i$ of (non-trivial) real groups embeds as a hyperbolic subgroup in $\mathscr{RF}(G)$ for some suitable group G. In this section we shall consider a modified embedding problem, in which some non-trivial group G is fixed in advance and the question is asked: which free products $\Lambda = \ast_{i\in I}\Lambda_i$ of non-trivial real groups embed as a hyperbolic subgroup in $\mathscr{RF}(G)$ *for this given group* G? Our result is as follows.[5]

Theorem 10.5 *Let G be a given non-trivial group, and let $\{\Lambda_i\}_{i\in I}$ be a family of non-trivial real groups. Then the free product $\Lambda = \ast_{i\in I}\Lambda_i$ embeds as a hyperbolic subgroup in $\mathscr{RF}(G)$ if and only if $|I| \leq |G|^{2^{\aleph_0}}$.*

We note that the answer afforded by Theorem 10.5 to the embedding problem just formulated is the best possible, in that Λ is realized as a hyperbolic subgroup of $\mathscr{RF}(G)$ whenever Λ can be embedded in $\mathscr{RF}(G)$ as a mere subset.

Proof of Theorem 10.5 Let G be a non-trivial group, and let $\{\Lambda_i\}_{i\in I}$ be a family of non-trivial real groups. If the group $\Lambda = \ast_{i\in I}\Lambda_i$ embeds in $\mathscr{RF}(G)$ then clearly, in view of Corollary 10.4,

$$|I| \leq \left|\bigcup_{i\in I}\Lambda_i\right| \leq |\Lambda| \leq |\mathscr{RF}(G)| = |G|^{2^{\aleph_0}}.$$

Suppose conversely that $|I| \leq |G|^{2^{\aleph_0}}$, let $\{\Lambda_{i_j}\}_{j\in J}$ be a system of representatives for the isomorphism types represented by the family $\{\Lambda_i\}$, where we may assume without loss of generality that $J\cap\mathbb{N} = \varnothing$ ($\mathbb{N}$ denoting the set of positive integers), and set

$$I_j := \{i\in I : \Lambda_i \cong \Lambda_{i_j}\}, \quad j\in J,$$

so that

$$\Lambda = \mathop{\ast}_{i\in I}\Lambda_i \cong \mathop{\ast}_{j\in J}\Lambda_{i_j}^{\ast|I_j|}.$$

Clearly $|J| \leq |I|$. By Corollary 10.3, each non-trivial real group can be realised

[5] This is Theorem 1.1 in Müller [36].

up to isomorphism as the centraliser of some test function, and it is in this way that we are going to think of the representatives Λ_{i_j}:

$$\Lambda_{i_j} \cong C_{\mathscr{RF}(G)}(g_j), \quad j \in J,$$

where each $g_j \in \mathscr{RF}(G)$ is some normalised test function, that is, $g_j(0) = 1_G$.

Next, let $\{f_\sigma\}_{\sigma \in S}$ be a family of pairwise locally incompatible test functions with bijective indexing and $|S| = |G|^{2^{\aleph_0}}$ (such a family exists by Corollary 10.2), let $\alpha_\sigma := L(f_\sigma) > 0$, and, for each $j \in J$, choose a subset $S_j \subseteq S$ with $|S_j| = |G|^{2^{\aleph_0}}$ and such that $S_{j_1} \cap S_{j_2} = \varnothing$ for $j_1, j_2 \in J$ and $j_1 \neq j_2$. This is possible, since $|I| \leq |G|^{2^{\aleph_0}}$, and thus

$$\left| \bigcup_{j \in J} S_j \right| = |G|^{2^{\aleph_0}} |J| \leq |G|^{2^{\aleph_0}} |I| \leq |G|^{2^{\aleph_0}} = |S|.$$

Given functions $g, g' \in \mathscr{RF}(G)$, we call g' a *segment* of g if there exist real numbers α, β such that $0 \leq \alpha < \beta \leq L(g)$, $L(g') = \beta - \alpha$, and

$$g'(\xi) = g(\xi + \alpha), \quad 0 < \xi < L(g').$$

Two segments g', g'' of g are identified if $L(g') = L(g'') = \alpha'$ and $g'|_{(0,\alpha')} = g''|_{(0,\alpha')}$.

Fix $j \in J$, and consider segments g' of elements $g \in C_{\mathscr{RF}(G)}(g_j) - \{\mathbf{1}_G\}$. Clearly, there are at most $2^{\aleph_0}$ essentially different choices for such a segment g' and, by local incompatibility, there can be at most two indices $\sigma \in S_j$ such that

$$f_\sigma(\alpha_\sigma - \eta) g'(\eta) = 1_G, \quad 0 < \eta < \varepsilon,$$

or

$$f_\sigma(\alpha_\sigma - \eta)(g')^{-1}(\eta) = 1_G, \quad 0 < \eta < \varepsilon,$$

for given g' and some $\varepsilon > 0$. Deleting the (at most $2^{\aleph_0}$) indices σ corresponding to such functions f_σ from S_j, we obtain a subset $S'_j \subseteq S_j$ with $|S'_j| = |S_j|$ and such that

$$f_\sigma g' f_\sigma^{-1} = f_\sigma \circ g' \circ f_\sigma^{-1}, \tag{10.14}$$

for each $\sigma \in S'_j$ and every segment g' of some element $g \in C_{\mathscr{RF}(G)}(g_j) - \{\mathbf{1}_G\}$.

Finally, for each $j \in J$, select a subset $S''_j \subseteq S'_j$ such that $|S''_j| = |I_j|$. Consider the system of subgroups

$$\mathscr{S} = \left\{ f_\sigma C_{\mathscr{RF}(G)}(g_j) f_\sigma^{-1} \right\}_{(j,\sigma)}$$

of $\mathscr{RF}(G)$, where (j,σ) runs through the set $\bigcup_{j\in J}\{j\}\times S''_j$, and let

$$\mathscr{H} := \Big\langle f_\sigma C_{\mathscr{RF}(G)}(g_j) f_\sigma^{-1} : (j,\sigma) \in \bigcup_{j\in J}\{j\}\times S''_j \Big\rangle$$

be the subgroup of $\mathscr{RF}(G)$ generated by $\mathscr{S}$. By (10.14) plus the fact that the functions f_σ are chosen to be pairwise locally incompatible, the family of subgroups $\mathscr{S}$ is independent in the sense of Definition 9.22, with each member of the family a hyperbolic subgroup since test functions are hyperbolic. Hence, by Lemma 9.23,

$$\mathscr{H} \cong \mathop{\ast}_{j\in J} \mathop{\ast}_{\sigma\in S''_j} f_\sigma C_{\mathscr{RF}(G)}(g_j) f_\sigma^{-1} \cong \mathop{\ast}_{j\in J} \Lambda_{i_j}^{*|I_j|} \cong \Lambda,$$

as required. We claim further that

$$\text{\textit{the cyclically reduced core } } c(h) \text{ \textit{of any element} } h\in\mathscr{H}-\{\mathbf{1}_G\} \text{ \textit{has positive length,}} \tag{10.15}$$

implying that $\mathscr{H}$ is hyperbolic by part (ii) of Proposition 3.13. An arbitrary element $h\in\mathscr{H}$ is of the form

$$h = f_{\sigma_1} g_1 f_{\sigma_1}^{-1} f_{\sigma_2} g_2 f_{\sigma_2}^{-1} \cdots f_{\sigma_r} g_r f_{\sigma_r}^{-1}, \tag{10.16}$$

where $g_\rho \in C_{\mathscr{RF}(G)}(g_{j_\rho}) - \{\mathbf{1}_G\}$, $\sigma_\rho \in S''_{j_\rho}$, and $\sigma_\rho \neq \sigma_{\rho+1}$ for $1\le\rho<r$. Further, in view of (10.14) and the fact that the functions f_σ are pairwise locally incompatible, we have

$$h = f_{\sigma_1}\circ g_1\circ f_{\sigma_1}^{-1}\circ f_{\sigma_2}\circ g_2\circ f_{\sigma_2}^{-1}\circ\cdots\circ f_{\sigma_r}\circ g_r\circ f_{\sigma_r}^{-1}; \tag{10.17}$$

see Lemma 9.13.

In proving our claim (10.15), we may clearly assume that h is cyclically reduced as a normal form element of the free product

$$\mathscr{H} = \mathop{\ast}_{j\in J} \mathop{\ast}_{\sigma\in S''_j} f_\sigma C_{\mathscr{RF}(G)}(g_j) f_\sigma^{-1};$$

that is, we may suppose that h as in (10.16) satisfies $\sigma_1\neq\sigma_r$ or $g_1 g_r \neq \mathbf{1}_G$. We now distinguish several cases.

Case 1: $r=1$. Here we have

$$h = f_{\sigma_1}\circ g_1\circ f_{\sigma_1}^{-1},$$

and g_1, being a non-trivial element of the centraliser $C_{\mathscr{RF}(G)}(g_{j_1})$, is a test function and hence cyclically reduced and of positive length; see Corollary 9.11. It follows that $c(h)=g_1$, and so $L(c(h))=L(g_1)>0$ as required.

Case 2: $\sigma_1 \neq \sigma_r$. Here $r \geq 2$ and, since $f_{\sigma_1}, f_{\sigma_r}$ are locally incompatible, we have $\varepsilon_0(h,h) = 0$, so $c(h) = h$ and $L(c(h)) = L(h) > 0$.

Case 3: $r > 1$ and $\sigma_1 = \sigma_r$. In this case we have $r \geq 3$, $j_1 = j_r$, and $g_1 g_r \neq \mathbf{1}_G$. Several subcases arise.

Case 3(a): $\varepsilon_0(g_r, g_1) = 0$. In this situation, since $L(g_1), L(g_r) > 0$ we have

$$c(h) = g_1 \circ f_{\sigma_1}^{-1} \circ \cdots \circ f_{\sigma_r} \circ g_r,$$

implying

$$L(c(h)) = \sum_{\rho=1}^{r-1} \left(2L(f_{\sigma_\rho}) + L(g_\rho)\right) + L(g_r) > 0.$$

For our remaining subcases we decompose g_1 as $c \circ g_1'$ and g_r as $g_r' \circ c^{-1}$ according to Lemma 2.15, so that $g_r g_1 = g_r' \circ g_1'$ and $L(c) = \varepsilon_0(g_r, g_1) > 0$.

Case 3(b): $L(g_1'), L(g_r') > 0$. Here, c, g_1' and g_r', c^{-1} are segments of g_1 and g_r respectively; hence, by (10.14), that is, by our choice of the sets S_j'',

$$\varepsilon_0(f_{\sigma_1}, c) = \varepsilon_0(g_1', f_{\sigma_1}^{-1}) = \varepsilon_0(f_{\sigma_r}, g_r') = 0$$

and so

$$h = (f_{\sigma_1} \circ c) \circ (g_1' \circ f_{\sigma_1}^{-1} \circ \cdots \circ f_{\sigma_r} \circ g_r') \circ (c^{-1} \circ f_{\sigma_r}^{-1})$$

with $g_1' \circ f_{\sigma_1}^{-1} \circ \cdots \circ f_{\sigma_r} \circ g_r'$ cyclically reduced. It follows that

$$L(c(h)) = L(g_1' \circ f_{\sigma_1}^{-1} \circ \cdots \circ f_{\sigma_r} \circ g_r') = \sum_{\rho=2}^{r-1} \left(2L(f_{\sigma_\rho}) + L(g_\rho)\right)$$
$$+ L(g_1') + 2L(f_{\sigma_1}) + L(g_r') > 0.$$

Case 3(c): $L(g_1') > 0$ and $L(g_r') = 0$. In this case we may assume without loss of generality that $g_r' = \mathbf{1}_G$, so that $g_r = c^{-1}$ and thus $\varepsilon_0(f_{\sigma_r}, c^{-1}) = 0$. Moreover, since the functions c and g_1' are segments of g_1 we have

$$\varepsilon_0(f_{\sigma_1}, c) = \varepsilon_0(g_1', f_{\sigma_1}^{-1}) = \varepsilon_0(f_{\sigma_r}, g_1') = 0,$$

and we find that

$$h = (f_{\sigma_1} \circ c) \circ (g_1' \circ f_{\sigma_1}^{-1} \circ \cdots \circ f_{\sigma_r}) \circ (c^{-1} \circ f_{\sigma_r}^{-1}),$$

where $g_1' \circ f_{\sigma_1}^{-1} \circ \cdots \circ f_{\sigma_r}$ is cyclically reduced. We conclude that

$$L(c(h)) = L(g_1' \circ f_{\sigma_1}^{-1} \circ \cdots \circ f_{\sigma_r}) = \sum_{\rho=2}^{r-1} \left(2L(f_{\sigma_\rho}) + L(g_\rho)\right) + 2L(f_{\sigma_1}) + L(g_1') > 0,$$

again as desired.

Case 3(d): $L(g'_1) = 0$ and $L(g'_r) > 0$. This case is handled analogously to the previous subcase.

Finally, it remains to consider the situation where $L(g'_1) = L(g'_r) = 0$; then, however, the element $g_1 g_r = g_r g_1 = g'_r g'_1$ is in $C_{\mathscr{RF}(G)}(g_{j_1})$ and is of length 0, implying $g_1 g_r = \mathbf{1}_G$. Hence, as we have assumed that h is cyclically reduced as a normal form element, this case does not arise in fact, and the proof of Theorem 10.5 is complete. □

Remark 10.6 Combining Lemma 9.23 with arguments adapted from the proof of Theorem 10.5, one can deduce the following *hyperbolicity criterion*, which generalises Theorem 9.21.

Theorem 10.7 *Let $\{\mathscr{H}_\sigma\}_{\sigma\in S}$ be a family of subgroups $\mathscr{H}_\sigma \leq \mathscr{RF}(G)$ with bijective indexing. Suppose that each $\mathscr{H}_\sigma$ is hyperbolic and that the family $\{\mathscr{H}_\sigma\}_{\sigma\in S}$ meets condition* (LI). *Then the join $\mathscr{H} = \langle \mathscr{H}_\sigma : \sigma \in S\rangle$ is a hyperbolic subgroup of $\mathscr{RF}(G)$ and is isomorphic to the free product $\ast_{\sigma\in S}\mathscr{H}_\sigma$.*

See Section 4 in Müller [36] for the proof of Theorem 10.7. Further, just as Corollary 9.24 follows from Theorem 9.21, Theorem 10.7 in turn allows us to deduce the following generalisation of Corollary 9.24.

Corollary 10.8 *Let $\{f_\sigma\}_{\sigma\in S} \subseteq \mathscr{RF}(G)$ be a family of pairwise locally incompatible cyclically reduced functions, with bijective indexing, and suppose that $|S| > 1$. Then the corresponding family $\{C_{\mathscr{RF}(G)}(f_\sigma)\}_{\sigma\in S}$ of centralisers generates a hyperbolic subgroup of $\mathscr{RF}(G)$ isomorphic to the free product $\ast_{\sigma\in S} C_{\mathscr{RF}(G)}(f_\sigma)$.*

10.6 The subgroup generated by a set of incompatible test functions

In what follows an overbar denotes abelianisation, and the homomorphisms

$$\pi^{ab} : \mathscr{RF}(G) \to \overline{\mathscr{RF}(G)}$$

$$\pi_0^{ab} : \mathscr{RF}(G) \to \mathscr{RF}(G)/\big(E(G)[\mathscr{RF}(G),\mathscr{RF}(G)]\big)$$

$$\pi : \mathscr{RF}(G) \to \mathscr{RF}_0(G) := \mathscr{RF}(G)/E(G)$$

are canonical. As a useful piece of general nonsense we note that, for a group Γ, a normal subgroup $\Delta \trianglelefteq \Gamma$, and a verbal functor $V_{\mathscr{W}}(\cdot) : \mathbf{Groups} \to \mathbf{Groups}$ on the category **Groups** of groups and homomorphisms associated with a set $\mathscr{W} = \{w_\mu(x_\nu)\}_{\mu \in M}$ of words in the variables x_ν, we have a canonical isomorphism

$$\Gamma/(\Delta V_{\mathscr{W}}(\Gamma)) \cong (\Gamma/\Delta)/V_{\mathscr{W}}(\Gamma/\Delta). \tag{10.18}$$

This follows immediately from the canonical isomorphism

$$\Gamma/(\Delta V_{\mathscr{W}}(\Gamma)) \cong (\Gamma/\Delta)/(\Delta V_{\mathscr{W}}(\Gamma)/\Delta)$$

plus the obvious fact that

$$\Delta V_{\mathscr{W}}(\Gamma)/\Delta = V_{\mathscr{W}}(\Gamma/\Delta).$$

As a special case of (10.18) we have a canonical isomorphism

$$\mathscr{RF}(G)/(E(G)[\mathscr{RF}(G), \mathscr{RF}(G)]) \cong \overline{\mathscr{RF}_0(G)},$$

which we shall tacitly use to view π_0^{ab} as a map

$$\pi_0^{ab} : \mathscr{RF}(G) \to \overline{\mathscr{RF}_0(G)}.$$

The purpose of this short section is to analyse the subgroup $\mathfrak{f}_S$ generated by a family $\{f_\sigma\}_{\sigma \in S}$ of pairwise locally incompatible test functions f_σ, as well as the images of $\mathfrak{f}_S$ under the projections π, π^{ab}, and π_0^{ab}. Proposition 10.9 below will turn out to be crucial when we are establishing the structural results concerning $\mathscr{RF}(G)$ and its quotient $\mathscr{RF}_0(G)$, which form the subject matter of the next section.

Proposition 10.9 *Let $\{f_\sigma\}_{\sigma \in S} \subseteq \mathscr{RF}(G)$ be a family of pairwise locally incompatible test functions, with bijective indexing. Then we have the following.*

(i) *The subgroup $\mathfrak{f}_S := \langle f_\sigma : \sigma \in S \rangle$ of $\mathscr{RF}(G)$ is free, with basis $\{f_\sigma\}_{\sigma \in S}$, and satisfies*

$$\mathfrak{f}_S \cap E(G) = \{\mathbf{1}_G\}. \tag{10.19}$$

In particular $\mathfrak{f}_S$ is hyperbolic and the image of $\mathfrak{f}_S$ under the projection π is free of rank $|S|$, with basis $\{\pi(f_\sigma)\}_{\sigma \in S}$.

(ii) *The image of $\mathfrak{f}_S$ under the projections π^{ab} and π_0^{ab} respectively is free abelian of rank $|S|$ with bases $\{\pi^{ab}(f_\sigma)\}_{\sigma \in S}$ and $\{\pi_0^{ab}(f_\sigma)\}_{\sigma \in S}$ respectively.*

Proof We first establish part (ii). Let $\mathfrak{N}$ denote either of the normal subgroups $[\mathscr{RF}(G), \mathscr{RF}(G)]$ and $E(G)[\mathscr{RF}(G), \mathscr{RF}(G)]$ of $\mathscr{RF}(G)$, and consider the relation

$$(f_{\sigma_1}\mathfrak{N})^{\gamma_1}(f_{\sigma_2}\mathfrak{N})^{\gamma_2}\cdots(f_{\sigma_r}\mathfrak{N})^{\gamma_r} = (f_{\sigma_1}^{\gamma_1} f_{\sigma_2}^{\gamma_2}\cdots f_{\sigma_r}^{\gamma_r})\mathfrak{N} = \mathfrak{N} \tag{10.20}$$

in $\pi^{ab}(\mathfrak{f}_S)$ or $\pi_0^{ab}(\mathfrak{f}_S)$ respectively, with $r \geq 0$, distinct indices $\sigma_1, \sigma_2, \ldots, \sigma_r$, and exponents $\gamma_1, \gamma_2, \ldots, \gamma_r \in \mathbb{Z} \setminus \{0\}$. If $r > 0$ then the function $f_{\sigma_1}^{\gamma_1} f_{\sigma_2}^{\gamma_2}\cdots f_{\sigma_r}^{\gamma_r}$ is a test function by Proposition 9.18; thus $f_{\sigma_1}^{\gamma_1} f_{\sigma_2}^{\gamma_2}\cdots f_{\sigma_r}^{\gamma_r} \notin \mathfrak{N}$ by Corollary 9.10, contradicting (10.20). Hence, we must have $r = 0$, so $\pi^{ab}(\mathfrak{f}_S)$ and $\pi_0^{ab}(\mathfrak{f}_S)$ are indeed free abelian groups, freely generated by the sets $\{\pi^{ab}(f_\sigma)\}_{\sigma \in S}$, respectively $\{\pi_0^{ab}(f_\sigma)\}_{\sigma \in S}$, of cardinality $|S|$.

In order to see that the group $\mathfrak{f}_S$ is free with basis $\{f_\sigma\}_{\sigma \in S}$, we invoke Lemma 9.23. Setting $\mathscr{H}_\sigma := \langle f_\sigma \rangle$ for $\sigma \in S$, we obtain a family of subgroups $\mathscr{H}_\sigma \leq \mathscr{RF}(G)$ that is bijectively indexed since $\langle f_{\sigma_1} \rangle = \langle f_{\sigma_2} \rangle$ forces $f_{\sigma_1} = f_{\sigma_2}^{\pm 1}$; this implies that $\sigma_1 = \sigma_2$, as the f_σ are bijectively indexed and pairwise locally incompatible. Moreover, each subgroup $\mathscr{H}_\sigma$ satisfies condition (T) of Definition 9.20 (in view of Corollary 9.4), and so $\mathscr{H}_\sigma \cap G_0 = \{\mathbf{1}_G\}$. Further, the family of subgroups $\{\mathscr{H}_\sigma\}_{\sigma \in S}$ is independent in the sense of Definition 9.22, since the functions f_σ are pairwise locally incompatible; see Lemma 9.13. Applying Lemma 9.23 we find that $\mathfrak{f}_S \cong \ast_{\sigma \in S}\langle f_\sigma \rangle$ is free, with basis $\{f_\sigma\}_{\sigma \in S}$, as claimed.

Finally, in order to establish equation (10.19) it clearly suffices to show that every subgroup of $\mathscr{RF}(G)$ generated by finitely many members of the family $\{f_\sigma\}_{\sigma \in S}$ meets $E(G)$ trivially. Let $\sigma_1, \sigma_2, \ldots, \sigma_k \in S$ be distinct indices, where $k \geq 1$. Then

$$\mathfrak{n} := \langle f_{\sigma_1}, f_{\sigma_2}, \ldots, f_{\sigma_k} \rangle \cap E(G)$$

is normal in the finitely generated free group

$$\mathfrak{f} := \langle f_{\sigma_1}, f_{\sigma_2}, \ldots, f_{\sigma_k} \rangle,$$

and the quotient

$$\mathfrak{f}/\mathfrak{n} \cong \mathfrak{f}E(G)/E(G)$$

projects onto

$$\tilde{\mathfrak{f}} := \langle \pi_0^{ab}(f_{\sigma_1}), \pi_0^{ab}(f_{\sigma_2}), \ldots, \pi_0^{ab}(f_{\sigma_k}) \rangle \leq \overline{\mathscr{RF}_0(G)}.$$

By part (ii), $\tilde{\mathfrak{f}}$ is free abelian of rank k; in particular, $(\mathfrak{f} : \mathfrak{n}) = \infty$, implying that $\mathfrak{n} = 1$ by a result of Greenberg [22]; see also Proposition 3.11 in Chapter I of Lyndon and Schupp [30]. Hence, assertion (10.19) is proved. It follows that $\mathfrak{f}_S$ is hyperbolic and that the restriction $\pi|_{\mathfrak{f}_S} : \mathfrak{f}_S \to \mathscr{RF}_0(G)$ is an embedding, so $\pi(\mathfrak{f}_S)$ is free of rank $|S|$ with basis $\{\pi(f_\sigma)\}_{\sigma\in S}$, as claimed. The proof of Proposition 10.9 is complete. □

10.7 A structure theorem for $\mathscr{RF}(G)$ and $\mathscr{RF}(G)/E(G)$

We are finally in a position to state and prove the following *structure theorem* concerning $\mathscr{RF}(G)$ and its quotient group $\mathscr{RF}_0(G)$.

Theorem 10.10 *Let G be a non-trivial group, set $\mathfrak{c}_G := |G|^{2^{\aleph_0}}$, and assume the axiom of choice. Then the following hold true.*

- (i) *The groups $\mathscr{RF}(G)$ and $\mathscr{RF}_0(G)$ contain a free subgroup of rank $\mathfrak{c}_G$ but are not free; in particular, $|\mathscr{RF}_0(G)| = \mathfrak{c}_G$.*
- (ii) *Every non-trivial torsion-free abelian group of rank at most $2^{\aleph_0}$ is realised (up to isomorphism) as the centraliser of a hyperbolic element in $\mathscr{RF}(G)$.*
- (iii) *The abelianised groups $\overline{\mathscr{RF}(G)}$ and $\overline{\mathscr{RF}_0(G)}$ contain a $\mathbb{Q}$-vector space of dimension $\mathfrak{c}_G$ as a direct summand; in particular, these groups contain a free abelian subgroup of rank $\mathfrak{c}_G$, but are not free abelian, and*

$$\left|\overline{\mathscr{RF}(G)}\right| = \mathfrak{c}_G = \left|\overline{\mathscr{RF}_0(G)}\right|.$$

- (iv) *Every non-trivial normal subgroup $\mathscr{N} \trianglelefteq \mathscr{RF}(G)$ contains a free subgroup of rank $\mathfrak{c}_G$; in particular, $|\mathscr{N}| = \mathfrak{c}_G$ and $\mathscr{N}$ is not soluble.*
- (v) *Let $\mathscr{N}_0 \trianglelefteq \mathscr{RF}_0(G)$ be a normal subgroup, and suppose that $\mathscr{N}_0$ has an element of infinite order. Then $\mathscr{N}_0$ contains a free subgroup of rank $\mathfrak{c}_G$; in particular, $|\mathscr{N}_0| = \mathfrak{c}_G$ and $\mathscr{N}_0$ is not soluble.*
- (vi) *If $\mathscr{N} \trianglelefteq \mathscr{RF}(G)$ has a non-trivial elliptic element then $\mathscr{N}$ contains a subgroup isomorphic to a free power of the form $U_0^{*\mathfrak{c}_G}$, where $U_0 := \mathscr{N} \cap G_0$.*

Proof Part (ii) of the theorem follows immediately from part (ii) of Proposition 3.13, Lemma 9.2, and Corollary 10.3.

Next, we establish parts (i) and (iii). Let $\{f_\sigma\}_{\sigma\in S}$ be a family of pairwise locally incompatible test functions with bijective indexing, such that $|S| = \mathfrak{c}_G$ and $C_{\mathscr{RF}(G)}(f_\sigma) \cong (\mathbb{Q},+)$ for all $\sigma \in S$ (such a family exists by Theorem 10.1). By part (i) of Proposition 10.9, the subgroup $\mathfrak{f}_S = \langle f_\sigma : \sigma \in S\rangle$ of $\mathscr{RF}(G)$ is free of rank $\mathfrak{c}_G$, and its image under the projection π is a free subgroup of

$\mathscr{R}\mathscr{F}_0(G)$ of the same rank. This last fact forces $|\mathscr{R}\mathscr{F}_0(G)| \geq \mathfrak{c}_G$, and equality follows (in the presence of the axiom of choice) from Corollary 10.4.

Now set

$$\mathscr{C} := \langle C_{\mathscr{R}\mathscr{F}(G)}(f_\sigma) : \sigma \in S \rangle \leq \mathscr{R}\mathscr{F}(G),$$

and consider the homomorphic images $\mathscr{C}' := \pi^{ab}(\mathscr{C})$ and $\mathscr{C}'_0 := \pi_0^{ab}(\mathscr{C})$ in $\overline{\mathscr{R}\mathscr{F}(G)}$ and $\overline{\mathscr{R}\mathscr{F}_0(G)}$ respectively. Both $\mathscr{C}'$ and $\mathscr{C}'_0$ are divisible (being generated by divisible subgroups) and each contains a free-abelian subgroup of rank $\mathfrak{c}_G$ by part (ii) of Proposition 10.9, namely the groups $\pi^{ab}(\mathfrak{f}_S)$ and $\pi_0^{ab}(\mathfrak{f}_S)$ respectively. By the structure theorem on divisible groups, $\mathscr{C}'$ and $\mathscr{C}'_0$ decompose as $\mathscr{C}' = V' \oplus T'$ and $\mathscr{C}'_0 = V'_0 \oplus T'_0$ respectively, where V' and V'_0 are $\mathbb{Q}$-vector spaces and T', T'_0 are the maximal torsion subgroups of $\mathscr{C}'$ and $\mathscr{C}'_0$ respectively; see, for instance, Theorem 19.1 in Fuchs [18]. Since $\pi^{ab}(\mathfrak{f}_S)$ and $\pi_0^{ab}(\mathfrak{f}_S)$ are torsion-free, they are embedded via the canonical projections $V' \oplus T' \to V'$ and $V'_0 \oplus T'_0 \to V'_0$ respectively into V' and V'_0 respectively, implying that $\dim_{\mathbb{Q}} V' \geq \mathfrak{c}_G$ and $\dim_{\mathbb{Q}} V'_0 \geq \mathfrak{c}_G$. Since $|\mathscr{R}\mathscr{F}(G)| = \mathfrak{c}_G$ by Corollary 10.4, we have

$$\dim_{\mathbb{Q}} V' = \mathfrak{c}_G = \dim_{\mathbb{Q}} V'_0.$$

Further, by a result of Baer, a divisible subgroup of an abelian group is a direct summand; see Baer [2] or Theorem 18.1 in Fuchs [18]. It follows that V' is a direct summand of $\overline{\mathscr{R}\mathscr{F}(G)}$ and that V'_0 is a direct summand of $\overline{\mathscr{R}\mathscr{F}_0(G)}$. The assertion concerning the cardinalities of $\overline{\mathscr{R}\mathscr{F}(G)}$ and $\overline{\mathscr{R}\mathscr{F}_0(G)}$ follows from the above plus Corollary 10.4.

Finally, the fact that $\mathscr{R}\mathscr{F}(G)$, for $G \neq \{\mathbf{1}_G\}$, is not a free group follows, for instance, from the existence of non-trivial elements with non-cyclic centraliser or from the (already established) fact that $\overline{\mathscr{R}\mathscr{F}(G)}$ is not free abelian. Similarly, the fact that $\overline{\mathscr{R}\mathscr{F}_0(G)}$ is not free abelian serves to show that $\mathscr{R}\mathscr{F}_0(G)$ itself is not free; alternatively Corollary 10.3, in conjunction with Corollaries 9.10 and 9.11, allows us to exhibit non-trivial elements with non-cyclic centraliser in $\mathscr{R}\mathscr{F}_0(G)$.

Our next task is to prove (iv) and (v). Let $\{f_\sigma\}_{\sigma \in S}$ be a family of test functions as described in Corollary 10.2 with $L(f_\sigma) = 1$ for all $\sigma \in S$, say. Since $\mathscr{N}$ is not trivial, it must (according to Proposition 3.25 and Corollary 3.27) contain a hyperbolic element h and, since $\mathscr{N}$ is normal, we must have $h_1 = c(h) \in \mathscr{N}$. By definition h_1 is cyclically reduced, and we have $L(h_1) > 0$ in view of part (ii) of Proposition 3.13 since h is hyperbolic. At this point, we observe that, for all but at most two indices $\sigma \in S$, we must have

$$f_\sigma h_1 f_\sigma^{-1} = f_\sigma \circ h_1 \circ f_\sigma^{-1}. \tag{10.21}$$

Indeed, suppose there are three distinct indices $\sigma_1, \sigma_2, \sigma_3 \in S$ such that

$$\varepsilon_0(f_{\sigma_j}, h_1) + \varepsilon_0(h_1, f_{\sigma_j}^{-1}) > 0, \quad j = 1,2,3.$$

Then there are two indices out of these three, σ_1 and σ_2, say, such that

$$\varepsilon_0(f_{\sigma_1}, h_1) > 0 \quad \text{and} \quad \varepsilon_0(f_{\sigma_2}, h_1) > 0$$

or

$$\varepsilon_0(h_1, f_{\sigma_1}^{-1}) > 0 \quad \text{and} \quad \varepsilon_0(h_1, f_{\sigma_2}^{-1}) > 0.$$

Since $\varepsilon_0(f,g) = \varepsilon_0(g^{-1}, f^{-1})$ by Lemma 2.12(i), we conclude in both cases that there exists a real number ε such that $0 < \varepsilon < 1$ and

$$f_{\sigma_1}(1-\eta) = f_{\sigma_2}(1-\eta), \quad 0 \le \eta \le \varepsilon.$$

It follows that

$$f_{\sigma_1}(\xi_0 + \eta') = f_{\sigma_2}(\xi_0 + \eta'), \quad |\eta'| \le \frac{\varepsilon}{2},$$

where $\xi_0 := 1 - \varepsilon/2$, contradicting the fact that f_{σ_1} and f_{σ_2} are locally incompatible. Hence, at most two of the test functions f_σ exhibit cancellation when conjugating h_1. Deleting these exceptional functions, we obtain a family $\{f_\sigma\}_{\sigma \in S'}$ of pairwise locally incompatible test functions with $|S'| = |S|$, such that (10.21) holds for all $\sigma \in S'$. Set

$$\mathcal{H}_\sigma := \langle f_\sigma h_1 f_\sigma^{-1} \rangle, \quad \sigma \in S',$$

and let

$$\mathcal{H} := \langle f_\sigma h_1 f_\sigma^{-1} : \sigma \in S' \rangle$$

be the subgroup generated by the $\mathcal{H}_\sigma$. In view of (10.21), and since h_1 is cyclically reduced and of positive length, an arbitrary non-trivial element h of $\mathcal{H}_\sigma$ is of the form

$$h = f_\sigma \circ \underbrace{h_1^{\operatorname{sgn}(\gamma)} \circ \cdots \circ h_1^{\operatorname{sgn}(\gamma)}}_{|\gamma| \text{ factors}} \circ f_\sigma^{-1},$$

where $\gamma \in \mathbb{Z} - \{0\}$. It follows from this that (1) each $\mathcal{H}_\sigma$ is hyperbolic (since $L(c(h)) = |\gamma| L(h_1) > 0$, see Proposition 3.13), in particular $\mathcal{H}_\sigma \cap G_0 = \{\mathbf{1}_G\}$, and that (2) the family $\{\mathcal{H}_\sigma\}_{\sigma \in S'}$ is independent in the sense of Definition 9.22 (see Lemma 9.13), as well as bijectively indexed. By construction $\mathcal{H}$ is contained in the normal subgroup $\mathcal{N}$ and, invoking Lemma 9.23, we find that $\mathcal{H}$ is free of rank $|S'| = |S| = \mathfrak{c}_G$, as required. The particular statements in (iv) are immediate now, in view of Corollary 10.4.

Now let $\mathcal{N}_0$ be as in (v), and let $h_0 \in \mathcal{N}_0$ be an element of infinite order. The preimage $\mathcal{N} = \pi^{-1}(\mathcal{N}_0)$ of $\mathcal{N}_0$ under π is a non-trivial normal subgroup of $\mathcal{RF}(G)$. Hence, by the proof of (iv), it contains a free subgroup $\mathfrak{f}$ of rank $\mathfrak{c}_G$, freely generated by a set of elements of the form $f_\sigma \circ c(h) \circ f_\sigma^{-1}$ that are constructed as in that proof, starting from an element h with $\pi(h) = h_0$. Let

$$\mathfrak{f}' = \langle f_{\sigma_1} \circ c(h) \circ f_{\sigma_1}^{-1}, \dots, f_{\sigma_k} \circ c(h) \circ f_{\sigma_k}^{-1} \rangle$$

be a subgroup of $\mathfrak{f}$ which is generated by $k \geq 1$ of the basis elements $f_\sigma \circ c(h) \circ f_\sigma^{-1}$. Then the quotient group

$$\mathfrak{f}'/\mathfrak{f}' \cap E(G) \cong \mathfrak{f}'E(G)/E(G) \leq \mathcal{RF}_0(G)$$

contains elements of infinite order, namely conjugates of h_0, and so is an infinite group. Thus, just as in the proof of Proposition 10.9(i), we conclude that $\mathfrak{f}' \cap E(G) = \{\mathbf{1}_G\}$ and hence that $\mathfrak{f} \cap E(G) = \{\mathbf{1}_G\}$. It follows that $\pi(\mathfrak{f})$ is a free group of rank $\mathfrak{c}_G$ contained in $\mathcal{N}_0$, as required.

Finally, we shall establish part (vi). Again, let $\{f_\sigma\}_{\sigma \in S}$ be a family of pairwise locally incompatible test functions with bijective indexing, $|S| = \mathfrak{c}_G$, and $L(f_\sigma) = 1$ for all $\sigma \in S$; the existence of such a family is guaranteed by Corollary 10.2. Set

$$\mathcal{H}_\sigma := f_\sigma U_0 f_\sigma^{-1}, \quad \sigma \in S,$$

where $U_0 = \mathcal{N} \cap G_0$ and let $\mathcal{H}$ be the subgroup of $\mathcal{RF}(G)$ generated by the groups $\mathcal{H}_\sigma$. An arbitrary non-trivial element h of $\mathcal{H}_\sigma$ has the form

$$h = f_\sigma \circ g \circ f_\sigma^{-1}$$

with some $g \in U_0 - \{\mathbf{1}_G\}$, so has length $L(h) = 2L(f_\sigma) = 2 > 0$. Hence $\mathcal{H}_\sigma \cap G_0 = \{\mathbf{1}_G\}$. Further, if $f_{\sigma_1} U_0 f_{\sigma_1}^{-1} = f_{\sigma_2} U_0 f_{\sigma_2}^{-1}$ then

$$(f_{\sigma_2}^{-1} f_{\sigma_1}) U_0 (f_{\sigma_2}^{-1} f_{\sigma_1})^{-1} = U_0; \tag{10.22}$$

and, for $\sigma_1 \neq \sigma_2$, the function $f_{\sigma_2}^{-1} f_{\sigma_1}$ is a product of two locally incompatible test functions and thus is also a test function, by part (ii) of Lemma 9.13 plus Lemma 9.14. Therefore, every non-trivial element on the left-hand side of (10.22) has positive length (and such elements exist since $U_0 \neq \{\mathbf{1}_G\}$), while the right-hand side is contained in G_0, a contradiction. It follows that $\sigma_1 = \sigma_2$, so that the family of subgroups $\{\mathcal{H}_\sigma\}_{\sigma \in S}$ is indeed bijectively indexed. Also, since the test functions f_σ are pairwise locally incompatible the family $\{\mathcal{H}_\sigma\}_{\sigma \in S}$ is independent. Hence, by Lemma 9.23,

$$\mathcal{H} \cong \mathop{\ast}_{\sigma \in S} f_\sigma U_0 f_\sigma^{-1} \cong U_0^{*\mathfrak{c}_G}$$

and, by construction, we have $\mathcal{H} \leq \mathcal{N}$, whence the result. □

10.8 Exercises

10.1. Show that every non-trivial torsion-free abelian group of rank at most $2^{\aleph_0}$ can be embedded as a proper subgroup into the additive reals.

10.2. Prove equation (10.1).

10.3. Establish assertion (10.5), that is, the test functions f_h with $h \in G^{\mathfrak{T}}$ are pairwise locally incompatible.

10.4. Given a proper real group Λ, show that there exists a subgroup $\Lambda' \leq \mathbb{R}$ such that $\Lambda < \Lambda' < \mathbb{R}$ and $(\Lambda' : \Lambda) \leq \aleph_0$.

10.5. Show that a proper real group Λ satisfies $(\mathbb{R} : \Lambda) = \infty$.

10.6. Show that a dense subgroup of the additive reals is generated by its intersection with any non-empty open interval.

10.7. Verify assertion (10.8), that is, establish the fact that the family $\mathfrak{F}$ of test functions constructed in case (ii) of the proof of Theorem 10.1 (see Section 10.3) consists of pairwise locally incompatible functions.

10.8. Complete the proof of Theorem 10.5 by dealing with the case where $L(g_1') = 0$ and $L(g_r') > 0$ (that is, case 3(d)).

10.9. Establish Theorem 10.7 and Corollary 10.8.

10.10. Show that $\mathscr{RF}(G)$ cannot be decomposed as the direct product of two proper normal subgroups.

10.11. Give an example of a hyperbolic element $f \in \mathscr{RF}(G)$ and a cyclically reduced function f_1 such that f is conjugate to f_1 and f_1 is not the core of f.

11

A generalisation to groupoids

11.1 Introduction

In Alperin and Moss [1] there is a construction of a family of groups acting freely on an $\mathbb{R}$-tree. We shall show that this construction and the construction of $\mathscr{RF}(G)$ have a common generalisation. This involves the idea of a *groupoid*, by which we mean a Brandt groupoid, that is, a small category in which every morphism is an isomorphism. We shall denote a groupoid by a pair (S,G), where S is the set of objects and G is the set of morphisms, ignoring the remaining structure on the groupoid (the notation for this will be given shortly).

In Section 11.2 we explain a general construction (the common generalization alluded to above), which associates with a given groupoid (S,G) a new groupoid $(S,\mathscr{ARF}(S,G))$ on the same set of objects. This construction is further analysed in Sections 11.3 and 11.4, mainly by the development of a cancellation theory generalising the corresponding proceedings in Sections 2.3 and 2.4. This involves straightforward generalisations of arguments in Chapter 2 but also makes use of the idea of an admissible function, adapted from [1].

In Section 11.5 we show that two major ideas from earlier chapters, namely cyclic reduction and exponent sums, generalise to our groupoid context. Next, in Section 11.6, we discuss the idea of a Lyndon length function on a groupoid, and use it to generalise the construction of the $\mathbb{R}$-tree $\mathbf{X_G}$ to $(S,\mathscr{ARF}(S,G))$. Finally, in Section 11.7, we consider the functoriality of our construction, but that is as far as we shall develop the theory in the present book.

We now return to matters of notation. Let (S,G) be a groupoid. For s, $t \in S$, we denote the set of morphisms from s to t by G_{st}. The identity element in G_{ss} will be denoted by 1_s, for $s \in S$. The groupoid (S,G) may be viewed as a directed graph with edge set G and vertex set S; if $f \in G_{st}$, we define its initial and terminal vertices via $\sigma(f) = s$ and $\tau(f) = t$ respectively, and we shall view f as an arc (a directed edge) from s to t. As is well known, this gives a forgetful functor from the category of groupoids to the category of directed graphs. Thus, objects will often be called vertices and the group G_{ss} will be termed the *vertex group at* s.

11.2 The construction

Let (S,G) be a groupoid. We begin with the ideas of left and right continuity from Alperin and Moss [1], suitably modified.

Two functions $f:(\alpha,\beta)\to S$ and $g:(\gamma,\delta)\to S$, where $\alpha,\beta,\gamma,\delta\in\mathbb{R}$, $\alpha<\beta$, and $\gamma<\delta$, are said to be *equivalent* if there exists $\varepsilon>0$ such that, for all η with $0<\eta<\varepsilon$, $f(\beta-\eta)=g(\delta-\eta)$. Let $\mathscr{S}$ be the set of equivalence classes of such functions. By considering constant functions $(\alpha,\beta)\to S$, it follows that the cardinalities of the sets S and $\mathscr{S}$ satisfy $|S|\le|\mathscr{S}|$; we shall fix an injective map $E: S\to\mathscr{S}$, denoted $s\mapsto E_s$.

Definition 11.1

(i) A mapping $f:[0,\alpha]\to S$, where $\alpha\in\mathbb{R}$ and $\alpha\ge 0$, is called *left continuous* if, for all $\xi\in\mathbb{R}$ with $0<\xi\le\alpha$, we have

$$f|_{(0,\xi)}\in E_{f(\xi)}.$$

(ii) The mapping f is *right continuous* if the function $f':[0,\alpha]\to S$ defined by $f'(x)=f(\alpha-x)$ is left continuous.

Thus, if f is left continuous then $f(\xi)$ is determined by $f|_{(\xi',\xi)}$, for any ξ' with $0<\xi'<\xi$; this motivates our terminology.

Definition 11.2

(i) Given a groupoid (S,G), we denote by $\mathscr{F}(S,G)$ the set of all maps $f:[0,\alpha]\to G$, where $\alpha\in\mathbb{R}$ and $\alpha\ge 0$.

(ii) If $f:[0,\alpha]\to G$ is in $\mathscr{F}(S,G)$, we call α the *length* of f, denoted by $L(f)$.

Next, we generalise the concept of a reduced function; this concept (after a considerable amount of work) allowed the transition from the monoid $\mathscr{F}(G)$ to the group $\mathscr{RF}(G)$ in Chapter 2.

Definition 11.3

(i) A function $f:[0,\alpha]\to G$ is called *reduced*, if for every interior point ξ_0 in the domain of f with $f(\xi_0)=1_s$ for some $s\in S$, and for every real ε satisfying $0<\varepsilon\le\min\{L(f)-\xi_0,\xi_0\}$, there exists δ such that $0<\delta\le\varepsilon$ and $f(\xi_0+\delta)\ne\big(f(\xi_0-\delta)\big)^{-1}$.

(ii) The set of all reduced functions in $\mathscr{F}(S,G)$ will be denoted by $\mathscr{RF}(S,G)$.

In order to ensure that the multiplication of two elements in the groupoid $(S,\mathscr{ARF}(S,G))$ to be constructed below is always defined when it should be, we shall need a slight modification of another idea from [1], namely that of an admissible function, which we introduce next.

Definition 11.4

(i) A function $f:[0,\alpha]\to G$ in $\mathscr{F}(S,G)$ is called *admissible* if the functions $f_1,f_2:[0,\alpha]\to S$, defined by $f_1(\xi)=\sigma(f(\xi))$ and $f_2(\xi)=\tau(f(\xi))$, are left and right continuous respectively.

(ii) The set of all functions in $\mathscr{RF}(S,G)$ which are admissible will be denoted by $\mathscr{ARF}(S,G)$.

The following observation will become important later.

Lemma 11.5 *If $f:[0,\alpha]\to G$ is admissible and if $\sigma(f(\xi))=\tau(f(\xi))$, where $0<\xi<\alpha$, then there exists a real number $\varepsilon>0$, depending on ξ, such that $\sigma(f(\xi-\delta))=\tau(f(\xi+\delta))$ for $0\le\delta<\varepsilon$.*

Proof For $\xi\in(0,\alpha)$, let $s_\xi=\sigma(f(\xi))=\tau(f(\xi))$. By the admissibility of f, the composite functions $\sigma\circ f|_{(0,\xi)}$ and $\tau\circ f'|_{(0,\alpha-\xi)}$, where $f'(x)=f(\alpha-x)$ for $x\in[0,\alpha]$, are both in E_{s_ξ}. Indeed,

$$\sigma\circ f|_{(0,\xi)}\in E_{\sigma(f(\xi))}=E_{s_\xi}$$

and

$$\tau\circ f'|_{(0,\alpha-\xi)}\in E_{\tau(f'(\alpha-\xi))}=E_{\tau(f(\xi))}=E_{s_\xi}.$$

The functions $\sigma\circ f|_{(0,\xi)}$ and $\tau\circ f'|_{(0,\alpha-\xi)}$ being equivalent, there exists a real number $\varepsilon>0$ such that

$$\sigma(f(\xi-\delta))=\tau(f'(\alpha-\xi-\delta))=\tau(f(\xi+\delta))$$

holds for all $0 < \delta < \varepsilon$. If we include the case where $\delta = 0$, which holds by assumption, the result follows. □

The construction of $\mathscr{ARF}(S,G)$ depends on the choice of an injective map $E : S \to \mathscr{S}$; we shall call such a map an *S-indexing for* $\mathscr{S}$ (short for 'injective indexing by S of a subset of $\mathscr{S}$'). In the rest of this section we shall assume that an S-indexing for $\mathscr{S}$ has been fixed.

Our initial aim is to make $\mathscr{ARF}(S,G)$ into a groupoid. We begin by setting $\sigma(f) := \sigma(f(0))$ and $\tau(f) := \tau(f(\alpha))$, where $f : [0,\alpha] \to G$ is any element of $\mathscr{F}(S,G)$, which gives $\mathscr{F}(S,G)$ the structure of a directed graph. We can then make $\mathscr{F}(S,G)$ into the set of morphisms of a small category with set of objects S. As in Chapter 2, partial multiplication, denoted by $*$, is defined via

$$(f*g)(\xi) = \begin{cases} f(\xi), & 0 \leq \xi < \alpha, \\ f(\alpha)g(0), & \xi = \alpha, \\ g(\xi - \alpha), & \alpha < \xi \leq \alpha + \beta, \end{cases}$$

where f, $g \in \mathscr{F}(S,G)$, are of lengths α, β respectively and $\tau(f) = \sigma(g)$, so that $f(\alpha)g(0)$ is defined. Thus, $f*g$ has length $\alpha + \beta$.

The identity element of $\mathscr{F}(S,G)$ at the vertex $s \in S$ is the function $\mathbf{1}_s$ defined on $[0,0] = \{0\}$ by $\mathbf{1}_s(0) = 1_s$, the identity element of the groupoid (S,G) at s. The proof of associativity is just the same as that in the proof of Proposition 2.1.

Note that $(S, \mathscr{F}(S,G))$ is a cancellative category (every morphism is epi and mono); again, the proof is the same as that of the corresponding fact for Proposition 2.1.

However, to obtain a groupoid we have to allow 'cancellation' in forming the product, just as we did when defining $\mathscr{RF}(G)$ in Chapter 2. In order to define the 'reduced product', it turns out to be necessary to restrict attention to the subset $\mathscr{ARF}(S,G)$ of $\mathscr{F}(S,G)$. This subset is the set of morphisms, and the set of objects is S, with the same definitions of $\sigma(f)$ and $\tau(f)$ as before, but with reduced multiplication (to be defined next) as the partial product.

Given $f, g \in \mathscr{F}(S,G)$, of lengths α and β respectively, set

$$\varepsilon_0 = \varepsilon_0(f,g) = \begin{cases} \sup \mathscr{E}(f,g) & \text{if } f(\alpha) = g(0)^{-1}, \\ 0 & \text{otherwise,} \end{cases}$$

where

$$\mathscr{E}(f,g) := \left\{\varepsilon \in [0,\min\{\alpha,\beta\}] : f(\alpha-\delta) = g(\delta)^{-1} \text{ for all } \delta \in [0,\varepsilon]\right\}.$$

Now let $f : [0,\alpha] \to G$ and $g : [0,\beta] \to G$ be admissible functions such that $\tau(f) = \sigma(g)$. Then we define the (reduced) product fg on the interval $[0,\alpha+\beta-2\varepsilon_0]$ via

$$(fg)(\xi) = \begin{cases} f(\xi), & 0 \le \xi < \alpha-\varepsilon_0 \\ f(\alpha-\varepsilon_0)g(\varepsilon_0), & \xi = \alpha-\varepsilon_0 \\ g(\xi-\alpha+2\varepsilon_0), & \alpha-\varepsilon_0 < \xi \le \alpha+\beta-2\varepsilon_0. \end{cases} \tag{11.1}$$

For this definition to make sense, we need to verify that $f(\alpha-\varepsilon_0)g(\varepsilon_0)$ is defined. If $\varepsilon_0 = 0$ then $\tau(f(\alpha)) = \sigma(g(0))$ by assumption, so $f(\alpha)g(0)$ is defined in that case. If $\varepsilon_0 > 0$ then $f(\alpha-\xi) = g(\xi)^{-1}$ for $0 \le \xi < \varepsilon_0$, so $\tau(f(\alpha-\xi)) = \sigma(g(\xi))$ for $0 \le \xi < \varepsilon_0$, thus

$$\sigma \circ g|_{(0,\varepsilon_0)} = \tau \circ f'|_{(0,\varepsilon_0)}.$$

Since the maps $\sigma \circ g$ and $\tau \circ f$ are left and right continuous respectively, by the admissibility of g and f, we have

$$E_{\sigma(g(\varepsilon_0))} \ni \sigma \circ g|_{(0,\varepsilon_0)} = \tau \circ f'|_{(0,\varepsilon_0)} \in E_{\tau(f'(\varepsilon_0))} = E_{\tau(f(\alpha-\varepsilon_0))}$$

implying that $\tau(f(\alpha-\varepsilon_0)) = \sigma(g(\varepsilon_0))$, as required, since E is injective. Hence, $f(\alpha-\varepsilon_0)g(\varepsilon_0)$ is always defined.

Further, it is easy to check that the product fg, if defined, is also admissible; see Exercise 11.1.

As our next result shows, reduced multiplication restricts to define a product on $\mathscr{ARF}(S,G)$.

Lemma 11.6 *If $f, g \in \mathscr{ARF}(S,G)$ and fg is defined then $fg \in \mathscr{ARF}(S,G)$.*

Proof It is enough to show that fg, as product of two reduced functions, is also reduced; however, the proof of this fact is the same as that of Lemma 2.7. □

We have $\mathbf{1}_s \in \mathscr{ARF}(S,G)$ for all $s \in S$ and $\mathbf{1}_s f = f$ for any $f \in \mathscr{ARF}(S,G)$ with $\sigma(f) = s$, as well as $f\mathbf{1}_s = f$ for any $f \in \mathscr{ARF}(S,G)$ with $\tau(f) = s$.

Definition 11.7 The (formal) *inverse* f^{-1} of an element $f \in \mathscr{F}(S,G)$ is the function defined on the same interval $[0,\alpha]$ as f by

$$f^{-1}(\xi) = \left(f(\alpha-\xi)\right)^{-1}, \quad 0 \le \xi \le \alpha.$$

11.3 Cancellation theory for $\mathscr{ARF}(S,G)$

The aim of this and the next section is to establish the following basic result.

Theorem 11.8 *With reduced multiplication as defined in the last section, $(S, \mathscr{ARF}(S,G))$ is a groupoid.*

It is easy to see that if $f \in \mathscr{ARF}(S,G)$ with $\sigma(f) = s$, $\tau(f) = t$ then $ff^{-1} = \mathbf{1}_s$ and $f^{-1}f = \mathbf{1}_t$; see Exercise 11.3. Thus, in order to establish Theorem 11.8 it remains only to show that (reduced) multiplication on $\mathscr{ARF}(S,G)$ is associative. For this purpose we introduce the analogue of the circle product defined in Chapter 2.

Definition 11.9 For $f,g \in \mathscr{F}(S,G)$, $f \circ g$ is defined if and only if $\tau(f) = \sigma(g)$ and $\varepsilon_0(f,g) = 0$, in which case $f \circ g = f * g$.

Thus, if $f,g \in \mathscr{ARF}(S,G)$ and $f \circ g$ is defined then we have

$$f \circ g = f * g = fg.$$

We shall establish a series of lemmas on the circle product, mostly referring to Chapter 2 for the proofs, which are essentially the same as in that chapter, although now more care is needed to ensure that endpoints of elements of $\mathscr{ARF}(S,G)$ match up so that various products are defined.

We begin by noting that Lemma 2.12 (inversion of star products) generalises straightforwardly as follows.

Lemma 11.10 **(Inversion of star products)**

(i) *Let $f_1, f_2 \in \mathscr{F}(S,G)$; then $\varepsilon_0(f_1,f_2) = \varepsilon_0(f_2^{-1}, f_1^{-1})$.*

(ii) *Let $f_1, f_2 \in \mathscr{F}(S,G)$, and let $f = f_1 * f_2$. Then $f^{-1} = f_2^{-1} * f_1^{-1}$; in particular, $f = f_1 \circ f_2$ implies $f^{-1} = f_2^{-1} \circ f_1^{-1}$.*

Next, we generalise Lemma 2.8.

Lemma 11.11 *For $f,g \in \mathscr{ARF}(G)$ such that $\tau(f) = \sigma(g)$, the following are equivalent:*

(i) $\varepsilon_0(f,g) = 0$;

(ii) $fg = f * g$;

(iii) $f * g$ *is reduced.*

Proof The proof that (i) $\Rightarrow$ (ii) is a straightforward computation, and the implication (ii) $\Rightarrow$ (iii) follows immediately from Lemma 11.6. The proof of (iii) $\Rightarrow$ (i) is similar to the proof of Lemma 2.8; one needs to replace 1_G by the appropriate identity element 1_s in the morphism set G. □

Remark 11.12 We note that the implication (iii) $\Rightarrow$ (i) in Lemma 11.11 in fact holds for $f,g \in \mathscr{F}(S,G)$.

Our next result generalises Lemma 2.14 (dissection of reduced functions) to our groupoid context.

Lemma 11.13 **(Dissection of reduced functions over a groupoid)** *Let $f : [0,\alpha] \to G$ be reduced, and let β be a real number such that $0 \le \beta \le \alpha$. Then there exist reduced functions $f_1 : [0,\beta] \to G$ and $f_2 : [0,\alpha-\beta] \to G$ such that $f = f_1 \circ f_2$. Moreover, f_1 and f_2 with these properties are uniquely determined once one of the values $f_1(\beta)$, $f_2(0)$ has been specified, and one of these values may be chosen arbitrarily in G, subject only to the condition that $\sigma(f_1(\beta)) = \sigma(f(\beta))$ if $f_1(\beta)$ is chosen, and to the condition that $\tau(f_2(0)) = \tau(f(\beta))$ if $f_2(0)$ is chosen. Moreover, if $f \in \mathscr{ARF}(S,G)$ then also f_1, $f_2 \in \mathscr{ARF}(S,G)$.*

Proof The equation $f = f_1 * f_2$ with $L(f_1) = \beta$ and $L(f_2) = \alpha - \beta$ is equivalent to the conjunction of the three statements

$$f_1(\xi) = f(\xi), \qquad 0 \le \xi < \beta,$$

$$f_2(\xi) = f(\xi+\beta), \quad 0 < \xi \le \alpha - \beta,$$

and

$$f_1(\beta) f_2(0) = f(\beta).$$

These equations can clearly be solved, and the fact that $f \in \mathscr{RF}(S,G)$ implies that f_1, $f_2 \in \mathscr{RF}(S,G)$ and that $f = f_1 * f_2 = f_1 \circ f_2$; the latter implication holds in view of Remark 11.12. The final equation implies that

$$\sigma(f_1(\beta)) = \sigma(f(\beta)), \quad \tau(f_2(0)) = \tau(f(\beta)), \quad \text{and} \quad \tau(f_1(\beta)) = \sigma(f_2(0)). \tag{11.2}$$

Together with the fact that (S,G) is a groupoid, it follows that f_1, f_2 are uniquely determined once one of the values $f_1(\beta), f_2(0)$ has been specified, and one of these values can be chosen subject to the constraints given in the lemma.

The final assertion of the lemma also follows from equations (11.2). □

Continuing with our program of establishing the associativity of reduced multiplication in the groupoid context, we next generalize Lemma 2.15 (visibility of cancellation) to the groupoid context.

Lemma 11.14 **(Visibility of cancellation over a groupoid)** *Let $f,g \in \mathscr{ARF}(S,G)$ be such that $\tau(f) = \sigma(g)$. Then there exist $f_1, g_1, u \in \mathscr{ARF}(S,G)$ such that $f = f_1 \circ u$, $g = u^{-1} \circ g_1$, and $fg = f_1 \circ g_1$.*

Proof Let $\varepsilon_0 = \varepsilon_0(f,g)$. If $\varepsilon_0 = 0$, then setting $f_1 := f$, $g_1 := g$, and $u := \mathbf{1}_{\tau(f)}$ satisfies the conditions of the lemma, so we may assume that $\varepsilon_0 > 0$. By Lemma 11.13 we can find functions
$f_1 : [0, L(f) - \varepsilon_0] \to G$, $g_1 : [0, L(g) - \varepsilon_0] \to G$, $u : [0, \varepsilon_0] \to G$, and $v : [0, \varepsilon_0] \to G$ in $\mathscr{ARF}(G)$ such that $f = f_1 \circ u$ and $g = v \circ g_1$. We have

$$f(L(f) - \delta) = (g(\delta))^{-1}, \quad 0 \leq \delta < \varepsilon_0. \tag{11.3}$$

Moreover, for these values of δ, we have

$$f(L(f) - \delta) = u(\varepsilon_0 - \delta)$$

as well as

$$g(\delta) = v(\delta),$$

so that equation (11.3) gives

$$\big(u(\varepsilon_0 - \delta)\big)^{-1} = v(\delta), \quad 0 \leq \delta < \varepsilon_0.$$

By (11.3) plus the admissibility of f and g, we conclude that

$$\tau(f(L(f) - \varepsilon_0)) = \sigma(g(\varepsilon_0)). \tag{11.4}$$

Further, by Lemma 11.13 we can choose $u(0)$ arbitrarily subject only to the condition that $\tau(u(0)) = \tau(f(L(f) - \varepsilon_0))$. In view of (11.4), we can therefore choose u, v such that $(u(0))^{-1} = v(\varepsilon_0)$, so that indeed $v = u^{-1}$.

At this stage, a straightforward calculation coupled with Lemmas 11.6 and 11.11 shows that $fg = f_1 \circ g_1$, as desired; see the proof of Lemma 2.15. □

Lemma 11.15 (Visible cancellation over a groupoid) *Suppose that $f = f_1 \circ u$ and $g = u^{-1} \circ g_1$, where $f_1, g_1, u \in \mathscr{ARF}(S,G)$. Then $fg = f_1 g_1$. Moreover, if $L(u) = \varepsilon_0(f,g)$ then $fg = f_1 \circ g_1$.*

Proof The proof is the same as that of Lemma 2.16; care needs to be taken that the various products of morphisms are defined, and 1_G needs to be replaced by the appropriate identity element in the morphism set G of the groupoid (S,G). □

Lemma 11.16 *Let $f,g,h \in \mathscr{ARF}(S,G)$, and suppose that $L(g) > 0$.*

(i) *If $\varepsilon_0(f,g) = 0$ then*

$$\varepsilon_0(fg,h) = 0 \iff \varepsilon_0(g,h) = 0.$$

(ii) *If $\varepsilon_0(g,h) = 0$ then*

$$\varepsilon_0(f,gh) = 0 \iff \varepsilon_0(f,g) = 0.$$

Proof The argument of Lemma 2.17 is easily adapted to prove this. □

Corollary 11.17 (Associativity of the circle product over a groupoid) *Let $f,g,h \in \mathscr{F}(S,G)$. Then $(f \circ g) \circ h$ is defined if and only if $f \circ (g \circ h)$ is defined, in which case*

$$(f \circ g) \circ h = f \circ (g \circ h).$$

Proof The argument of Corollary 2.18 can be used to prove this. □

11.4 Proof of Theorem 11.8

It remains only to establish the associativity of reduced multiplication. Let $f,g,h \in \mathscr{ARF}(S,G)$ be reduced and admissible functions. Clearly $f(gh)$ is defined if and only if $(fg)h$ is defined, both statements being equivalent to the conjunction of $\tau(f) = \sigma(g)$ and $\tau(g) = \sigma(h)$. Assume that this is the case; then we need to show that $f(gh) = (fg)h$.

By Lemma 11.14 (visibility of cancellation over a groupoid), there exist elements $f_1, g_1, g_2, h_1, u, v \in \mathscr{ARF}(S,G)$, such that

$$f = f_1 \circ u,\ g = u^{-1} \circ g_1,\ fg = f_1 \circ g_1,$$

as well as

$$g = g_2 \circ v,\ h = v^{-1} \circ h_1,\ gh = g_2 \circ h_1.$$

Case 1: $L(u) < L(g_2)$. By Lemma 11.13 (dissection of reduced functions over a groupoid), we can write $g_2 = w \circ g_3$ with $w, g_3 \in \mathscr{ARF}(S,G)$ and $L(w) = L(u)$, so $L(g_3) > 0$. By Corollary 11.17 (associativity of the circle product over a groupoid), we have

$$g = (w \circ g_3) \circ v = w \circ (g_3 \circ v) = u^{-1} \circ g_1.$$

Comparing values in the last equation, we find that

$$w(\xi) = u^{-1}(\xi), \quad 0 \leq \xi < L(u).$$

Further, since $L(u) < L(g_2)$ by our case assumption, we have

$$g_2(L(u)) = g(L(u)) = u^{-1}(L(u)) g_1(0),$$

and thus

$$\sigma(u^{-1}(L(u))) = \sigma(g(L(u))) = \sigma(g_2(L(u))).$$

By Lemma 11.13, we can therefore choose w such that

$$w(L(u)) = u^{-1}(L(u)),$$

so that $w = u^{-1}$ and $g_2 = u^{-1} \circ g_3$.

By Corollary 11.17,

$$g = g_2 \circ v = (u^{-1} \circ g_3) \circ v = u^{-1} \circ (g_3 \circ v) = u^{-1} \circ g_1,$$

hence $g_1 = g_3 \circ v$, since the category $(S, \mathscr{F}(S,G))$ is cancellative.

Again by Corollary 11.17,

$$fg = f_1 \circ g_1 = f_1 \circ (g_3 \circ v) = (f_1 \circ g_3) \circ v;$$

hence, by Lemma 11.15,

$$(fg)h = ((f_1 \circ g_3) \circ v)(v^{-1} \circ h_1) = (f_1 \circ g_3) h_1.$$

We note that, by the computation of fg just given, $\varepsilon_0(f_1, g_3) = 0$. Next, making use of Corollary 11.17 once more, we find that

$$gh = g_2 \circ h_1 = (u^{-1} \circ g_3) \circ h_1 = u^{-1} \circ (g_3 \circ h_1),$$

implying $\varepsilon_0(f_1g_3,h_1)=0$, and thus

$$(fg)h=(f_1\circ g_3)\circ h_1$$

by part (i) of Lemma 11.16 applied to the functions f_1,g_3,h_1.

Similarly, using Corollary 11.17 and Lemma 11.15,

$$\begin{aligned}f(gh)&=f(g_2\circ h_1)\\&=(f_1\circ u)((u^{-1}\circ g_3)\circ h_1)\\&=(f_1\circ u)(u^{-1}\circ(g_3\circ h_1))\\&=f_1(g_3\circ h_1)\\&=f_1\circ(g_3\circ h_1),\end{aligned}$$

where we have employed Lemma 11.16(ii) in the last step. The fact that $(fg)h=f(gh)$ follows now from Corollary 11.17.

Case 2: $L(u)\geq L(g_2)$. As in case 1, making use of the fact that

$$\sigma(u^{-1}(L(g_2)))=\sigma(g(L(g_2)))=\sigma(g_2(L(g_2)))$$

and Lemma 11.13, we can write $u^{-1}=g_2\circ u_1$ for some $u_1\in\mathscr{ARF}(S,G)$. We now distinguish two subcases.

Case 2(a): $L(u_1)>0$. Applying Lemma 11.10 together with Corollary 11.17, we find that

$$f=f_1\circ u=f_1\circ(u_1^{-1}\circ g_2^{-1})=(f_1\circ u_1^{-1})\circ g_2^{-1};$$

in particular, $\varepsilon_0(f_1,u_1^{-1})=0$. Also,

$$g=u^{-1}\circ g_1=(g_2\circ u_1)\circ g_1=g_2\circ(u_1\circ g_1)=g_2\circ v,$$

hence $v=u_1\circ g_1$. Now

$$\begin{aligned}(fg)h&=(f_1\circ g_1)(v^{-1}\circ h_1)\\&=(f_1\circ g_1)((g_1^{-1}\circ u_1^{-1})\circ h_1)\\&=(f_1\circ g_1)(g_1^{-1}\circ(u_1^{-1}\circ h_1))\\&=f_1(u_1^{-1}\circ h_1)\\&=f_1\circ(u_1^{-1}\circ h_1),\end{aligned}$$

where we have made use of Lemma 11.16(ii) in the last step. Similarly,

$$\begin{aligned} f(gh) &= (f_1 \circ u)(g_2 \circ h_1) \\ &= (f_1 \circ (u_1^{-1} \circ g_2^{-1}))(g_2 \circ h_1) \\ &= ((f_1 \circ u_1^{-1}) \circ g_2^{-1})(g_2 \circ h_1) \\ &= (f_1 \circ u_1^{-1})h_1 \\ &= (f_1 \circ u_1^{-1}) \circ h_1, \end{aligned}$$

using Lemma 11.16(i) for the final step. By Corollary 11.17, we again conclude that $(fg)h = f(gh)$, as required.

Case 2(b): $L(u_1) = 0$. Here, if necessary we can change the values $g_2(L(u))$, $v(0)$, and $h_1(0)$ in such a way that $g = g_2 \circ v$ and $h = v^{-1} \circ h_1$ still hold, and additionally $g_2 = u^{-1}$. Then $g = u^{-1} \circ v = u^{-1} \circ g_1$, so $v = g_1$. Consequently, by Lemma 11.15,

$$(fg)h = (f_1 \circ g_1)(v^{-1} \circ h_1) = (f_1 \circ v)(v^{-1} \circ h_1) = f_1 h_1,$$

and

$$f(gh) = (f_1 \circ u)(g_2 \circ h_1) = (f_1 \circ u)(u^{-1} \circ h_1) = f_1 h_1,$$

completing the proof.

11.5 Cyclic reduction and exponent sums

The purpose of this short section is to show that two key ideas concerning $\mathscr{RF}(G)$, namely that of cyclic reduction and that of the exponent sum, generalise to our present context. This is done for possible further applications; we shall not use these ideas here.

Definition 11.18 A function $f \in \mathscr{ARF}(S,G)$ is called *cyclically reduced* if $\varepsilon_0(f,f) = 0$, or equivalently, if either f^2 is undefined or $L(f^2) = 2L(f)$.

Lemma 11.19

(i) *Let $f \in \mathscr{ARF}(S,G)$. Then there exist $t, f_1 \in \mathscr{ARF}(S,G)$ with f_1 cyclically reduced, such that $f = t \circ f_1 \circ t^{-1}$.*

(ii) *If $f = t \circ f_1 \circ t^{-1} = s \circ f_2 \circ s^{-1}$, where $t, s, f_1, f_2 \in \mathscr{ARF}(S,G)$ and f_1, f_2 are cyclically reduced, then $s = tg$ and $f_2 = g^{-1} f_1 g$ for some g with $L(g) = 0$.*

Proof (i) For $f \in \mathscr{ARF}(S,G)$, let $\varepsilon_0(f) := \min\{\varepsilon_0(f,f), L(f)/2\}$. By Lemma 11.13 (dissection of reduced functions over a groupoid), there exist $t, f_0 \in \mathscr{ARF}(S,G)$ such that $f = t \circ f_0$ and $L(t) = \varepsilon_0(f)$. Then

$$L(f_0) = L(f) - L(t) = L(f) - \varepsilon_0(f) \geq L(f)/2 \geq \varepsilon_0(f),$$

so, again by Lemma 11.13, there exist $u, f_1 \in \mathscr{ARF}(S,G)$ such that $f_0 = f_1 \circ u$, where $L(u) = \varepsilon_0(f)$. By Corollary 11.17 (associativity of the circle product over a groupoid) $t \circ f_1$ is defined, so $\varepsilon_0(t, f_1) = 0$ and

$$L(t \circ f_1) = L(t) + L(f_1) = L(f) - \varepsilon_0(f).$$

For $0 \leq \xi < \varepsilon_0(f) \leq \varepsilon_0(f,f)$ we have

$$f(L(f) - \xi) = f(\xi)^{-1}$$

as well as

$$f(\xi) = t(\xi);$$

also,

$$L(f) - \xi > L(f) - \varepsilon_0(f) = L(t \circ f_1)$$

so

$$f(L(f) - \xi) = u(\varepsilon_0(f) - \xi)$$

and hence $t(\xi) = u^{-1}(\xi)$. By the admissibility of f, we have

$$\sigma(t(\varepsilon_0(f))) = \tau(u(0)).$$

According to Lemma 11.13, we can choose t such that $t(\varepsilon_0(f)) = u(0)^{-1}$; this only requires the value of $f_0(0)$ to be changed, which can be done by changing the value of $f_1(0)$ without altering the value of $u(0)$, even if $L(f_1) = 0$. Thus $t = u^{-1}$, so $f = t \circ f_1 \circ t^{-1}$.

It remains to show that f_1 is cyclically reduced; we may assume that $L(f_1) > 0$ and that f_1^2 is defined. Then, since $L(f_1) = L(f) - 2\varepsilon_0(f)$, we have $\varepsilon_0(f) < L(f)/2$ so that

$$L(t) = \varepsilon_0(f) = \varepsilon_0(f,f).$$

By Lemma 11.15 (visible cancellation over a groupoid) and repeated use of Corollary 11.17, we have

$$\begin{aligned} f^2 &= ((t \circ f_1) \circ t^{-1})(t \circ (f_1 \circ t^{-1})) \\ &= (t \circ f_1) \circ (f_1 \circ t^{-1}) \\ &= t \circ (f_1 \circ f_1) \circ t^{-1}; \end{aligned}$$

in particular, $\varepsilon_0(f_1, f_1) = 0$ as required.

(ii) Assume without loss of generality that $L(t) \geq L(s)$. Then by Lemma 11.13, we can write $t = s \circ u$ for some $u \in \mathscr{ARF}(S,G)$. By part (ii) of Lemma 11.10 we have $t^{-1} = u^{-1} \circ s^{-1}$, and so

$$s \circ f_2 \circ s^{-1} = f = s \circ u \circ f_1 \circ u^{-1} \circ s^{-1}.$$

Hence $f_2 = u \circ f_1 \circ u^{-1}$ and since f_1 and f_2 are cyclically reduced,

$$L(f_2^2) = 2L(f_2) = 2\big(2L(u) + L(f_1)\big),$$

and also

$$L(f_2^2) = L(uf_1^2u^{-1}) \leq 2L(u) + L(f_1^2) = 2\big(L(u) + L(f_1)\big).$$

Therefore $L(u) = 0$ and, putting $g = u^{-1}$, it follows that $s = tg$ and $f_2 = g^{-1}f_1g$, as claimed. □

Let G_0 be the set of morphisms of $\mathscr{F}(S,G)$ of length 0. These are in one-to-one correspondence with G via $f \mapsto f(0)$, and are automatically in $\mathscr{ARF}(S,G)$. In fact, (S, G_0) is a subgroupoid of $(S, \mathscr{ARF}(S,G))$ isomorphic to (S,G) via this bijective map $G_0 \to G$ and the identity map on S.

Next, we observe that the idea of 'exponent sum' in Chapter 5 generalises routinely.

Definition 11.20 Let (S,G) be a groupoid, $g \in G$ and $f \in \mathscr{ARF}(S,G)$. We define

$$\mu_g(f) = m^*\big(\{\xi \in [0, L(f)] : f(\xi) = g\}\big),$$

where m^* denotes outer Lebesgue measure.

Thus, $\mu_g(f)$ is a real number less than or equal to $L(f)$.

Lemma 11.21 *For $g \in G$, and $f, f_1, f_2 \in \mathscr{ARF}(G)$ with $\varepsilon_0(f_1, f_2) = 0$, we have*

(i) $\mu_g(f^{-1}) = \mu_{g^{-1}}(f)$;

(ii) $\mu_g(f_1 \circ f_2) = \mu_g(f_1) + \mu_g(f_2)$.

Proof The proof of Lemma 5.5 applies without modification. □

Definition 11.22 Let $g \in G$. We define a mapping $e_g : \mathscr{ARF}(S,G) \to \mathbb{R}$, called the *exponent sum of* $\mathscr{ARF}(S,G)$ *relative to* $g \in G$, by

$$e_g(f) = \mu_g(f) - \mu_{g^{-1}}(f).$$

Proposition 11.23 *The mapping e_g is a groupoid map.*

Proof The proof is just the same as that of Proposition 5.7, except that Lemma 11.14 and Lemma 11.21 are used instead of Lemma 2.15 and Lemma 5.5. □

We shall show in the next section that the vertex groups of $(S, \mathscr{ARF}(S,G))$ have a canonical action on an $\mathbb{R}$-tree. In the special case where S has one element, so that G is a group, the admissibility condition is automatic, the construction reduces to that in Chapter 2, and $\mathscr{ARF}(S,G)$ is the group $\mathscr{RF}(G)$. In Alperin and Moss [1], the group Γ_S with a free action on an $\mathbb{R}$-tree, where S is any set, is the vertex group of $(S, \mathscr{ARF}(S,G))$ with respect to a chosen vertex $\hat{s} \in S$, where (S,G) is the simplicial groupoid on S. Thus $G = S \times S$, with multiplication $(s,t)(t,u) = (s,u)$ and $1_s = (s,s)$.

11.6 Lyndon length functions on groupoids

Let (G,S) be a groupoid, let Λ be an ordered abelian group, and let $L : G \to \Lambda$ be a (set-theoretic) mapping. For $g, h \in G$ with $\sigma(g) = \sigma(h)$, put

$$c(g,h) = \tfrac{1}{2}(L(g) + L(h) - L(g^{-1}h))$$

(in general, this is an element of $\mathbb{Q} \otimes_{\mathbb{Z}} \Lambda$).

Definition 11.24 A function $L : G \to \Lambda$ is a *Lyndon length function* (or a *length function* for short), if the following hold.

(i) For all $s \in S$, $L(1_s) = 0$.

(ii) For all $g \in G$, $L(g^{-1}) = L(g)$.

(iii) We have $c(g,h) \geq \min\{c(g,k), c(h,k)\}$ whenever g, h, $k \in G$ and $\sigma(g) = \sigma(h) = \sigma(k)$.

The following are easy consequences of the axioms.

(iv) For all $g \in G$, $L(g) \geq 0$.

(v) $L(gh) \le L(g) + L(h)$ if $\tau(g) = \sigma(h)$.

(vi) $0 \le c(g,h) \le \min\{L(g), L(h)\}$ if $\sigma(g) = \sigma(h)$.

The function c is symmetric in the sense that $c(g,h) = c(h,g)$ if $\sigma(g) = \sigma(h)$. Also, $c(g,g) = L(g)$ for all $g \in G$.

For $s \in S$, let $G_{*s} = \{g \in G \mid \tau(g) = s\} = \bigcup_{t \in S} G_{ts}$. There is an equivalence relation $\approx$ on G_{*s} defined by $g \approx h$ if and only if $L(gh^{-1}) = 0$. (This relation is transitive by properties (iv) and (v).) Let $X_s = G_{*s}/\approx$ be the set of equivalence classes, and let $\langle g \rangle$ denote the equivalence class of $g \in G_{*s}$. We can define $d_s : X_s \times X_s \to \Lambda$ by $d_s(\langle g \rangle, \langle h \rangle) = L(gh^{-1})$; it is easy to verify that d_s is a Λ-metric on X_s (see Section A.1 for the definition).

For $g \in G_{st}$ there is a metric isomorphism $\psi_g : X_s \to X_t$ given by $\langle h \rangle \mapsto \langle hg \rangle$, with inverse $\psi_{g^{-1}}$.

Let $x_0 = \langle 1_s \rangle$. Then $(\langle g \rangle \cdot \langle h \rangle)_{x_0} = c(g^{-1}, h^{-1})$, hence (X_s, d_s) is 0-hyperbolic (see Section A.2). Assume additionally that

(vii) $c(g,h) \in \Lambda$ for all g, $h \in G$ for which $c(g,h)$ is defined.

By Theorem A.23, there exist a Λ-tree (X'_s, d'_s) and an isometric embedding $\phi_s : X_s \to X'_s$. Taking $\phi = \phi_s$, $Z = X'_t$ and $\psi : X_s \to X'_t$ to be the composite of ψ_g and ϕ_t in Theorem A.23, ψ_g extends to a metric isomorphism $\mu_g : X'_s \to X'_t$ making the diagram

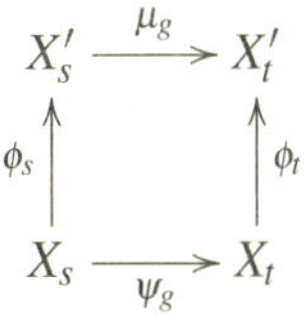

$$\begin{array}{ccc} X'_s & \xrightarrow{\mu_g} & X'_t \\ {\scriptstyle \phi_s}\uparrow & & \uparrow{\scriptstyle \phi_t} \\ X_s & \xrightarrow[\psi_g]{} & X_t \end{array}$$

commutative. We have defined a functor R from (S,G) to the category $\mathscr{T}_\Lambda$, whose objects are all Λ-trees and whose morphisms are the Λ-isometries between them, written on the right, where $R(s) = X'_s$ for $s \in S$ and $R(g) = \mu_g$. In particular, the vertex group G_{ss} acts (on the right) on the Λ-tree X'_s as isometries.

Remark 11.25 Note that, for $g \in G_{st}$, $L(g) = d_t(x_t, x_s g)$. Conversely, given a functor $R : (S,G) \to \mathscr{T}_\Lambda$, where (S,G) is a groupoid, and a choice of base-point x_s in $R(s)$, for every $s \in S$, the map $L : G \to \Lambda$ defined by $L(g) = d_t(x_t, x_s g)$ for $g \in G_{st}$ is a Lyndon length function (d_s being the metric on $R(s)$). We shall not elaborate on this here.

Suppose that (S,G) is a groupoid and we have chosen an S-indexing for

$\mathscr{S}$. The idea of a Lyndon length function can be applied to $(S, \mathscr{ARF}(S,G))$, because of the following proposition. Note that, in the case $\Lambda = \mathbb{R}$, condition (vii) above is automatically satisfied by any Lyndon length function.

Proposition 11.26 *The map $L : \mathscr{ARF}(S,G) \to \mathbb{R}$, where $L(f)$ is the length of the domain of f, is a Lyndon length function.*

Proof The proof of Proposition 3.1 works here, on taking suitable care over endpoints. In particular, if $c(f,g)$ is defined then it is equal to $\varepsilon_0(f^{-1}, g)$. □

Thus, there is a functor R from $\mathscr{ARF}(S,G)$ to the category $\mathscr{T}_{\mathbb{R}}$ of $\mathbb{R}$-trees and isometric maps; that is, an $\mathbb{R}$-tree action of $\mathscr{ARF}(S,G)$. In particular, the vertex group of $\mathscr{ARF}(S,G)$ at a vertex s acts as isometries on the $\mathbb{R}$-tree $R(s)$ (and the restriction of L to this vertex group is a Lyndon length function in the usual sense). Unfortunately, this is a right action; such an action is necessary to ensure R is covariant. Therefore it does not reduce to the construction in Chapter 2 when (S,G) is a group. However, it is easy to modify the construction to obtain a left action generalising that in Chapter 2. One has to change the equivalence relation $\approx$ by defining $f \approx g$ to hold if and only if $L(f^{-1}g) = 0$, and defining the metric d_s by $d_s(\langle f \rangle, \langle g \rangle) := L(f^{-1}g)$. The rest of the construction then proceeds in a similar manner to that in Chapter 2, using Theorem A.29 to obtain an $\mathbb{R}$-tree with $\mathscr{ARF}(S,G)_{ss}$ acting as isometries on the left. We denote this $\mathbb{R}$-tree by $\mathbf{X}_s$. The construction in Theorem A.29 gives $\mathbf{X}_s$ a canonical base-point x_0, and the stabiliser of x_0 is the set of elements of $\mathscr{ARF}(S,G)_{ss}$ of length 0, which is the vertex group at s of the subgroupoid (S, G_0) of $\mathscr{ARF}(S,G)$. Further, $\mathbf{X}_s$ is spanned by the orbit of x_0.

The idea of a strongly regular Lyndon length function is given in Definition A.31.

Lemma 11.27 *If (S,G) is connected and $s \in S$ then L, when restricted to $\mathscr{ARF}(S,G)_{ss}$, is a strongly regular Lyndon length function. Consequently the action of $\mathscr{ARF}(S,G)_{ss}$ on $\mathbf{X}_s$ is transitive.*

Proof Suppose that $f \in \mathscr{ARF}(S,G)_{ss}$ and let α be a real number with $0 \le \alpha \le L(f)$. By Lemma 11.13, there exist f_1, $f_2 \in \mathscr{ARF}(S,G)$ such that $f = f_1 \circ f_2$ and $L(f_1) = \alpha$. This implies that $\tau(f_1(\alpha)) = \sigma(f_2(\alpha)) = t$, say. Choose $g \in G$ with $\sigma(g) = t$ and $\tau(g) = s$, and let h be the element of $\mathscr{ARF}(S,G)$ with $L(h) = 0$ and $h(0) = g$. Put $f_1' = f_1 h$ and $f_2' = h^{-1} f_2$. Then f_1', $f_2' \in \mathscr{ARF}(S,G)_{ss}$ and $f = f_1' \circ f_2'$, which proves strong regularity, since

$L(f) = L(f_1') + L(f_2')$. It follows now from Proposition A.37 that the action is transitive. □

Thus, if (S,G) is connected, then the stabilisers of points for the action in Lemma 11.27 are the conjugates of the vertex group at s of (S,G_0).

11.7 Functoriality

Let $F : (S,G) \to (S',G')$ be a groupoid map, that is, a functor, which we shall view as a mapping $S \cup G \to S' \cup G'$ sending S to S' and G to G'. Recall that F preserves inverses: $F(g^{-1}) = F(g)^{-1}$ for $g \in G$. If f is a mapping with codomain S or G, we denote the composite of f and F by $\hat{F}(f)$. Let $\mathscr{S}$ be the set of equivalence classes of functions defined on open real intervals with values in S, as defined in Section 11.1, and let $\mathscr{S}'$ be the corresponding set for S'. Assume that an S-indexing for $\mathscr{S}$ has been chosen, and suppose that the following condition is satisfied:

$$\left.\begin{array}{c}\textit{if } f:(a,b)\to S,\ g:(c,d)\to S \textit{ are two functions with } f\in E_s \textit{ and } g\in E_t\\ \textit{such that } \hat{F}(f) \textit{ and } \hat{F}(g) \textit{ are equivalent then } F(s)=F(t).\end{array}\right\}(**)$$

Then we can (and do) choose an S'-indexing $S' \to \mathscr{S}'$ for $\mathscr{S}'$ such that

$$\{\hat{F}(f) \mid f \in E_s\} \subseteq E_{F(s)}, \quad s \in S.$$

It follows that if f is left continuous then so is $\hat{F}(f)$, hence if f is right continuous then so is $\hat{F}(f)$. Since F is a groupoid map, it follows that if $f : [0,m] \to G$ is admissible then so is $\hat{F}(f)$.

Circumstances under which the condition $(**)$ on F is satisfied are:

(i) $F : S \to S'$ is one-to-one; or

(ii) S' has one element, that is, G' is a group.

Note that $L(\hat{F}(f)) = L(f)$ for $f \in \mathscr{ARF}(S,G)$.

For $s \in S$, set $G_{s*} = \bigcup_{t\in S} G_{st} = \{g \in G \mid \sigma(g) = s\}$. We call F *locally injective* if, for every $s \in S$, the restriction of F to a map $G_{s*} \to G'_{F(s)*}$ is injective. (Higgins [25] uses the term *star injective* for this property.)

Lemma 11.28 *Suppose that F is locally injective and satisfies the condition $(**)$. If $f \in \mathscr{ARF}(S,G)$ then $\hat{F}(f) \in \mathscr{ARF}(S',G')$.*

Proof We need to show that $\hat{F}(f)$ is reduced. Suppose that $f(x) \in G_{st} = \mathrm{Hom}(s,t)$ and that $\hat{F}(f)(x) = 1_{F(s)}$; then $f(x) = 1_s$ by local injectivity. By Lemma 11.5, there exists a real number $\varepsilon' > 0$ such that $\sigma(\hat{F}(f)(x+\delta)) = \tau(\hat{F}(f)(x-\delta) = \sigma(\hat{F}(f)(x-\delta)^{-1})$ for $0 \le \delta < \varepsilon'$.

Since f is reduced, if $0 < \varepsilon \le \min\{L(f)-x_0, x_0\}$ there exists δ such that $0 < \delta \le \varepsilon$ and $f(x_0+\delta) \ne \big(f(x_0-\delta)\big)^{-1}$. Replacing ε by $\min\{\varepsilon, \varepsilon'/2\}$, we may assume that $\delta < \varepsilon'$. But then $\hat{F}(f)(x+\delta) \ne \big(\hat{F}(f)(x-\delta)\big)^{-1}$; for otherwise, since $\sigma(\hat{F}(f)(x+\delta)) = \sigma\big((\hat{F}(f)(x-\delta))^{-1}\big)$, we would havc that $f(x_0+\delta) = \big(f(x_0-\delta)\big)^{-1}$ by local injectivity, a contradiction. Therefore, $\hat{F}(f)$ is reduced. □

Lemma 11.29 *Suppose that F is locally injective and satisfies the condition $(**)$. If f, $g \in \mathscr{ARF}(S,G)$ and fg is defined, then $\varepsilon_0(f,g) = \varepsilon_0(\hat{F}(f),\hat{F}(g))$.*

Proof Since F is a locally injective groupoid map preserving L,

$$f(L(f)) = (g(0))^{-1} \iff (\hat{F}(f))(L(\hat{F}(f))) = (\hat{F}(g))(0))^{-1}.$$

Hence, if $f(L(f)) \ne (g(0))^{-1}$ then $\varepsilon_0(f,g) = 0 = \varepsilon_0(\hat{F}(f),\hat{F}(g))$.

Suppose that $f(L(f)) = (g(0))^{-1}$; then $(\hat{F}(f))(L(\hat{F}(f))) = ((\hat{F}(g))(0))^{-1}$ and hence

$$\varepsilon_0(f,g) = \sup \mathscr{E}(f,g) \;\text{ and }\; \varepsilon_0(\hat{F}(f),\hat{F}(g)) = \sup \mathscr{E}(\hat{F}(f),\hat{F}(g)).$$

Also, for $\varepsilon \in [0, \min\{L(f), L(g)\}]$,

$$\begin{aligned} \varepsilon \in \mathscr{E}(f,g) &\implies f(L(f)-\eta) = (g(\eta))^{-1} \;\text{ for }\; 0 \le \eta \le \varepsilon \\ &\implies (\hat{F}(f))(L(\hat{\varphi}(f_1)) - \eta) = ((\hat{F}(g))(\eta))^{-1} \;\text{ for }\; 0 \le \eta \le \varepsilon \\ &\implies \varepsilon \in \mathscr{E}(\hat{F}(f),\hat{F}(g)). \end{aligned}$$

Hence $\varepsilon_0(f,g) \le \varepsilon_0(\hat{F}(f),\hat{F}(g))$. Suppose that $\varepsilon_0(f,g) < \varepsilon_0(\hat{F}(f),\hat{F}(g))$, and abbreviate $\varepsilon_0(f,g)$ to ε_0. Then $(\hat{F}(f))(L(f)-\varepsilon_0) = (\hat{F}(g))(\varepsilon_0)^{-1}$ and, as noted when the reduced product was defined, $\tau(f(L(f)-\varepsilon_0)) = \sigma(g(\varepsilon_0)) = \tau(g(\varepsilon_0)^{-1})$ by admissibility. Hence $f(L(f)-\varepsilon_0) = g(\varepsilon_0)^{-1}$ since F is locally injective; in particular, $\sigma(f(L(f)-\varepsilon_0)) = \tau(g(\varepsilon_0))$.

It follows from the admissibility of f and g that there exists $\delta > 0$ such that $\sigma(f(L(f)-\varepsilon_0-x)) = \tau(g(\varepsilon_0+x))$ for $0 \le x < \delta$, and we may choose δ such that $\delta < \varepsilon_0(\hat{F}(f),\hat{F}(g)) - \varepsilon_0(f,g)$. Therefore $\sigma(f(L(f)-\eta)) = \tau(g(\eta)) = \sigma(g(\eta)^{-1})$ for $\varepsilon_0 \leq \eta < \delta + \varepsilon_0$ and, for such values of η, we have $\eta < \varepsilon_0(\hat{F}(f),\hat{F}(g))$, so that $(\hat{F}(f))(L(f)-\eta)) = (\hat{F}(g))(g(\eta))^{-1} = \hat{F}(g))(g(\eta)^{-1})$. By the local injectivity of F, $(\hat{F}(f))(L(f)-\eta) = g(\eta)^{-1}$ for $0 \leq \eta < \delta$. But then $\theta \in \mathscr{E}(f,g)$ for $\varepsilon_0 < \theta < \delta$, contradicting the fact that $\varepsilon_0 = \sup \mathscr{E}(f,g)$. Hence $\varepsilon_0(f,g) = \varepsilon_0(\hat{F}(f),\hat{F}(g))$ as required. □

Lemma 11.30 *Suppose that F is locally injective and satisfies the condition $(**)$. If f, $g \in \mathscr{ARF}(S,G)$ and fg is defined then $\hat{F}(fg) = (\hat{F}(f))(\hat{F}(g))$.*

Proof Since F is a groupoid map, it is easy to see that $\hat{F}(f*g) = \hat{F}(f)*\hat{F}(g)$, so that if $fg = f \circ g$ then $F(fg) = F(f) \circ F(g)$ by Lemma 11.29. In general, by Lemma 11.14 we can write $f = f_1 \circ u$, $g = u^{-1} \circ g_1$, where $fg = f_1 \circ g_1$. Then

$$\begin{aligned} F(f) &= F(f_1) \circ F(u), \\ F(g) &= F(u)^{-1} \circ F(g), \\ F(fg) &= F(f_1) \circ F(g_1) = F(f_1)F(g_1). \end{aligned}$$

Hence

$$F(f)F(g) = F(f_1)F(g_1) = F(fg).$$

□

We have shown that if F is locally injective and satisfies the condition $(**)$ then F defines a groupoid map $\hat{F} : \mathscr{ARF}(S,G)$ to $\mathscr{ARF}(S',G')$, where $\hat{F}(s) = F(s)$ for $s \in S$. Further, $\hat{F}$ preserves L; that is, $L(\hat{F}(f)) = L(f)$ for $f \in \mathscr{ARF}(S,G)$.

Lemma 11.31 *Suppose that F is locally injective and satisfies the condition $(**)$. Then the groupoid map $\hat{F}$ is locally injective.*

Proof Suppose that f, $g \in \mathscr{ARF}(S,G)$, $\sigma(f) = \sigma(g)$, and $\hat{F}f = \hat{F}g$. Then $L(f) = L(g) = \alpha$, say, and $F(f(0)) = F(g(0))$, whence $f(0) = g(0)$ since F is locally injective. This proves the lemma in the case $\alpha = 0$, so assume that $\alpha > 0$. Let $E = \{\xi \in [0,\alpha] \mid f(\eta) = g(\eta) \text{ for all } \eta \in [0,\xi)\}$, a non-empty set as $0 \in E$, and let $\gamma = \sup E$. Then γ satisfies the following four assertions, which are followed by brief justifications.

(1) $\gamma > 0$. For τf and τg are right continuous and $\tau(f(0)) = \tau(g(0))$, hence there exists $\delta > 0$ such that $\tau(f(\xi)) = \tau(g(\xi))$ for $0 \leq \xi < \delta$. Since F is locally injective, it follows that $f(\xi) = g(\xi)$ for $0 \leq \xi < \delta$, so that $\delta \in E$, hence $\gamma \geq \delta > 0$.

(2) $\gamma \in E$. For if $\xi \in [0,\gamma)$, there exists $\eta \in E$ such that $\xi < \eta \leq \gamma$, so that $\xi \in [0,\eta)$, which implies that $f(\xi) = g(\xi)$, as required.

(3) $f(\gamma) = g(\gamma)$. For $\sigma(f(\xi)) = \sigma(g(\xi))$ for $0 \leq \xi < \gamma$, and σf, σg are left continuous, so that $\sigma(f(\gamma)) = \sigma(g(\gamma))$, hence $f(\gamma) = g(\gamma)$ as F is locally injective.

(4) $\gamma = \alpha$. For suppose that $\gamma < \alpha$. By (3), $\tau(f(\gamma)) = \tau(g(\gamma))$ and, arguing as in (1), there exists $\delta' > 0$ such that $f(\xi) = g(\xi)$ for $\gamma \leq \xi < \gamma + \delta'$. But then $\gamma + \delta' \in E$, contradicting the fact that $\gamma = \sup E$.

From (2), (3), and (4) we obtain $f = g$, proving the lemma. □

In particular, if $s \in S$, then $\hat{F}$ restricted to the vertex group of $(S, \mathscr{ARF}(S,G))$ at s is a monomorphism to the vertex group of $(S', \mathscr{ARF}(S',G'))$ at $F(s)$.

We remark that if F is an isomorphism, so that the condition $(**)$ is satisfied, then $\hat{F}$ is an isomorphism.

Let G be a group acting freely on an $\mathbb{R}$-tree (X,d), and let S be the set constructed in Section 4.3. Recall that S was given an arbitrary group structure and the resulting group was called H. Define $p : (S, S \times S) \to H$, where $(S, S \times S)$ is the simplicial groupoid on S, by $(g,h) \mapsto g^{-1}h$. Then p is a universal covering map (see Example 4, p. 98 in Higgins [25]); in particular p is locally injective and satisfies the condition $(**)$ since H is a group. Therefore, there is an induced groupoid map $\hat{p} : \mathscr{ARF}(S, S \times S) \to \mathscr{RF}(H)$ which is injective on vertex groups and length-preserving.

Let L_{y_0} be the Lyndon length function corresponding to a point $y_0 \in X$. According to Theorem 4.2 in Alperin and Moss [1], there is a length-preserving embedding from G to a vertex group of $\mathscr{ARF}(S, S \times S)$. The argument is as follows. By Theorem 4.1 in the present text, it can be assumed that the action of G on X is transitive, which simplifies the argument in [1]. For $g \in G$, define F_g by

$$F_g(\xi) := (s_{g_\xi}, s_{g^{-1}g_\xi}), \quad (g \in G, \xi \in [0, L_{y_0}(g)])$$

where (as in Section 4.3) g_ξ is the unique element of G such that $g_\xi y_0$ is the point of $[y_0, gy_0]$ at distance ξ from y_0. Then, for $g \neq 1$, $s_{g_\xi} \neq s_{g^{-1}g_\xi}$ for all $\xi \in [0, L_{y_0}(g)]$ (see Lemma 7.1 in Chiswell and Müller [12]), so that $F_g(\xi) \neq 1_s$ for any $\xi \in [0, L_{y_0}(g)]$ and $s \in S$. Consequently, $F_g \in \mathscr{RF}(S, S \times S)$ for all $g \in G$.

If g, $h \in G$ and $\varepsilon_0 = \varepsilon_0(g,h)$, simple modifications of the proof of Lemma 4.14 show that $F_g(L(f) - \varepsilon_0)F_h(\varepsilon_0)$ is defined. Therefore Formula (11.1) can be used to define the product F_gF_h, and we have $F_gF_h = F_{gh}$, again by modification of the proof of Lemma 4.14. Then $\theta : G \to \mathscr{RF}(S, S\times S)$, $g \mapsto F_g$, becomes a homomorphism onto its image, which is clearly length-preserving ($L(\theta(g)) = L_{y_0}(g)$); hence θ is injective. Also $\sigma(\theta(g)) = \tau(\theta(g)) = s_{1_G}$ for all $g \in G$.

If the image of θ were contained in $\mathscr{ARF}(S, S\times S)_{ss}$, where $s = s_{1_G}$, then θ would be the desired embedding; then composing θ with $\hat{p}$ would give a length-preserving embedding of G into $\mathscr{RF}(H)$. This is the idea behind the argument in Section 4.3. However, it is difficult to see how to define the S-indexing needed to construct $\mathscr{ARF}(S, S\times S)$ in such a way that it is an injective map. Fortunately, this is not needed; indeed, defining $\psi(g) = p \circ F_g$ gives, in the present notation, the embedding of Theorem 4.9.

However, if S is any set, and $s \in S$ then the vertex group $\mathscr{ARF}(S, S\times S)_{ss}$ acts transitively on the $\mathbb{R}$-tree $\mathbf{X}_s$, by Lemma 11.27. As noted before Lemma 11.27, the stabiliser of the canonical base-point x_0 of $\mathbf{X}_s$ is isomorphic to the vertex group $(S, S\times S)_{ss}$, which is trivial. Thus $\mathscr{ARF}(S, S\times S)_{ss}$ acts freely on $\mathbf{X}_s$. This was used in an example in [1].

This suggests two possible areas of further study. First, one could try to find more general conditions on functions in $\mathscr{RF}(S, S\times S)$, where S is a set, which give rise to subsets of $\mathscr{RF}(S, S\times S)$, such that a (partial) binary operation can be defined by formula (11.1), so that the subset becomes a groupoid and the vertex groups act freely on corresponding $\mathbb{R}$-trees. This might lead to a class of groups which are universal for free $\mathbb{R}$-tree actions and which themselves act freely on $\mathbb{R}$-trees. Of course, the action of $\mathscr{RF}(H)$ on its canonical $\mathbb{R}$-tree $\mathbf{X}_H$ is not free for any non-trivial group H.

Second, looking at the vertex groups of $\mathscr{ARF}(S, S\times S)$ for suitable sets S and a suitable choice of S-indexing might yield interesting new examples of groups having a free action on an $\mathbb{R}$-tree.

11.8 Exercises

11.1. Show that the (reduced) product of two admissible functions, when defined, is also admissible.

11.2. Convince yourself that the proof of Lemma 2.7 can be used to establish Lemma 11.6.

11.3. Show that, if $f \in \mathscr{ARF}(S, G)$ with $\sigma(f) = s$ and $\tau(f) = t$, then $ff^{-1} = \mathbf{1}_s$ and $f^{-1}f = \mathbf{1}_t$.

11.4. Adapt the proof of Lemma 2.16 to obtain a proof of Lemma 11.15.

11.5. Establish Lemma 11.16.

11.6. Convince yourself that the proof of Corollary 2.18 can be used to establish Corollary 11.17.

11.7. Show that the map $d_s : X_s \times X_s \to \Lambda$ introduced in Section 11.6 is a well-defined Λ-metric.

Appendix A
The basics of Λ-trees

A.1 The definition

The concept of a Λ-tree was first defined by Morgan and Shalen [34]. A Λ-tree is a special kind of metric space, where the metric takes values in some (totally) ordered abelian group Λ; the usual axioms for a metric make sense in this context. Thus, a Λ-*valued metric* on a set X is a mapping $d : X \times X \to \Lambda$ satisfying $d(x,y) \geq 0$, $d(x,y) = 0$ if and only if $x = y$, $d(x,y) = d(y,x)$ for all x, $y \in X$, and the triangle inequality: $d(x,y) \leq d(x,z) + d(z,y)$ for all x, y, $z \in X$. We call a pair (X,d), where d is a Λ-valued metric on X, a Λ-*metric space*. Thus, a metric space in the usual sense is an $\mathbb{R}$-metric space, that is, the special case when Λ is the additive group of the real numbers with its usual ordering. Two examples of Λ-metric spaces are of particular interest here.

Example A.1 Let X be the set of vertices of a connected graph Y, and let d be the path metric on X, that is, $d(x,y)$ is the minimum length of a path in Y with endpoints x, y. (The *length* of a path is the number of edges in the path.) Then (X,d) is a $\mathbb{Z}$-metric space.

Example A.2 There is a Λ-valued metric on Λ itself, for any ordered abelian group Λ, defined by $d(x,y) = |x-y|$, where, for $a \in \Lambda$, $|a| = \max\{a, -a\}$.

If (X,d) and (X',d') are Λ-metric spaces, an *isometry* from (X,d) to (X',d') is a mapping $f : X \to X'$ such that $d(x,y) = d'(f(x), f(y))$ for all $x, y \in X$. Such a mapping is automatically one-to-one but need not be onto. If it is onto, we call it a *(metric) isomorphism*, and a metric isomorphism $f : X \to X$ is called a *metric automorphism* of (X,d). Two Λ-metric spaces (X,d), (X',d') are said to be *metrically isomorphic* if there exists a metric isomorphism $f : X \to X'$.

We define closed intervals in Λ just as for $\mathbb{R}$;

$$[a,b]_\Lambda = \{x \in \Lambda \mid a \le x \le b\}$$

for a, $b \in \Lambda$ with $a \le b$, and we set $[b,a]_\Lambda = [a,b]_\Lambda$.

Definition A.3 A *segment* in a Λ-metric space (X,d) is the image of an isometry $\alpha : [a,b]_\Lambda \to X$ for some a, $b \in \Lambda$ with $a \le b$. Note that $a = b$ is allowed. The *endpoints* of the segment are $\alpha(a)$, $\alpha(b)$.

The endpoints of the segment are characterised as the points of the segment at maximum distance apart. A Λ-metric space (X,d) is called *geodesic* if, for all x, $y \in X$, there is a segment in X with endpoints x, y. Note that a geodesic Λ-metric space need not be path connected. Indeed it can be discrete. For it is easy to see that a $\mathbb{Z}$-metric space (X,d) is geodesic if and only if there exists a graph Y with $X = V(Y)$ and d the path metric on X; see Exercise A.1.

This leads to the following question: given a geodesic $\mathbb{Z}$-metric space (X,d), when is X the set of vertices of a tree, such that d is the path metric? One answer is given by the following proposition, which was used by Morgan and Shalen as a model for their definition of a Λ-tree.

Proposition A.4 *Let (X,d) be a geodesic $\mathbb{Z}$-metric space. Then there exists a tree Γ, such that $X = V(\Gamma)$ and such that d is the path metric of Γ if and only if*

$$\left.\begin{array}{c}\textit{whenever two segments of } (X,d) \textit{ intersect in a single point, which is an}\\ \textit{endpoint of both, then their union is again a segment.}\end{array}\right\}(*)$$

Proof We refer to Lemma 3.5 in Chapter 1 of Chiswell [10] for the proof of this. □

Proposition A.4 motivates the following definition.

Definition A.5 A Λ-*tree* is a Λ-metric space (X,d) such that:

(i) (X,d) is geodesic;

(ii) the condition $(*)$ in Proposition A.4 holds;

(iii) The intersection of two segments with a common endpoint is also a segment.

It follows easily from axiom (iii) that if x, y are points of a Λ-tree (X,d) then there is a unique segment whose set of endpoints is $\{x,y\}$; this segment is denoted by $[x,y]$. In axiom (ii), if $[x,y] \cap [x,z] = \{x\}$ then $\sigma = [x,y] \cup [x,z]$ is a

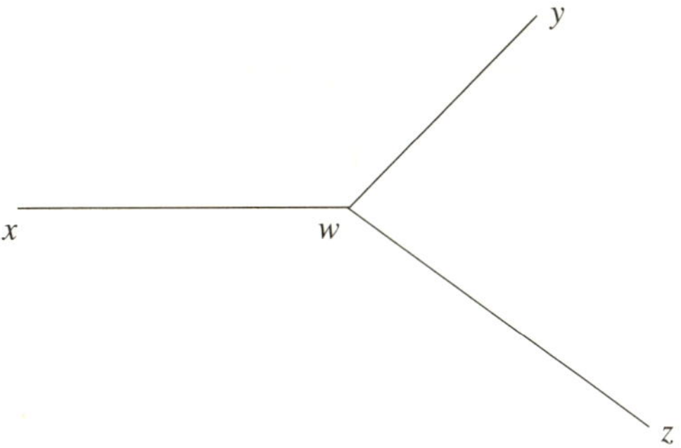

Figure A.1 The segments joining three points in a Λ-tree.

segment. It is not difficult to see that y, z are the points at maximum distance apart in σ, and so are the endpoints of σ; that is, $\sigma = [y,z]$. Similarly, in axiom (iii), one can show that if x, y, $z \in X$ then x is an endpoint of $[x,y] \cap [x,z]$, so that $[x,y] \cap [x,z] = [x,w]$ for some unique $w \in X$. Further, the latter is symmetric in x, y and z; that is, $[x,y] \cap [y,z] = [y,w]$ and $[x,z] \cap [y,z] = [z,w]$. This follows because $[y,z] = [y,w] \cup [w,z]$ by axiom (ii). Also,

$$[x,w] \cap [y,w] = [y,w] \cap [z,w] = [z,w] \cap [x,w] = \{w\}.$$

(For more details, see Lemma 1.1 in Chapter 2 of [10].) All this is illustrated by Figure A.1. Because of the shape of the diagram, we make the following definition.

Definition A.6 The point w described above is denoted $Y(x,y,z)$, and is called the *median* of the triple x,y,z.Y(x,y,z)

Remark A.7 If x, $v \in X$ (where (X,d) is a Λ-tree) and x_m denotes the point at distance m from v in $[v,x]$, where $0 \leq m \leq d(v,x)$, one can show that

$$d(x_m, y_n) = m + n - 2\min\{m, n, d(v,w)\},$$

where $[v,x] \cap [v,y] = [v,w]$; see Exercise A.2.

For later use, if (X,d) is a Λ-tree and x, $y \in X$, we define $(x,y) := [x,y] - \{x,y\}$, $[x,y) := [x,y] - \{y\}$, and $(x,y] := [x,y] - \{x\}$.

We continue with a number of examples of Λ-trees.

Example A.8 The metric space (Λ, d), where d is given by $d(a,b) = |a-b|$ is a Λ-tree. It is easy to see that a segment in Λ must be of the form $[a,b]_\Lambda$ for some a, $b \in \Lambda$.

Example A.9 A $\mathbb{Z}$-metric space (X,d) is a $\mathbb{Z}$-tree if and only if there is a simplicial tree Γ such that $X = V(\Gamma)$ and d is the path metric of Γ. This follows from Proposition A.4 and the remarks preceding it (axiom (iii) is automatically satisfied in any $\mathbb{Z}$-metric space).

Example A.10 Our next example is an $\mathbb{R}$-tree, the geometric realisation $\mathrm{real}(Y)$ of a simplicial tree Y, in which we view the unoriented edges as part of the tree, each such edge being metrically isomorphic to the unit interval $I = [0,1]_\mathbb{R}$, with endpoints identified appropriately. The metric is defined in a similar way to the path metric. For a full discussion of this example, we refer to Section 2 in Chapter 2 of [10]. A *polyhedral* $\mathbb{R}$-tree is one that is homeomorphic to $\mathrm{real}(Y)$ for some simplicial tree Y.

Example A.11 Let $X = \mathbb{R}^2$ be the plane, but with the metric d defined by

$$d((x_1,y_1),(x_2,y_2)) = \begin{cases} |y_1| + |y_2| + |x_1 - x_2| & \text{if } x_1 \neq x_2, \\ |y_1 - y_2| & \text{if } x_1 = x_2. \end{cases}$$

Thus, to measure the distance between two points not on the same vertical line, we take their projections onto the horizontal axis, and add their distances to these projections and the distance between the projections (the usual euclidean distance). If they are on the same vertical line, their distance is the euclidean distance. This is illustrated in Figure A.2. For a proof that this is an $\mathbb{R}$-tree see Section 2 in Chapter 2 of [10].

We shall see shortly that there is a notion of direction at a point in a Λ-tree, and we can therefore define the degree or valency of a point in a Λ-tree. If Y is a simplicial tree, the points of $\mathrm{real}(Y)$ of degree greater than 2 are a subset of $V(Y)$ and form a discrete set. However, at each point of the real axis $\mathbb{R} \times \{0\}$ in the present example, one can show that there are four directions (corresponding to up, down, left and right in the picture). The metric restricted to the real axis is the usual euclidean distance, so the real axis is not a discrete subspace of X. We shall also see that homeomorphisms of $\mathbb{R}$-trees preserve the degree of points. It follows that this example is not polyhedral.

There is a natural concept of subtree of a Λ-tree.

Definition A.12 A *subtree* of a Λ-tree (X,d) is a subset Y of X such that x, $y \in Y$ implies $[x,y] \subseteq Y$.

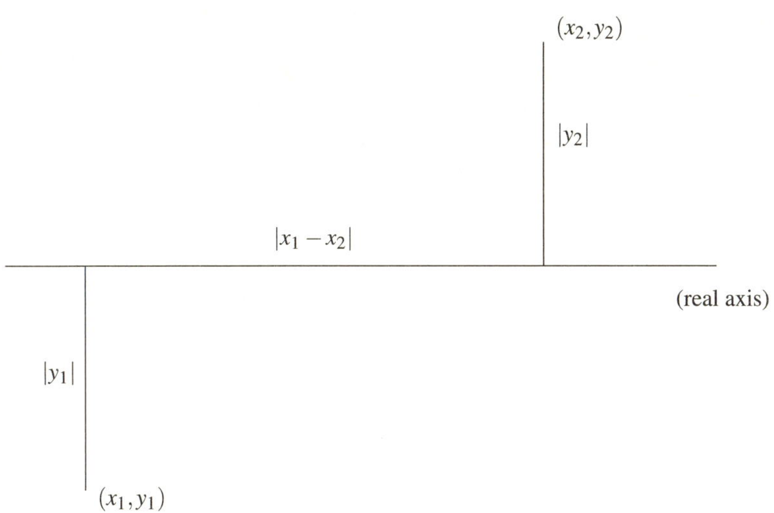

Figure A.2 A non-polyhedral $\mathbb{R}$-tree.

A subtree Y is itself a Λ-tree with metric equal to the restriction of d to $Y \times Y$. The intersection of a family of subtrees of a Λ-tree is also a subtree. If S is a subset of a Λ-tree (X,d), we define the *subtree spanned by S* to be the intersection of all subtrees of X containing S. If the subtree spanned by S is X, we say that (X,d) is spanned by S.

Remark A.13 If $v \in S$, where S is a subset of a Λ-tree, one can show that the subtree spanned by S is equal to $\bigcup_{x \in S} [v, x]$; see Exercise A.3.

Definition A.14 A subtree Y is called *closed* if it is convex-closed, that is, if the intersection of Y with any segment of X is either empty or a segment of X.

For example, a segment is a closed subtree (see Lemma 1.7 in Chapter 2 of [10]). An example of a subtree which is not closed is the interval $(0,1)$ in $\mathbb{R}$. The intersection of any collection of subtrees is a subtree, and the intersection of finitely many closed subtrees is a closed subtree. However, the intersection of an arbitrary family of closed subtrees need not be a closed subtree. For example, take $\Lambda = \mathbb{Q}$ and $X = \Lambda$, and let Y_r be the segment joining r and 2, where $r \in \mathbb{Q}$ and $0 \leq r < \sqrt{2}$. Then the intersection of $\bigcap_r Y_r$ with the segment $[0,2]_{\mathbb{Q}}$ is not a segment in our sense. In the case $\Lambda = \mathbb{Z}$, $X = V(\Gamma)$ for some simplicial tree Γ, the subtrees of X correspond to subtrees of Γ and they are all closed.

The main result concerning closed subtrees is the following lemma.

Lemma A.15 *Let A and B be non-empty closed subtrees of a Λ-tree (X,d) such that $A \cap B = \varnothing$. Then there exist unique points $a \in A$, $b \in B$ such that $[a,b] \cap A = \{a\}$, $[a,b] \cap B = \{b\}$ and if $a_1 \in A$, $b_1 \in B$, then $[a,b] \subseteq [a_1,b_1]$.*

Proof See Lemma 1.9 in Chapter 2 of [10]. □

We call $[a,b]$ in Lemma A.15 the *bridge* between A and B. The lemma applies in particular when $B = \{b\}$ is a single point and A is a closed subtree. Also, if A is a closed subtree and $a \in A$, we call $[a,a] = \{a\}$ the bridge between A and a.

A Λ-metric space (X,d) has a topology with a basis consisting of the open balls

$$B(x,r) = \{y \in X \mid d(x,y) < r\},$$

where $x \in X$ and $r \in \Lambda$, $r > 0$, just as in the case $\Lambda = \mathbb{R}$. Any closed subtree of a Λ-tree is closed in this topology (see Exercise A.4), but the converse is false in general, because of the example above, where the intersection of a family of closed subtrees is not a closed subtree. However, the converse is true in the case $\Lambda = \mathbb{R}$, since the intersection of a subtree closed in this topology with a segment is a compact connected subset of the segment, hence a segment.

A.2 Λ-trees as hyperbolic spaces

Suppose that (X,d) is a Λ-metric space. Choose a point $v \in X$ and, for x, $y \in X$, define

$$(x \cdot y)_v = \tfrac{1}{2}\big(d(x,v) + d(y,v) - d(x,y)\big).$$

This is an element of the ordered abelian group $\frac{1}{2}\Lambda$, which is a subgroup of the ordered abelian group $\mathbb{Q} \otimes_{\mathbb{Z}} \Lambda$. The elements of $\mathbb{Q} \otimes_{\mathbb{Z}} \Lambda$ can be described as fractions a/m, where $a \in \Lambda$, $m \in \mathbb{Z}$ and $m \neq 0$, and $a/m = b/n$ if and only if $na = bm$. Addition is formally the same as the usual addition of fractions, and the ordering is defined by $a/m > 0$ if and only if $ma > 0$. (This is enough to specify the ordering.) The group $\frac{1}{2}\Lambda$ is the subgroup $\{a/2 \mid a \in \Lambda\}$ of $\mathbb{Q} \otimes_{\mathbb{Z}} \Lambda$.

Remark A.16 If (X,d) is a Λ-tree, then $[v,x] \cap [v,y] = [v,w]$ for some w, and $(x \cdot y)_v = d(v,w)$ (see Exercise A.6); in particular $(x \cdot y)_v \in \Lambda$ for all $x, y, v \in X$.

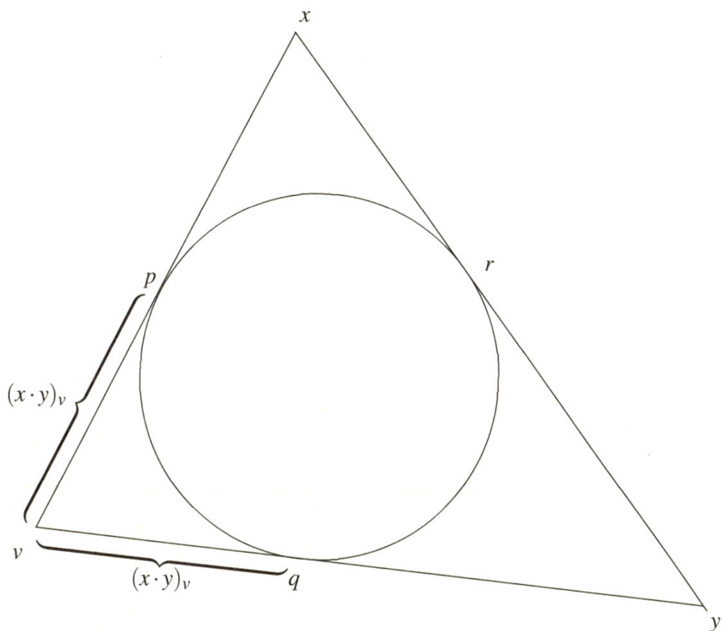

Figure A.3 The product $(x \cdot y)_v$ in euclidean or hyperbolic space.

There is also a simple geometric interpretation of $(x \cdot y)_v$ in real euclidean and hyperbolic space, illustrated in Figure A.3. Since $d(x, p) = d(x, r)$, $d(y, r) = d(y, q)$, $d(v, p) = d(v, q)$ and $d(x, v) = d(x, p) + d(p, v)$, etc., it follows easily that $(x \cdot y)_v = d(v, p) = d(v, q)$.

Definition A.17 Let (X, d) be a Λ-metric space, let $v \in X$, and let $\delta \in \Lambda$ with $\delta \geq 0$. Then (X, d) is called *δ-hyperbolic with respect to v* if

$$x \cdot y \geq \min\{x \cdot z, y \cdot z\} - \delta, \quad x, y, z \in X,$$

where $x \cdot y$ means $(x \cdot y)_v$, etc.

The notion of δ-hyperbolic given in this definition depends on the choice of the reference point v; however, we have the following result.

Lemma A.18 *If (X, d) is δ-hyperbolic with respect to v, and t is any other point of X, then (X, d) is 2δ-hyperbolic with respect to t.*

Proof First note that if t is a point of X, we have, by direct calculation, the identity

$$(x \cdot y)_t = d(t,v) + (x \cdot y)_v - (x \cdot t)_v - (y \cdot t)_v \qquad (**)$$

for all $x, y \in X$. We need to show that $(x \cdot y)_t \geq \min\{(x \cdot z)_t, (y \cdot z)_t\} - 2\delta$, for all $x, y, z \in X$. On adding $(x \cdot t)_v + (y \cdot t)_v + (z \cdot t)_v - d(t,v)$ to both sides and using $(**)$ above, the required inequality becomes

$$(x \cdot y)_v + (z \cdot t)_v \geq \min\left\{(x \cdot z)_v + (y \cdot t)_v, (y \cdot z)_v + (x \cdot t)_v\right\} - 2\delta$$

for all x, y, z and $t \in X$. Now interchanging x and y if necessary, and also interchanging z and t if appropriate, we can assume that $x \cdot z$ is greater than or equal to $x \cdot t$, $y \cdot z$ and $y \cdot t$, where $x \cdot z$ means $(x \cdot z)_v$, etc. Then by assumption

$$x \cdot y \geq (y \cdot z) - \delta \text{ and } z \cdot t \geq (x \cdot t) - \delta.$$

The lemma follows on adding these inequalities. □

Definition A.19 The Λ-metric space (X,d) is δ-hyperbolic if it is δ-hyperbolic with respect to all points of X, and is hyperbolic if it is δ-hyperbolic for some $\delta \geq 0$.

This definition was originally given by Gromov [23] for $\mathbb{R}$-metric spaces, but it works for any Λ-metric space and has the advantage that it applies to any such space. However, in a geodesic $\mathbb{R}$-metric space, it is equivalent to saying that geodesic triangles are, in various senses, 'thin' (see Coornaert, Delzant, and Papadopoulos [16], Ghys and de la Harpe [20] or Short [46]). These ideas of thin triangles are easier to follow intuitively than the definition just given. We shall consider one such idea of thinness, which works to some extent for arbitrary Λ-metric spaces.

We shall use $[x,y]$ to denote a segment with endpoints x and y in a Λ-metric space (X,d), even if this segment is not unique. A *geodesic triangle* in X consists of three segments $[x,y]$, $[y,z]$ and $[z,x]$, where x, y, $z \in X$; these points are called the *vertices* of the triangle. Having chosen three such segments, we make the convention that $[x,y] = [y,x]$, etc.

For example, Figure A.1 is a picture of a geodesic triangle in a Λ-tree and, of course, Figure A.3 illustrates a geodesic triangle in euclidean or hyperbolic space.

Definition A.20 Let (X,d) be a Λ-metric space, let $x,y,z \in X$ be points, and let $\delta \in \Lambda$ with $\delta \geq 0$. Consider the triangle Δ consisting of the segments $[x,y]$, $[y,z]$, and $[z,x]$.

(i) The triangle Δ is called *δ-thin with respect to the vertex x* if, for all $u \in [x,y]$ and $v \in [x,z]$ such that $d(u,x) = d(v,x) \leq (y \cdot z)_x$, we have $d(u,v) \leq \delta$.

(ii) The triangle Δ is called *δ-thin* if it is δ-thin with respect to each of its vertices.

This definition is equivalent to that in [16] and [20] when $\Lambda = \mathbb{R}$, but does not need the mapping f_Δ (which can be defined only for $\Lambda = \mathbb{R}$) used in these books.

Lemma A.21 *Suppose that (X,d) is a Λ-metric space, and let $\delta \in \Lambda$ with $\delta \geq 0$.*

(i) *If (X,d) is δ-hyperbolic, then all geodesic triangles are 4δ-thin.*

(ii) *If (X,d) is geodesic, if $(x \cdot y)_z \in \Lambda$ for all $x,y,z \in X$, and if all geodesic triangles in X are δ-thin then (X,d) is δ-hyperbolic.*

Proof The proof of Proposition 3.1 in Chapter 1 of [16] for $\mathbb{R}$-metric spaces works in general. □

Referring to Figure A.1, we have $(x \cdot y)_z = d(z,w)$ and, symmetrically, $(y \cdot z)_x = d(x,w)$, $(z \cdot x)_y = d(y,w)$. It follows that, in a Λ-tree, geodesic triangles are 0-thin, hence Λ-trees are 0-hyperbolic; cf. Remark A.16. In fact, one can easily prove directly that Λ-trees are 0-hyperbolic, and this is part of the proof of Proposition A.22 below.

By Lemma A.18, if (X,d) is 0-hyperbolic with respect to one point then it is 0-hyperbolic with respect to any other point. Also, (X,d) is 0-hyperbolic with respect to v if and only if for all $x,y,z \in X$, at least two of $(x \cdot y)_v$, $(y \cdot z)_v$ and $(z \cdot x)_v$ are equal, and not greater than the third. This can easily be reinterpreted as follows: (X,d) is 0-hyperbolic if and only if for all $x,y,z,t \in X$, at least two of $d(x,y) + d(z,t)$, $d(x,z) + d(y,t)$ and $d(y,z) + d(x,t)$ are equal, and not less than the third. (This is known as the *four point condition*; a geometric interpretation for Λ-trees is provided after Lemma 1.6 in Chapter 2 of Chiswell [10].) The following result characterises Λ-trees as a subclass of the class of all 0-hyperbolic Λ-metric spaces.

Proposition A.22 *Assume that (X,d) is a Λ-metric space, and let v be a point of X. Then (X,d) is a Λ-tree if and only if the following three conditions hold:*

(i) *(X,d) is geodesic;*

(ii) *(X,d) is 0-hyperbolic with respect to v;*

(iii) *For all $x,y \in X$, we have $(x \cdot y)_v \in \Lambda$.*

Proof See Lemmas 1.6 and 4.3 in Chapter 2 of [10]. □

We have noted that if condition (ii) holds then (X,d) is 0-hyperbolic with respect to any point of X. It is also true that if condition (iii) holds then $(x \cdot y)_t \in \Lambda$ for all x, y and $t \in X$. This follows from Equation $(**)$ in the proof of Lemma A.18.

We now come to a basic theorem, which states that if conditions (ii) and (iii) in Propositon A.22 are satisfied by the Λ-metric space (X,d) with respect to the point $v \in X$ then (X,d) can be embedded in a Λ-tree.

Theorem A.23 *Let (X,d) be a Λ-metric space, and let v be a point of X such that:*

(i) *for all $x,y \in X$, we have $(x \cdot y)_v \in \Lambda$;*

(ii) *(X,d) is 0-hyperbolic with respect to v.*

Then there exist a Λ-tree (X',d') and an isometry $\phi : X \to X'$ such that, if $\psi : X \to Z$ is any isometry of X into a Λ-tree Z, then there is a unique isometry $\mu : X' \to Z$ such that $\mu \circ \phi = \psi$.

Outline of proof. We shall abbreviate $(x \cdot y)_v$ to $x \cdot y$. Let

$$Y = \{(x,m) : \ x \in X, \ m \in \Lambda \text{ and } 0 \leq m \leq d(v,x)\}.$$

Define

$$(x,m) \sim (y,n) :\iff m = n \leq x \cdot y, \quad (x,m),(y,n) \in Y.$$

This is an equivalence relation on Y (it is transitive by Assumption (ii)). Let $X' = Y/\sim$, and let $\langle x,m \rangle$ denote the equivalence class of (x,m). We define the metric by

$$d'(\langle x,m \rangle, \langle y,n \rangle) = m + n - 2\min\{m,n,x \cdot y\}, \quad \langle x,m \rangle, \langle y,n \rangle \in X'$$

(this makes sense because of assumption (i)), and the mapping $\phi : X \to X'$ is given by

$$\phi(x) := \langle x, d(x,v) \rangle, \quad x \in X.$$

See the proof of Theorem 4.4 in Chapter 2 of Chiswell [10] for the details of why this works. The idea of the proof is that $\langle x,m \rangle$ should represent the point on the segment $[v,x]$ (more accurately $[\phi(v),\phi(x)]$) at a distance m from v. The definition of the distance is motivated by Remark A.7 above. Knowing this, the last part is easy to prove. Given an isometry $\psi : X \to Z$ and $x \in X$, denote by x_m the point on the segment in Z with endpoints $\psi(v)$ and $\psi(x)$, at a distance m from $\psi(v)$. Then the mapping $\mu : X' \to Z$ defined by $\mu(\langle x,m \rangle) = x_m$ is an isometry by Remark A.7 and is the only one with the required factorization property, since isometries of Λ-trees have to preserve segments. □

Remarks A.24

(i) If $\theta : X \to X$ is an isometry, then in Theorem A.23 there is a unique isometry $\mu : X' \to X'$ such that $\mu \circ \phi = \phi \circ \theta$ (take $\psi = \phi \circ \theta$ in Theorem A.23). It follows that, if G is a group acting as metric automorphisms on X, there is an induced action on X', such that $g\phi(x) = \phi(gx)$ for all $g \in G$ and $x \in X$.

(ii) In Theorem A.23, X' is spanned by $\phi(X)$. This follows directly from the proof, or one can apply the last part of the theorem with Z equal to the subtree spanned by $\phi(X)$ and $\psi = \phi$. The mapping μ is then the identity on $\phi(X)$ and so on Z (because isometries of Λ-trees map segments to segments) and is one-to-one, hence $Z = X'$.

We can now introduce the idea of a *direction at a point* in a Λ-tree.

Let (X,d) be a Λ-tree, and let $v \in X$. Put

$$D = \{ [v,x] : x \in X - \{v\} \}.$$

Define $[v,x] \equiv [v,y]$ to mean that $[v,x] \cap [v,y] \neq \{v\}$. This means that, if $[v,x] \cap [v,y] = [v,w]$ (as in the axioms for a Λ-tree), then $w \neq v$. Since $(x \cdot y)_v = d(v,w)$ (see Remark A.16), this is equivalent to $(x \cdot y)_v > 0$. It follows from the fact that (X,d) is 0-hyperbolic that, if $(x \cdot y)_v > 0$ and $(y \cdot z)_v > 0$, then $(x \cdot z)_v > 0$. It now follows easily that $\equiv$ is an equivalence relation on D, and the equivalence classes are called *directions at* v. We can then define the *degree* of v to be the cardinality of the set $D/\equiv$ of directions at v.

A point x in a Λ-tree (X,d) is called an *endpoint* if it has degree 1, an *edge point* if it has degree 2, and a *branch point* if it has degree greater than 2.

There is an alternative way of looking at directions in $\mathbb{R}$-trees and, in order to explain this, we shall prove some lemmas.

Lemma A.25 *If (X,d) is a Λ-tree and $x_0,\ldots,x_n \in X$, then*

$$[x_0,x_n] \subseteq \bigcup_{i=1}^{n} [x_{i-1},x_i].$$

Proof If $n=2$, $[x_0,x_1] \cap [x_1,x_2] = [x_1,w]$ for some w. By the discussion after the definition of a Λ-tree,

$$[x_0,x_2] = [x_0,w] \cup [w,x_2] \subseteq [x_0,x_1] \cup [x_1,x_2].$$

If $n \geq 2$ then $[x_0,x_n] \subseteq [x_0,x_{n-1}] \cup [x_{n-1},x_n]$ by the case $n=2$, and the result follows by induction on n. □

Lemma A.26 *Let (X,d) be an $\mathbb{R}$-tree, and let $\alpha : [a,b]_{\mathbb{R}} \to X$ be a continuous map. If $x = \alpha(a)$ and $y = \alpha(b)$ then $[x,y] \subseteq \mathrm{Im}(\alpha)$.*

Proof Let $A = \mathrm{Im}(\alpha)$. Since A is a closed subset of X (being compact), it is enough to show that every point of $[x,y]$ is within distance ε of A, for all real $\varepsilon > 0$.

Given $\varepsilon > 0$, we can partition the domain of α, say $a = t_0 < \cdots < t_n = b$ so that, for $1 \leq i \leq n$, we have $d(\alpha(t_{i-1}),\alpha(t_i)) < \varepsilon$. Then all points of $[\alpha(t_{i-1}),\alpha(t_i)]$ are at distance less than ε from A, for $1 \leq i \leq n$. Finally, we have

$$[x,y] \subseteq \bigcup_{i=1}^{n} [\alpha(t_{i-1}),\alpha(t_i)]$$

by Lemma A.25. □

Lemma A.27 *Let (X,d) be an $\mathbb{R}$-tree, and let $v \in X$ be any point. Then for $x,y \in X - \{v\}$, we have $[v,x] \equiv [v,y]$ if and only if x and y are in the same path component of $X - \{v\}$.*

Proof If $v \in [x,y]$, then $[x,v] \cap [v,y] = \{v\}$, so $[v,x] \equiv [v,y]$ implies $[x,y] \subseteq X - \{v\}$, hence x and y lie in the same path component of $X - \{v\}$.

Conversely, if $\alpha : [a,b]_{\mathbb{R}} \to X - \{v\}$ is a continuous map, with $x = \alpha(a)$ and $y = \alpha(b)$, then $[x,y] \subseteq \mathrm{Im}(\alpha)$ by Lemma A.26, so $v \notin [x,y]$, and $[v,x] \cap [v,y] \neq \{v\}$ by Axiom (c) for a Λ-tree, hence $[v,x] \equiv [v,y]$. □

Thus, the directions at a point v in an $\mathbb{R}$-tree (X,d) are in one-to-one correspondence with the path components of $X - \{v\}$. It follows that homeomorphisms of $\mathbb{R}$-trees preserve degrees of points.

The reader can now establish that Example A.11 in §A.1 is not a polyhedral $\mathbb{R}$-tree.

A.3 Lyndon length functions

Let G be a group, and let Λ be an ordered abelian group. A mapping $L : G \to \Lambda$ is called a (Λ-valued) *Lyndon length function* if the following axioms hold.

(i) $L(1) = 0$.

(ii) For all $g \in G$, $L(g) = L(g^{-1})$.

(iii) For all g, h, $k \in G$, we have $c(g,h) \geq \min\{c(h,k), c(k,g)\}$, where $c(g,h)$ is defined to be $\frac{1}{2}(L(g) + L(h) - L(g^{-1}h))$.

Axiom (iii) is equivalent to the statement that, for all g, h, $k \in G$, at least two of $c(g,h)$, $c(h,k)$, $c(k,g)$ are equal, and not greater than the third. Note that c is symmetric by axiom (ii). It is not hard to verify that these axioms imply the following properties; cf. Exercise A.7.

(iv) For all $g \in G$, $L(g) \geq 0$.

(v) For all g, $h \in G$, $L(gh) \leq L(g) + L(h)$.

(vi) For all g, $h \in G$, $0 \leq c(g,h) \leq \min\{L(g), L(h)\}$.

Property (v) is called the triangle inequality; (vi) is a consequence of (ii) and (v).

Example A.28 Let G be a group acting as isometries on a Λ-tree (X,d) and let $v \in X$. Then the *displacement function* $L_v : G \to \Lambda$ defined by $L_v(g) = d(v, gv)$ is a Lyndon length function on G. By applying the isometry g^{-1} to $[v, gv] \cup [v, hv]$, we see that $c(g,h)$ is equal to $d(v,w)$, where $[v, gv] \cap [v, hv] = [v, w]$, i.e. $w = \mathrm{Y}(v, gv, hv)$; hence $c(g,h) = (gv \cdot hv)_v$ (cf. Remark A.16). It follows from the fact that X is 0-hyperbolic that L_v satisfies axiom (iii) above, and clearly it satisfies (i) and (ii). It also follows that L_v satisfies

(vii) $c(g,h) \in \Lambda$ for all g, $h \in G$.

A special case occurs when G is a free group with basis S acting on its Cayley graph with respect to S, which is a tree, so that G acts on the corresponding $\mathbb{Z}$-tree. (See Lemma 1 in Chapter 8 of Cohen [14] or Section 3.2 in Chapter I of Serre [45].) The points of this $\mathbb{Z}$-tree are just the elements of G and, taking v to be the point 1_G, $L_v(g)$ is the length of the reduced word in $S \cup S^{-1}$ representing g, and $c(g,h)$ is the length of the largest common initial segment of the reduced words representing the elements g and h.

In fact, the next theorem shows that all Λ-valued Lyndon length functions satisfying (vii) arise as displacement functions from isometric actions on Λ-trees. Also, if L is any Lyndon length function then $2L$ is a Lyndon length function satisfying (vii).

Theorem A.29 *Let G be a group and let $L : G \to \Lambda$ be a Lyndon length function satisfying property* (vii). *Then there are a Λ-tree (X',d'), an action of G on X' and a point $v' \in X'$ such that $L = L_{v'}$ and (X',d') is spanned by the orbit Gv'.*

Proof Defining $g \approx h$ if and only if $L(g^{-1}h) = 0$ gives an equivalence relation on G (it is transitive by the triangle inequality). Denote the set of equivalence classes by X and the equivalence class of g by $\langle g \rangle$. We obtain a Λ-metric space (X,d) by defining $d(\langle g \rangle, \langle h \rangle) = L(g^{-1}h)$ (the triangle inequality for d follows from that for L). Also, G acts as isometries on X by defining $g\langle h \rangle = \langle gh \rangle$ for $g, h \in G$. Direct computation shows that

$$(\langle g \rangle \cdot \langle h \rangle)_{\langle 1_G \rangle} = c(g,h),$$

so, by axiom (iii) and property (vii), (X,d) satisfies the hypotheses of Theorem A.23 (with $v = \langle 1_G \rangle$), and we obtain a corresponding Λ-tree (X',d') and an isometry $\phi : X \to X'$. We can denote the point $\langle \langle g \rangle, m \rangle$ of X' simply by $\langle g, m \rangle$ and take $v' = \langle 1, 0 \rangle$ as base-point. Thus, from the proof of Theorem A.23, $\phi(\langle g \rangle) = \langle g, L(g) \rangle$ for $g \in G$, and $v' = \phi(v)$.

By Remark A.24(i), the action of G on X induces an action of G on X' such that $g\phi(u) = \phi(gu)$ for all $u \in X$ and $g \in G$. Hence $gv' = \langle g, L(g) \rangle$ for $g \in G$ and, by the definition of d' in the proof of Theorem A.23, $d'(v', gv') = L(g)$, hence $L = L_{v'}$.

Finally, for $g \in G$ we have

$$\phi(\langle g \rangle) = \phi(g\langle 1 \rangle) = g\phi(\langle 1 \rangle) = gv',$$

thus $\phi(X) = Gv'$. It follows from Remark A.24(ii) that X' is spanned by Gv', as claimed. □

Next, we show that two transitivity properties of group actions on Λ-trees, which involve the geometry of Λ-trees, can be expressed by certain conditions on the length function, called regularity and strong regularity. The concept of a regular length function has already been used in the literature (see, for instance, Myasnikov, Remeslennikov, and Serbin [40]). We begin with the definitions.

Definition A.30 A length function $L : G \to \Lambda$ on a group G is termed *regular* if, for every choice of $x, y \in G$, there exist $u, x_1, y_1 \in G$, such that $x = u \circ x_1$, $y = u \circ y_1$ and $L(u) = c(x, y)$.

Here, $x = u \circ x_1$ means that $x = ux_1$ and $L(x) = L(u) + L(x_1)$ (by Lemma 2.8, this is consistent with our use of the circle notation for $\mathscr{RF}(G)$).

Definition A.31 A length function $L : G \to \Lambda$ is termed *strongly regular* if, for each $g \in G$ and every $a \in \Lambda$ such that $0 \leq a \leq L(g)$, there exist elements $g_1, g_2 \in G$ such that $g = g_1 \circ g_2$ and $L(g_1) = a$.

We shall justify this terminology by showing that, assuming property (vii) above (which is obviously necessary and is automatically satisfied for $\Lambda = \mathbb{R}$), a strongly regular length function is regular. First we shall give a direct argument. Then we shall establish criteria for regularity and strong regularity for length functions arising from actions on Λ-trees, and these criteria in turn will provide an alternative proof.

To begin, we need a simple lemma.

Lemma A.32 *Let $L : G \to \Lambda$ be a Lyndon length function. Then, for all $x, y, z \in G$, we have the following:*

(i) *if $c(x, y) = 0$ and $c(y^{-1}, z) < L(y)$ then $c(x, yz) = 0$;*

(ii) *if $L(x) = 0$ then $L(xy) = L(y)$.*

Proof (i) By direct computation,

$$c(y, yz) + c(y^{-1}, z) = L(y);$$

hence, we have $c(y, yz) > 0$, since by assumption $c(y^{-1}, z) < L(y)$. As it is also assumed that $c(x, y) = 0$, it follows that $c(x, yz) = 0$ by axiom (iii) for length functions.

(ii) Since

$$0 \leq c(x^{-1}, y) \leq L(x^{-1}) = L(x)$$

by property (vi) of length functions, we have $c(x^{-1},y)=0$. Hence $L(x)+L(y)=L(xy)$ and so $L(y)=L(xy)$, as claimed. □

Proposition A.33 *Let $L : G \to \Lambda$ be a strongly regular length function such that $c(x,y) \in \Lambda$ for all $x,y \in G$. Then L is regular.*

Proof Let x, $y \in G$. Since

$$0 \leq c(x,y) \leq \min\{L(x),\ L(y)\},$$

by strong regularity we can write $x = u \circ x_1$, $y = v \circ y_1$, where $L(u) = L(v) = c(x,y)$. Then

$$c(x,y) = \tfrac{1}{2}\left(L(u)+L(v)+L(x_1)+L(y_1)-L(x^{-1}y)\right),$$

which implies

$$L(x_1)+L(y_1) = L(x^{-1}y) = L(x_1^{-1}(u^{-1}v)y_1). \tag{$*$}$$

Suppose for a contradiction that $L(u^{-1}v) \neq 0$. Then $c(u,v) < L(u) = L(v)$ and $c(x_1,u^{-1}) = c(u^{-1},x_1) = 0$ (since $x = u \circ x_1$); hence, by part (i) of Lemma A.32, we have $c(x_1,u^{-1}v) = 0$.

Also, $c(y_1,v^{-1}) = 0$ (as $y = v \circ y_1$) and $c(v,u) = c(u,v) < L(v)$, so, again using part (i) of Lemma A.32, we deduce that $c(y_1,v^{-1}u) = 0$.

Thus we have $c(x_1,u^{-1}v) = 0$, and $0 = c((u^{-1}v)^{-1},y_1) < L(u^{-1}v)$ holds by assumption; hence, using part (i) of Lemma A.32 once more, we find that

$$c(x_1,(u^{-1}v)y_1) = 0.$$

This means that

$$L(x_1)+L((u^{-1}v)y_1) = L(x_1^{-1}(u^{-1}v)y_1)$$

and, since $c(y_1,(u^{-1}v)^{-1}) = 0$, we have

$$L(x_1)+L(u^{-1}v)+L(y_1) = L(x_1^{-1}(u^{-1}v)y_1),$$

which implies $L(x_1)+L(y_1) < L(x_1^{-1}(u^{-1}v)y_1)$, contradicting $(*)$. Thus, we have $L(u^{-1}v) = 0$.

Let $a = u^{-1}v$, so that $L(a) = 0$. Then $y = u(ay_1)$ and, by part (ii) of Lemma A.32, we have

$$L(u)+L(ay_1) = L(u)+L(y_1) = L(v)+L(y_1) = L(y),$$

so that $y = u \circ (ay_1)$. Hence L is regular, as claimed. □

Remark A.34 A regular length function need not be strongly regular. For example, $L(x) = |x|$ defines a length function on the additive group $\mathbb{R}$; the restriction $L: \mathbb{Q} \to \mathbb{R}$ is regular, but not strongly regular. Another example is that of a free group $F(S)$ on a basis S; defining $L(w)$ to be the length of the reduced word on $S^{\pm 1}$ representing w gives one of the prototype examples of a length function. This length function is easily seen to be regular by considering initial segments of the word representing w. Hence $2L : F(S) \to \mathbb{Z}$ is also a regular length function but is clearly not strongly regular, as there are no elements of odd length.

The next result gives our geometric interpretation of a regular length function. Specifically, it uses the ideas of *direction* and *branch point* in a Λ-tree, defined in Section A.2.

Some notation is needed for the proof. If (X,d) is a Λ-tree and $x_0, \ldots, x_n$ are points in X, we write

$$[x_0, x_n] = [x_0, x_1, \ldots, x_n]$$

to mean that, if $\alpha : [0, d(x_0, x_n)]_\Lambda \to X$ is the unique isometry with $\alpha(0) = x_0$ and $\alpha(d(x_0, x_n)) = x_n$, then $x_i = \alpha(a_i)$, where $0 = a_0 \leq a_1 \leq \cdots \leq a_n = d(x_0, x_n)$. (The existence and uniqueness of α follow by the remark after Lemma 2.3 in Chapter 1 of Chiswell [10].) This means that $x_0, x_1, \ldots, x_n$ are elements of $[x_0, x_n]$ listed in order of increasing distance from x_0.

Proposition A.35 *Let G be a group acting on a Λ-tree (X,d), and let $x_0 \in X$. Let Y be the subtree of X spanned by the orbit of x_0. Then the following are equivalent.*

(i) *Every branch point of Y belongs to the orbit Gx_0.*

(ii) *The Lyndon length function L_{x_0} is regular.*

Proof First, note that Y is clearly a G-invariant subtree of X and, by Remark A.13, we have $Y = \bigcup_{g \in G} [x_0, gx_0]$; see also Exercise A.3.

Assume (i). Let g, $h \in G$ and let $w = Y(x_0, gx_0, hx_0)$, (see Definition A.6). Then $w \in Gx_0$. This is clear if w is one of x_0, gx_0, or hx_0. Otherwise $[w, x_0]$, $[w, gx_0]$, and $[x_0, hx_0]$ define three different directions at x_0 in Y, by the discussion after Definition A.5. Thus w is a branch point of Y, so $w \in Gx_0$ by assumption.

Let $w = kx_0$, where $k \in G$. Then (as noted after property (vi) of length functions), we have

$$L_{x_0}(k) = d(x_0, w) = c(g, h).$$

Also, $w \in [x_0, gx_0]$. Thus

$$\begin{aligned} d(x_0, gx_0) &= d(x_0, kx_0) + d(kx_0, gx_0) \\ &= d(x_0, kx_0) + d(x_0, k^{-1}gx_0); \end{aligned}$$

that is,

$$L_{x_0}(g) = L_{x_0}(k) + L_{x_0}(k^{-1}g),$$

so that $g = k \circ (k^{-1}g)$. Similarly, since $w \in [x_0, hx_0]$ we have $h = k \circ (k^{-1}h)$. Hence L_{x_0} is regular.

Conversely, assume (ii), and let w be a branch point of Y. If $w = x_0$ then $w \in Gx_0$, so we may assume that $w \neq x_0$. Then $[w, x_0]$ defines a direction at w, and we can choose two more directions in Y at w, defined by $[w, y]$ and $[w, z]$, say. Now $y, z \in Y$, so that $y \in [x_0, gx_0]$ and $z \in [x_0, hx_0]$ for some $g, h \in G$. Since $w \in [x_0, y]$ it follows that $[x_0, gx_0] = [x_0, w, y, gx_0]$, so that $y \in [w, gx_0]$ and $[x_0, gx_0] = [x_0, w] \cup [w, gx_0]$ (using the notation established just before the proposition). Similarly, $[x_0, hx_0] = [x_0, w, z, hx_0]$, $z \in [w, hx_0]$ and $[x_0, hx_0] = [x_0, w] \cup [w, hx_0]$.

It follows from Lemma 1.5 in Chapter 2 of Chiswell [10] that $[gx_0, hx_0] = [gx_0, y, w, z, hx_0]$; in particular, $[w, gx_0] \cap [w, hx_0] = \{w\}$. This implies that

$$[x_0, gx_0] \cap [x_0, hx_0] = [x_0, w],$$

so $w = Y(x_0, gx_0, hx_0)$. The situation is illustrated by the following figure.

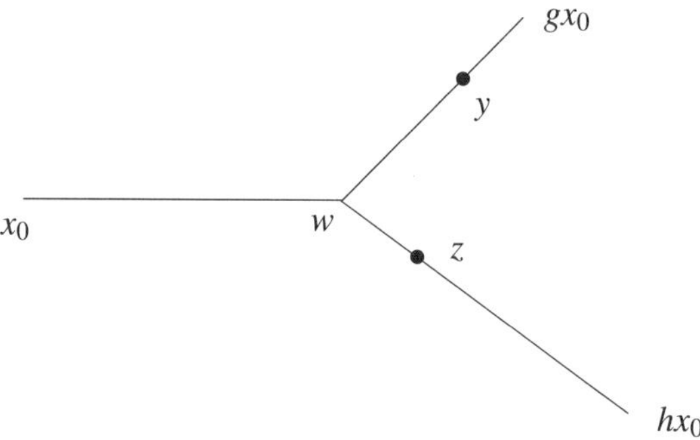

From the comments after the statement of property (vi) of length functions, it follows that $c(g,h) = d(x_0, w)$. By assumption there exists $k \in G$ such that $c(g,h) = L_{x_0}(k)$, $g = k \circ (k^{-1}g)$ and $h = k \circ (k^{-1}h)$. Thus $d(x_0, w) = d(x_0, kx_0)$. Now

$$L_{x_0}(k) + L_{x_0}(k^{-1}g) = L_{x_0}(g),$$

that is,

$$d(x_0,kx_0)+d(kx_0,gx_0)=d(x_0,gx_0),$$

hence $kx_0 \in [x_0,gx_0]$. Similarly, $kx_0 \in [x_0,hx_0]$, so

$$kx_0 \in [x_0,gx_0] \cap [x_0,hx_0] = [x_0,w].$$

It follows that $w = kx_0 \in Gx_0$. □

Remark A.36 The action of G on Y preserves the degrees of points; thus, if one point of Gx_0 is a branch point then they all are. Hence, Proposition A.35(i) is equivalent to the assertion that

either Y has no branch points or the set of branch points of Y is Gx_0.

This in turn is equivalent to:

either Y is linear (that is, isomorphic to a subtree of Λ)
or the set of branch points of Y is Gx_0;

see Exercise 2 after Lemma 3.2 in Chapter 2 of [10].

Our next result spells out the precise connection between the strong regularity of a length function on a group G and the associated action of G on the corresponding Λ-tree.

Proposition A.37 *Suppose that a group G acts by isometries on a Λ-tree (X,d), and let $x_0 \in X$ be any point. Then the following assertions are equivalent.*

(i) *The group G is transitive on the subtree of X spanned by the orbit of x_0.*

(ii) *The displacement function L_{x_0} is strongly regular.*

Proof (i) ⇒ (ii). Let $g \in G$, and let $a \in \Lambda$ be such that $0 \le a \le L_{x_0}(g)$. Let y be the point on the segment $[x_0,gx_0]$ at a distance a from x_0. Then $y = g_1x_0$ for some $g_1 \in G$ by (i), and we have

$$a = d(x_0,y) = d(x_0,g_1x_0) = L_{x_0}(g_1).$$

Set $b := L_{x_0}(g) - a$ and let $g_2 := g_1^{-1}g$, so that $g = g_1g_2$. Then we have

$$L_{x_0}(g_2) = d(x_0,g_2x_0) = d(g_1x_0,g_1g_2x_0) = d(y,gx_0) = b,$$

hence

$$L_{x_0}(g_1)+L_{x_0}(g_2) = a+b = L_{x_0}(g);$$

we conclude that L_{x_0} is strongly regular.

(ii) $\Rightarrow$ (i). Let y be a point in Y, where

$$Y := \bigcup_{g\in G} [x_0, gx_0]$$

is the subtree of (X,d) spanned by the orbit of x_0 (cf. Remark A.13). Then $y \in [x_0, g_0x_0]$ for some $g_0 \in G$. Let $a := d(x_0, y)$, so that $0 \le a \le L_{x_0}(g_0)$. Since L_{x_0} is strongly regular by assumption, there exist elements $g_1, g_2 \in G$ such that $g_0 = g_1g_2$, $L_{x_0}(g_1) = a$, and $L_{x_0}(g_1) + L_{x_0}(g_2) = L_{x_0}(g_0)$. Then

$$\begin{aligned} c(g_1^{-1}, g_2) &= \tfrac{1}{2}\left[L_{x_0}(g_1^{-1}) + L_{x_0}(g_2) - L_{x_0}(g_1g_2)\right] \\ &= \tfrac{1}{2}\left[L_{x_0}(g_1) + L_{x_0}(g_2) - L_{x_0}(g_0)\right] \\ &= 0. \end{aligned}$$

Also, as noted after the statement of property (vi) of length functions, we have $0 = c(g_1^{-1}, g_2) = d(x_0, w)$, where $[x_0, g_1^{-1}x_0] \cap [x_0, g_2x_0] = [x_0, w]$; that is,

$$[x_0, g_1^{-1}x_0] \cap [x_0, g_2x_0] = \{x_0\},$$

so by axiom (ii) in the definition A.5 of a Λ-tree (see the remarks following the definition), we have $x_0 \in [g_1^{-1}x_0, g_2x_0]$. It follows that $g_1x_0 \in [x_0, g_0x_0]$ and that it is at distance a from x_0, hence $g_1x_0 = y$, and G acts transitively on the span of Gx_0, as claimed. □

The last two propositions provide an alternative proof of Proposition A.33. If $L: G \to \Lambda$ is a strongly regular Lyndon length function, with $c(g,h) \in \Lambda$ for all $g, h \in G$ then, by Theorem A.29, there are an action on a Λ-tree and a point x_0 in the tree such that $L = L_{x_0}$. By Proposition A.37 the action of G is transitive, and so, by Proposition A.35, L is regular.

A.4 Some special properties of $\mathbb{R}$-trees

We begin with some alternative ways of defining the concept of an $\mathbb{R}$-tree which have been used in the literature and then follow with a discussion of completeness for $\mathbb{R}$-trees.

An *arc* in an $\mathbb{R}$-metric space is defined to be a homeomorphic image of a compact interval (that is, a segment) in $\mathbb{R}$ with more than one point; its *endpoints* are the images of the endpoints of the interval under the homeomorphism. Thus, any segment with more than one point is an arc. The proof of the next result combines arguments in Morgan and Shalen [34] and Tits [50].

Proposition A.38 *Let (X,d) be an $\mathbb{R}$-metric space. Then the following are equivalent.*

(i) *(X,d) is an $\mathbb{R}$-tree;*

(ii) *given two points of X, there is a unique arc having them as endpoints, and it is a segment;*

(iii) *(X,d) is geodesic and contains no subspace homeomorphic to the circle S^1.*

Proof (i) $\Rightarrow$ (ii). Let α be a homeomorphism from a compact interval $[a,b]_{\mathbb{R}}$ into X, with image A, and let $x = \alpha(a)$, $y = \alpha(b)$. By Lemma A.26, $[x,y] \subseteq A$. Hence $\alpha^{-1}([x,y])$ is a connected subset of $[a,b]$ containing a and b and so is equal to $[a,b]$. It follows that $A = [x,y]$.

(ii) $\Rightarrow$ (iii). If X contains a homeomorphic image of a circle then the images of two semicircles give two distinct arcs with the same endpoints.

(iii) $\Rightarrow$ (i). We have to verify axioms (ii) and (iii) in Definition A.5. First, we show that, if x, $y \in X$, there is a unique segment in X whose set of endpoints is $\{x,y\}$. For otherwise, using appropriate reflections and translations (see Section 2 in Chapter 1 of Chiswell [10]), there are isometries $\alpha, \beta : [0,a] \to X$ with $\alpha(0) = \beta(0) = x$, $\alpha(a) = \beta(a) = y$, say, and $\alpha(b) \neq \beta(b)$ for some $b \in [0,a]$. Let l be the largest element of $[0,b]$ on which α and β agree, and let m be the smallest element of $[b,a]$ on which α and β agree. Then $\alpha([l,m])$, $\beta([l,m])$ are two arcs meeting only at their endpoints, so by the 'gluing lemma' of elementary topology we can define a continuous bijection from their union onto S^1; since their union is compact and S^1 is Hausdorff, this map is a homeomorphism, contrary to our assumption.

Next, we show that axiom (iii) of Definition A.5 is satisfied. Let σ and τ be segments with a common endpoint v. Again we can find surjective isometries $\alpha : [0,a]_{\mathbb{R}} \to \sigma$, $\beta : [0,b]_{\mathbb{R}} \to \tau$ with $\alpha(0) = \beta(0) = v$. Define

$$I = \{c \in [0, \min\{a,b\}] : \alpha(c) = \beta(c)\}.$$

Then $\alpha(I) = \beta(I) = \sigma \cap \tau$. For if $u \in \sigma \cap \tau$ then $u = \alpha(t) = \beta(t)$, where $t = d(v,u)$, and the reverse inclusion is obvious. Also, since $\mathbb{R}$ is Hausdorff,

I is closed and so compact. If x, $y \in I$ and $z \in \Lambda$ with $x \leq z \leq y$ then $z \in I$, for otherwise the images of $\alpha|_{[x,y]_\Lambda}$ and $\beta|_{[x,y]_\Lambda}$ are distinct segments in X with the same endpoints, contrary to what has been proved. Thus I is a connected subspace of $\mathbb{R}$. Hence I is $[0,c]_{\mathbb{R}}$ for some c, and it follows that $\sigma \cap \tau$ is a segment.

It remains to verify axiom (ii) of Definition A.5. Suppose that $[x,y] \cap [x,z] = \{x\}$. If $x \in [y,z]$, then it is easy to see that $[x,y] \cup [x,z] = [y,z]$. Suppose $x \notin [y,z]$, so that

$$[y,z] \cap [y,x] = [y,w], \qquad \text{where } w \neq x,$$
$$[y,z] \cap [z,x] = [z,v], \qquad \text{where } v \neq x.$$

It follows that each pair of the segments $[x,w]$, $[w,v]$ and $[v,x]$ meet only at a single endpoint. For example, since $w \in [x,y]$ and w, $v \in [y,z]$ we have

$$\begin{aligned}[x,w] \cap [w,v] &= [x,w] \cap [x,y] \cap [w,v] \\ &\subseteq [x,w] \cap [x,y] \cap [y,z] \\ &= [x,w] \cap [y,w] = \{w\}.\end{aligned}$$

It now follows as before that the union of the three segments is homeomorphic to S^1, a contradiction. This completes the proof of the implication (iii)$\Rightarrow$(i), and hence of the proposition. □

Condition (iii) of Proposition A.38 may be viewed as an analogue of the definition of an ordinary tree (a connected circuit-free graph), where circuits are replaced by homeomorphic images of a circle. In Section A.2, we characterised Λ-trees using the idea of a 0-hyperbolic space. For $\mathbb{R}$-trees, we have the following.

Proposition A.39 *An $\mathbb{R}$-metric space (X,d) is an $\mathbb{R}$-tree if and only if*

(i) *it is connected and*

(ii) *it is* 0*-hyperbolic.*

Proof Since an $\mathbb{R}$-tree is geodesic, it is path connected, hence connected, and it is 0-hyperbolic by Proposition A.22. Conversely, assume conditions (i) and (ii). By Theorem A.23 there is an embedding of (X,d) in an $\mathbb{R}$-tree (X',d'). Let x, $y \in X$, suppose that $v \in X' - X$ and that $v \in [x,y]$. Then $[v,x] \not\equiv [v,y]$, and the points x, y are in different path components of $X' - \{v\}$ by Lemma A.27. Let C be the path component of $X' - \{v\}$ containing x. We show that C is open and closed in $X' - \{v\}$. For if u, $w \in X' - \{v\}$ and $d(u,w) < d(u,v)$, then $d(u,z) < d(u,v)$ for all $z \in [u,w]$, so that $v \notin [u,w]$. Hence the open ball B with centre u and radius $d(u,v)$ in $X' - \{v\}$ is path connected. Thus if $u \in C$ then $B \subseteq C$, and if $u \notin C$ then $B \cap C = \varnothing$. It follows that C and its complement are open in $X' - \{v\}$, as required.

Since $X \subseteq X' - \{v\}$, it follows that $X \cap C$ is open and closed in X. Since $x \in X \cap C$ and $y \notin X \cap C$, this contradicts the connectedness of X. Thus, we have $[x,y] \subseteq X$, and (X,d) is geodesic. It follows that (X,d) is an $\mathbb{R}$-tree by Proposition A.22. □

Next, we consider completeness in $\mathbb{R}$-trees. As usual, an $\mathbb{R}$-metric space is called *complete* if every Cauchy sequence converges. Recall that every $\mathbb{R}$-metric space (X,d) has a *completion*, that is, a complete metric space $(\hat{X},\hat{d})$, together with an isometry $\varphi : X \to \hat{X}$, such that $\varphi(X)$ is dense in $\hat{X}$. This characterises $(\hat{X},\hat{d})$ up to a metric isomorphism. Further, there is a canonical construction of $(\hat{X},\hat{d})$ as equivalence classes of Cauchy sequences in X, the mapping φ sending x to the equivalence class of the constant sequence with value x. Slightly more detail is provided at the end of Section 2 in Chapter 2 of [10]; it is also noted there (Lemma 2.11) that if (X,d) is 0-hyperbolic, then so is $(\hat{X},\hat{d})$. Using this, it is easy to prove the following result of Imrich [26].

Proposition A.40 *If (X,d) is an $\mathbb{R}$-tree then so is its completion $(\hat{X},\hat{d})$.*

Proof By Proposition A.39, (X,d) is connected and 0-hyperbolic; also, X is dense in $\hat{X}$. Hence $(\hat{X},\hat{d})$ is connected (being the closure of a connected subspace) and 0-hyperbolic. Therefore $(\hat{X},\hat{d})$ is an $\mathbb{R}$-tree by Proposition A.39. □

Our main concern now is to establish a geometric criterion for an $\mathbb{R}$-tree to be complete. This criterion appears to be well known among experts, although we have not been able to trace its proof in the literature. It involves the idea of an end in a Λ-tree, and we begin with a brief description of this. For more detail and proofs of the assertions below, see Section 3 in Chapter 2 of [10].

Let (X,d) be a Λ-tree. Let $a \in \Lambda$, and let A be a subtree of Λ such that $A \subseteq \{x \in \Lambda \mid a \leq x\}$ and $a \in A$. Let $\alpha : A \to X$ be an isometry; then $L := \alpha(A)$ is a linear subtree of (X,d) with $x := \alpha(a)$ as an endpoint. We call such a subtree a *linear subtree from x*. The linear ordering on Λ induces a linear ordering on L via α and, if $y \in L$, it is easy to see that the set

$$L_y = \{z \in L \mid y \leq z\}$$

is a linear subtree from y, with $L = [x,y] \cup L_y$ and $[x,y] \cap L_y = \{y\}$.

A linear subtree from x which is maximal with respect to inclusion is called an *X-ray* from x. Given X-rays L, L' with specified endpoints (not necessarily the same), we define $L \equiv L'$ to mean that $L \cap L' = L_v = L'_v$ for some $v \in X$. (It is necessary to specify the endpoint because an X-ray could be a segment, which has two endpoints.)

One can show that $\equiv$ is an equivalence relation on the set of X-rays with specified endpoints. The equivalence classes are called *ends* of X. If ε is an end of X, and $L \in \varepsilon$, we say that L *represents* ε. If ε is an end of X and $x \in X$, there is a unique ray from x representing ε, which is denoted $[x, \varepsilon\rangle$.

If y is an endpoint of X then there is an end ε_y whose X-rays are the segments $[x, y]$ for $x \in X, x \neq y$. Also, if X has just a single point x then there is an end ε_x whose only ray is $[x, x] = \{x\}$. Ends of the form ε_y for some $y \in X$ are called *closed ends* of X, and ends not of this form are called *open ends* of X.

If ε is an end of a Λ-tree (X, d) and x, y are points of X, then the rays $[x, \varepsilon\rangle$ and $[y, \varepsilon\rangle$ intersect in a ray $[z, \varepsilon\rangle$, and we have

$$\begin{aligned} [x, \varepsilon\rangle &= [x, z] \cup [z, \varepsilon\rangle, \\ [y, \varepsilon\rangle &= [y, z] \cup [z, \varepsilon\rangle. \end{aligned} \tag{$*$}$$

We are interested in the case $\Lambda = \mathbb{R}$; in this case an open end is an isometric image of an interval of the form either $[a, b)$, with $b \in \mathbb{R}$ and $a < b$, or $[a, \infty)$.

Suppose that x, $y \in X$ and let $\Lambda = \mathbb{R}$. Then $[x, \varepsilon\rangle$ is bounded (in the usual sense for $\mathbb{R}$-metric spaces) if and only if $[y, \varepsilon\rangle$ is bounded. This follows from equations $(*)$, both statements being equivalent to the assertion that $[z, \varepsilon\rangle$ is bounded. In this situation, we say that ε *has bounded rays*.

Our criterion for an $\mathbb{R}$-tree to be complete is based on the following simple lemma, which is implicit in Proposition 1.6 of Mayer, Nikiel, and Oversteegen [33].

Lemma A.41 *Let (X, d) be an $\mathbb{R}$-tree, and let Y be a dense subset of X. Choose a base-point $x_0 \in X$. Then:*

(i) *if $x \in X$ and x is not an endpoint of X then $x \in [x_0, y]$ for some $y \in Y$;*

(ii) *if Y is a subtree of X then every point of $X - Y$ is an endpoint of X.*

Proof (i) We may assume that $x_0 \neq x$, since otherwise any point y in Y will satisfy $x \in [x_0, y]$. Then $[x, x_0]$ defines a direction at x, and there is at least one more direction at x, defined by $[x, z]$, say, where $z \in X$. Then $x \in [x_0, z]$, so that $[x, z] \subseteq [x_0, z]$. The open ball

$$B(z, d(z, x)) := \{w \in X : d(z, w) < d(z, x)\}$$

contains a point of Y, say y. Since $p := Y(y,x_0,z) \in [x_0,z]$, we have $p \in [x,z]$; moreover $p \neq x$, for otherwise

$$d(y,z) = d(y,p) + d(p,x) + d(x,z) \geq d(x,z),$$

contradicting the fact that $y \in B(z, d(z,x))$. By Lemma 1.5 in Chapter 2 of [10], it follows that $[x_0,y] = [x_0,x,p,y]$, in particular, that $x \in [x_0,y]$.

(ii) We can choose $x_0 \in Y$. Then, by part (i), every point $x \in X$ which is not an endpoint is in $[x_0,y]$ for some $y \in Y$, and $[x_0,y] \subseteq Y$, whence (ii). □

Before giving our completeness criterion for ℝ-trees, another definition is needed.

Definition A.42 Let (X,d) be an R-tree, and let x_0 be a base-point in X. A sequence $\{y_i\}$ in X is called *monotone increasing with respect to* x_0 if $i < j$ implies $y_i \in [x_0, y_j]$.

We now have the following criterion for the completeness of an ℝ-tree.

Proposition A.43 *Let (X,d) be an ℝ-tree. Then the following assertions are equivalent:*

(i) *(X,d) is complete;*

(ii) *every monotone increasing Cauchy sequence with respect to some base-point converges;*

(iii) *(X,d) has no open ends with bounded rays.*

Proof We shall denote the completion of X by $\hat{X}$. Obviously (i) implies (ii). Assume (ii), and fix a base-point x_0. Suppose that ε is an open end with bounded rays in X. Then $[x_0,\varepsilon\rangle$ is the image of an isometry $\alpha : [0,a) \to X$ for some $a \in \mathbb{R}$ with $a > 0$. Since isometries of metric spaces extend to isometries of their completions, α extends to an isometry $[0,a] \to \hat{X}$, also denoted by α. Let $y = \alpha(a)$, so that $[x_0,\varepsilon\rangle \subseteq [x_0,y]$. Let $\{a_n\}$ be a monotone increasing sequence in $[0,a)$ converging to a. Then $\alpha(a_n)$ is monotone increasing with respect to x_0 in X and converges to y, so $y \in X$ by assumption. Then $[x_0,y] \subseteq X$, so $[x_0,\varepsilon\rangle$ is not a ray in X, a contradiction. Hence (iii) holds.

Now assume (iii). If x, $y \in \hat{X}$ then no point of (x,y) is an endpoint of $\hat{X}$, so $(x,y) \subseteq X$ by part (ii) of Lemma A.41. Hence, if $x \in X$ and $y \in \hat{X} - X$ then $[x,y) \subseteq X$, and it is a ray in X. For otherwise, $[x,y) \subseteq [x,z]$ for some $z \in X$. We can find a sequence $\{y_i\}$ in $[x,y)$ converging to y and, since $[x,z]$

is (sequentially) compact, the Cauchy sequence (y_i) converges in $[x,z]$. Hence $y \in [x,z]$, so that $y \in X$, a contradiction. By assumption no such ray exists, so no such point y exists, hence $X = \hat{X}$; that is, X is complete as required. □

Remark A.44 The proof of Proposition A.43 shows that in fact the points of $\hat{X} - X$ are in bijective correspondence with the open ends of X with bounded rays; if $y \in \hat{X} - X$ and ε is the corresponding end of X, then $[x,y] = [x,\varepsilon\rangle \cup \{y\}$ for all $x \in X$. (See Theorem 1.11 in [33].)

As an application of Proposition A.43, we shall obtain a generalisation of a theorem of Wilkens. Suppose that a group G acts by isometries on an $\mathbb{R}$-tree (X,d). Then there is an induced action by isometries on the completion $\hat{X}$. If $x \in \hat{X} - X$ is fixed by G, then from the proof of Proposition A.43 we see that if $y \in X$ then $[y,x)$ is a bounded ray defining an open end ε of X. If $g \in G$ then $g[y,x) = [gy,x)$ is a ray defining ε. (Since x is an endpoint of $\hat{X}$, $[y,x] \cap [gy,x] = [w,x]$ for some $w \neq x$, so $[y,x) \cap [gy,x) = [w,x)$ is a ray in X.) Thus G fixes ε.

An action by isometries of a group G on a metric space (X,d) is said to be *bounded* if, for some point $x_0 \in X$, there is a positive constant K such that $d(x_0, gx_0) \leq K$ for all $g \in G$. (Note that if this is true for some point x_0, it is true for every point in X by a simple application of the triangle inequality.)

Wilkens [54] showed that if a group G has a bounded action by isometries on a complete $\mathbb{R}$-tree then there is a global fixed point. Now if G has a bounded action on an $\mathbb{R}$-tree, then the induced action on the completion is easily seen to be bounded as well. This gives the following.

Proposition A.45 *If a group has a bounded action by isometries on an $\mathbb{R}$-tree then either there is a global fixed point or there exists a fixed open end with bounded rays.*

A.5 Isometries of Λ-trees

We shall now consider the behaviour of an isometry of a Λ-tree onto itself, that is, a metric automorphism. There is a classification of such isometries similar to that for hyperbolic space (in fact there is such a classification for any proper geodesic hyperbolic $\mathbb{R}$-metric space; see Coornaert, Delzant, and Papadopoulos [16] or Ghys and de la Harpe [20]). There are three possible kinds of isometry in a Λ-tree: elliptic, hyperbolic, and inversions. Inversions do not occur in the case $\Lambda = \mathbb{R}$, or indeed whenever $\Lambda = 2\Lambda$, and, unlike hyperbolic space, there are no parabolic isometries.

Let (X,d) be a Λ-tree, where Λ is an arbitrary ordered abelian group, and let g be an isometry of X onto X. We call g *elliptic* if it has a fixed point. The behaviour of such isometries is described in the next lemma, where $\langle g \rangle$ denotes the cyclic subgroup generated by g in the group of metric automorphisms of X. If $[x,y]$ is a segment, there is a unique point p such that $d(x,p) = d(p,y)$ if and only if $d(x,y) \in 2\Lambda$. This point, when it exists, is called the *midpoint* of $[x,y]$.

Lemma A.46 *Let g be an elliptic isometry, and let X^g denote the set of fixed points of g. Then we have the following.*

(i) *The fixed-point set X^g is a closed, non-empty, $\langle g \rangle$-invariant subtree of X.*

(ii) *If $x \in X$ and $[x,p]$ is the bridge between x and X^g then p is the midpoint of $[x,gx]$ and, for all $a \in X^g$, we have $[a,x] \cap [a,gx] = [a,p]$. Hence $d(x,gx) = 2d(x,p)$.*

Proof See Lemma 1.1 in Chapter 3 of Chiswell [10]. □

Lemma A.47 *The following are equivalent:*

(i) *there is a segment of X that is invariant under g, and the restriction of g to this segment has no fixed points;*

(ii) *there is a segment $[x,y]$ in X such that $gx = y$, $gy = x$, and $d(x,y) \notin 2\Lambda$;*

(iii) *g^2 has a fixed point, but g does not;*

(iv) *g^2 has a fixed point and, for all $x \in X$, we have $d(x,gx) \notin 2\Lambda$.*

Proof See Lemma 1.2 in Chapter 3 of [10]. □

Definition A.48 If g satisfies the equivalent conditions of Lemma A.47, it is called an *inversion*.

Remark A.49 If Λ is such that $\Lambda = 2\Lambda$, in particular if $\Lambda = \mathbb{R}$, then g cannot be an inversion, by criterion (ii) or criterion (iv) of Lemma A.47.

Examples of inversions are provided by certain reflections, for example the reflection $g : x \mapsto 1 - x$ in the case where $X = \Lambda = \mathbb{Z}$. In fact inversions in $\mathbb{Z}$-trees correspond to graph automorphisms of the corresponding tree which fix an (unoriented) edge but interchange its endpoints.

Definition A.50 A metric automorphism of (X,d) that is not an inversion and not elliptic is called a *hyperbolic* isometry.

Given a metric automorphism g of the Λ-tree (X,d), we define

$$A_g = \{p \in X : [g^{-1}p, p] \cap [p, gp] = \{p\}\}$$

and call A_g the *characteristic subset* of g. If g is elliptic, it follows from Lemma A.46 that $A_g = X^g$. If g is an inversion then $A_g = \varnothing$. For, suppose that $p \in A_g$ and that g has no fixed point but g^2 does. It follows from Lemma A.46 that the midpoint p of $[g^{-1}p, gp]$ is fixed by g^2. But then $g^{-1}p = gp$, and since $p \in [g^{-1}p, gp]$ we have $p = gp$, a contradiction.

Our next result summarises properties of the characteristic set A_g in the case where g is hyperbolic.

Theorem A.51 *Suppose that g is a hyperbolic automorphism of the Λ-tree (X,d). Then the following hold.*

(i) *The characteristic set A_g is a non-empty, closed, $\langle g \rangle$-invariant subtree of X.*

(ii) *The set A_g is a linear tree, and g restricted to A_g is equivalent to a translation $a \mapsto a + \ell(g)$ for some $\ell(g) \in \Lambda$ with $\ell(g) > 0$.*

(iii) *If $x \in X$, and if $[x, p]$ is the bridge between x and A_g, then we have*

$$\begin{aligned} [x, gx] \cap A_g &= [p, gp], \\ [x, gx] &= [x, p, gp, gx], \\ p &= Y(g^{-1}x, x, gx), \end{aligned}$$

and $d(x, gx) = \ell(g) + 2d(x, p)$.

Proof See Theorem 1.4 in Chapter 3 of [10]. □

When g is hyperbolic, A_g is called the *axis* of g (which explains the notation). When g is elliptic or an inversion, we put $\ell(g) = 0$. Thus $\ell(g)$ is defined for any g, and it is called the *hyperbolic length* of g.

Corollary A.52 *Let g be a metric automorphism of a Λ-tree (X,d).*

(i) *For any $x \in X$,*

$$\ell(g) = \max\{L_x(g^2) - L_x(g), 0\}.$$

(ii) *Assume that g is not an inversion. Then*

$$\ell(g) = \min\{d(x,gx) : x \in X\} = \min_{x\in X} L_x(g)$$

and

$$A_g = \{p \in X : d(p,gp) = \ell(g)\}.$$

Proof (i) Suppose that g is elliptic. From Lemma A.46 we have $L_x(g) = 2d(x,p)$, where p is a point fixed by g, hence by g^2. By the triangle inequality,

$$L_x(g^2) \le d(x,p) + d(p,g^2x) = d(x,p) + d(g^2p, g^2x) = 2d(x,p)$$

and so

$$\max\{L_x(g^2) - L_x(g), 0\} = 0 = \ell(g),$$

as required.

Next, assume that g is an inversion, let $x \in X$ and let p be a point of X fixed by g^2. Put $w = Y(x,p,gp)$. Then $gw = Y(gx,gp,p) \in [p,gp]$ and $[x,w] \cap [p,gp] = \{w\}$, so $[x,w] \cap [w,gw] = \{w\}$. Also $g^2w \in [gp,g^2p] = [p,gp]$ and $d(w,p) = d(g^2w, g^2p) = d(g^2w,p)$, hence $w = g^2w$. It follows that

$$[gx,gw] \cap [w,gw] = [gx,gw] \cap [g^2w,gw] = \{gw\}.$$

Since g has no fixed points, $w \neq gw$, and it follows by Lemma 1.5 in Chapter 2 of [10] that $[x,gx] = [x,w,gw,gx]$. By Lemma 1.4 in Chapter 2 of [10] we have

$$d(x,gx) = d(x,w) + d(w,gw) + d(gw,gx) = 2d(x,w) + d(w,gw) > 2d(x,w).$$

By the triangle inequality,

$$L_x(g^2) = d(x,g^2x) \le d(x,w) + d(w,g^2x) = d(x,w) + d(g^2w,g^2x) = 2d(x,w),$$

hence $L_x(g^2) < L_x(g)$ and we conclude again that

$$\max\{L_x(g^2) - L_x(g), 0\} = 0 = \ell(g).$$

Finally, if g is hyperbolic, let $[p,x]$ be the bridge between x and A_g. By Theorem A.51, $L_x(g) = \ell(g) + 2d(p,gp)$. It is a consequence of Theorem 4.3 that $A_{g^2} = A_g$, that $\ell(g^2) = 2\ell(g)$, and that g^2 is hyperbolic (for the details see Lemma 1.7 in Chapter 3 of [10]). Hence $L_x(g^2) = 2\ell(g) + 2d(p,gp)$, so that $L_x(g^2) - L_x(g) = \ell(g) > 0$ and (i) follows again.

(ii) This is clear if g is elliptic and follows from Theorem A.51(iii) if g is hyperbolic. □

Remark A.53 If g and h are isometries of a Λ-tree onto itself, then g is an inversion if and only if $A_g = \varnothing$, and g is an inversion if and only if hgh^{-1} is an inversion. It therefore follows from part (ii) of Corollary A.52 that, for all such g, h, $A_{hgh^{-1}} = hA_g$ and $\ell(hgh^{-1}) = \ell(g)$.

In the main text, two further observations on hyperbolic length are needed; these are collected together in the next result.

Lemma A.54 *Let g and h be isometries of a Λ-tree (X,d).*

(i) *If n is an integer then $\ell(g^n) = |n|\ell(g)$ and $A_{g^n} \subseteq A_g$; if moreover g is hyperbolic then $A_{g^n} = A_g$.*

(ii) *Suppose that g and h commute. If g,h are both hyperbolic then $A_g = A_h$. If g is elliptic and h is hyperbolic, then $A_h \subseteq A_g$ (that is, g fixes A_h pointwise).*

Proof We refer to Lemma 1.7(3) and Lemma 3.9(2) in Chapter 3 of [10] for the proofs of parts (i) and (ii), respectively. □

If G is a group acting on a Λ-tree (X,d), then there is a mapping $\ell : G \to \Lambda$, where $\ell(g)$ is the hyperbolic length of the isometry of X induced by g. We call this the *hyperbolic length function* for the action. The relation between ℓ and the Lyndon length functions L_x, for $x \in X$, is given by Corollary A.52.

A.6 Free actions

An action by isometries of a group G on a Λ-tree (X,d) is free (in the usual sense) if 1_G is the only element of G which acts as an elliptic isometry. For arbitrary Λ this assumption is often inadequate, and we have to assume that the action is *free and without inversions*, which means that each $g \in G - \{1\}$ acts as a hyperbolic isometry on (X,d). Equivalently, $\ell(g) > 0$ for all $g \in G - \{1\}$. To express this property of an action in terms of Lyndon length functions, we make the following definition.

Definition A.55 A Lyndon length function $L: G \to \Lambda$ is called *free* if it satisfies $L(g^2) > L(g)$ for all $g \in G - \{1\}$.

This is the term used in Myasnikov, Remeslennikov, and Serbin [40]; Lyndon originally used the term *archimedean* for this property. By Corollary A.52(2) we immediately obtain the following criterion.

Lemma A.56 *Suppose that a group G acts on a Λ-tree (X,d), and let $x \in X$ be any point. Then the action is free and without inversions if and only if L_x is a free Lyndon length function.*

If Λ is an ordered abelian group, we call a group Λ-*free* if it has a free action without inversions on some Λ-tree; and a group is called *tree-free* if it is Λ-free for some Λ. Note that every subgroup of a Λ-free group is also Λ-free. Tree-free groups are automatically torsion-free, by part (ii) of Theorem A.51. Any ordered abelian group Λ is Λ-free, since Λ acts on itself by translations. It follows that any torsion-free abelian group is tree-free, for such a group can be made into an ordered abelian group.[1] Torsion-free abelian groups of rank at most $2^{\aleph_0}$ are $\mathbb{R}$-free since they embed into the additive group of the reals.

As a consequence of Bass–Serre theory, a group is $\mathbb{Z}$-free if and only if it is a free group; cf Section 3 of Chapter I in [45]. In fact it is easy to see, without recourse to the general theory, that a free group acts freely on its Cayley graph with respect to a basis, this action giving rise to a free action without inversions on the corresponding $\mathbb{Z}$-tree. Let us denote by $\mathfrak{TF}(\Lambda)$ the class of Λ-free groups. Apart from the case $\Lambda = \mathbb{Z}$, no characterization in purely group-theoretic terms is known for $\mathfrak{TF}(\Lambda)$.

So far, we have seen that $\mathfrak{TF}(\Lambda)$ is subgroup closed and consists only of torsion-free groups. We shall now show that it is closed under the formation of free products via the next lemma, which is essentially due to Harrison.

Lemma A.57 *Let $\{G_i : i \in I\}$ be a family of groups, and suppose that $L_i : G_i \to \Lambda$ is a Lyndon length function, for $i \in I$. For $g \in G := \ast_{i \in I} G_i$, define $L(g)$ to be $\sum_{j=1}^{n} L_{i_j}(g_j)$, where $g = g_1 \cdots g_n$ is the reduced decomposition of g, with $g_j \in G_{i_j}$ and $L(1_G) = 0$. Then L is a Lyndon length function on G. Moreover, if all L_i are free then so is L.*

[1] See Levi [28] or Section IV.5 in Kertész [27].

Proof Axiom (i) for a Lyndon length function ($L(1)=0$) is satisfied by definition, and if g is expressed in normal form as in the lemma, then the normal form of g^{-1} is $g^{-1}=g_n^{-1}\cdots g_1^{-1}$; hence axiom (ii) ($L(g)=L(g^{-1})$ for all $g\in G$) is also satisfied.

Note that L restricted to G_i is L_i, so we can omit the subscript on L_i. We next observe that L satisfies the triangle inequality, that is, $L(gh)\le L(g)+L(h)$ for all g, $h\in G$. This is clear if one of g, h is 1; otherwise, we have expressions in normal form, $g=g_n\cdots g_1$ and $h=h_1\cdots h_m$. Let p be the least integer $j\ge 1$ such that $g_j\ne h_j^{-1}$. Then we can write $gh=g_n\cdots g_p h_p\cdots h_m$ (note that $p=n+1$ is allowed, when $gh=h_{n+1}\cdots h_m$, as is $p=m+1$, and both can happen simultaneously, in which case $gh=1$). There are two cases.

Case 1: Either g_p, h_p are in different free factors, or $p=n+1$, or $p=m+1$. In this case, $L(gh)=\sum_{j=p}^n L(g_j)+\sum_{j=p}^m L(h_j)\le L(g)+L(h)$. (If $p>n$ then $\sum_{j=p}^n L(g_j)$ is to be interpreted as 0; similar conventions are used elsewhere in the proof.)

Case 2: Otherwise, g_p, h_p are in the same free factor G_i, but $g_p\ne h_p^{-1}$. In this case, $L(gh)=\sum_{j=p+1}^n L(g_j)+\sum_{j=p+1}^m L(h_j)+L(g_ph_p)$, and $L(g_ph_p)\le L(g_p)+L(h_p)$ since $L|_{g_i}=L_i$ satisfies the triangle inequality (Property (v) of Lyndon length functions, given at the start of Section A.3). Thus we again obtain $L(gh)\le L(g)+L(h)$.

We have to verify axiom (iii) for Lyndon length functions; that is, we have to show that, for all $g,h,k\in G$,

$$c(g,h)\ge\min\{c(g,k),c(h,k)\},\tag{A.1}$$

where $c(g,h)=\frac{1}{2}(L(g)+L(h)-L(g^{-1}h))$. (Note that, by axiom (ii), c is symmetric, that is, $c(g,h)=c(h,g)$.) Inequality (A.1) is clear if (at least) one of g, h, k is 1; otherwise we can write down normal forms

$$g=g_1\cdots g_m,\quad h=h_1\cdots h_n,\quad k=k_1\cdots k_p.$$

Let q be the least integer $j\ge 1$ such that $k_j\ne g_j$, and let r be the least integer $j\ge 1$ such that $k_j\ne h_j$. Interchanging g and h if necessary, we can assume that $q\le r$.

Assume first that k^{-1}, g fall into case 1 above. Then it is easily seen that $c(k,g)=\sum_{j=1}^{q-1}L(k_j)$. Now $k_j=g_j$ for $j<q$ and $k_j=h_j$ for $j<r$, so $g_j=h_j$ for $j<q$. Hence, using the properties of L already established,

$$L(g^{-1}h)=L(g_m^{-1}\cdots g_q^{-1}h_q\cdots h_n)\le\sum_{j=q}^m L(g_j)+\sum_{j=q}^n L(h_j)$$

from which we obtain

$$c(g,h) \geq \frac{1}{2}\left(\sum_{j=1}^{q-1} L(g_j) + \sum_{j=1}^{q-1} L(h_j)\right) = \sum_{j=1}^{q-1} L(k_j) = c(k,g),$$

establishing inequality (A.1) in this case.

Now assume that k^{-1}, g fall into case 2 above. Then

$$L(k^{-1}g) = \sum_{j=q+1}^{p} L(k_j) + \sum_{j=q+1}^{m} L(g_j) + L(k_q^{-1}g_q)$$

from which we obtain $c(k,g) = \sum_{j=1}^{q-1} L(k_j) + c(k_q, g_q)$. If $q < r$ then $k_q = h_q$, but $k_q \neq g_q$, so that g_q, h_q are in the same free factor but $g_q \neq h_q$, hence

$$c(g,h) = \sum_{j=1}^{q-1} L(g_j) + c(g_q,h_q) = \sum_{j=1}^{q-1} L(k_j) + c(g_q,k_q) = c(g,k),$$

and (A.1) holds in this case.

Assume that $q = r$. If k^{-1} and h fall into case 1 above, then so do g^{-1} and h, hence

$$c(g,h) = \sum_{j=1}^{q-1} L(g_j) = \sum_{j=1}^{q-1} L(k_j) = c(h,k),$$

and so inequality (A.1) holds again. If k^{-1}, h fall into case 2 above, then $c(h,k) = \sum_{j=1}^{q-1} L(k_j) + c(k_q,h_q)$. Also,

$$L(g^{-1}h) = L(g_m^{-1}\cdots g_q^{-1}h_q\cdots h_n) \leq \sum_{j=q+1}^{m} L(g_j) + \sum_{j=q+1}^{n} L(h_j) + L(g_q^{-1}h_q);$$

hence

$$c(g,h) \geq \frac{1}{2}\left(\sum_{j=1}^{q} L(g_j) + \sum_{j=1}^{q} L(h_j) - L(g_q^{-1}h_q)\right) = \sum_{j=1}^{q-1} L(k_j) + c(g_q,h_q).$$

Now since g_q, h_q, k_q are all in the same free factor, and L restricted to this factor is a Lyndon length function, we have $c(g_q,h_q) \geq \min\{c(g_q,k_q), c(h_q,k_q)\}$, hence again $c(g,h) \geq \min\{c(g,k), c(h,k)\}$, and L is a Lyndon length function.

Finally, suppose that every L_i is free. Then $L_i(g) \neq 0$ for all $1 \neq g \in G_i$ and all $i \in I$. For, by property (vi) of Lyndon length functions, if $L_i(g) = 0$ then $L_i(g^2) = c(g,g^2) = 0$, which implies $g = 1$ by freeness. If $1 \neq g \in G$, we can write g in reduced form:

$$g = h_1\cdots h_m(g_1\cdots g_n)h_m^{-1}\cdots h_1^{-1},$$

where $m \geq 0$, $n \geq 1$, and either $n = 1$ or $g_1 \neq g_n^{-1}$. It now follows easily that $L(g^2) > L(g)$, as required. □

Proposition A.58 *If $\{G_i : i \in I\}$ is any family of Λ-free groups, then the free product $\ast_{i\in I} G_i$ is also Λ-free.*

Proof Let (X_i, d_i) be a Λ-tree on which G_i acts, and choose $x_i \in X_i$. Then $L_i := L_{x_i}$ is a free Lyndon length function on G_i, and Lemma A.57 gives us a free Lyndon length function L on $G := \ast_{i\in I} G_i$. Replacing L by $2L$ gives a Lyndon length function which is still free and satisfies $c(g,h) \in \Lambda$ for all g, $h \in G$. By Theorem A.29 there is an action of G on a Λ-tree (X,d) such that $2L = L_x$ for some $x \in X$. Since L_x is free, the action of G on X is free and without inversions. □

It follows from Proposition A.58 that free products of copies of $(\mathbb{R},+)$ are $\mathbb{R}$-free, as are free groups. The latter assertion also follows from the general fact that, if Λ, Λ' are ordered abelian groups such that $\Lambda \leq \Lambda'$ then $\mathfrak{TF}(\Lambda) \subseteq \mathfrak{TF}(\Lambda')$; see Lemma 2.1 in Chapter 3 of [10].

We now note some further properties of tree-free groups. A subgroup H of a group G is termed *malnormal* if, for all $g \in G$:

$$H \cap gHg^{-1} \neq \{1\} \implies g \in H.$$

A group is called a CSA *group*, if all its maximal abelian subgroups are malnormal.[2] Further, a group G is called *commutative transitive*, if the following equivalent conditions are satisfied.

(i) Given $a,b,c \in G$ such that $[a,b] = 1$, $[b,c] = 1$, and $b \neq 1$, then $[a,c] = 1$.

(ii) Centralisers of non-identity elements of G are abelian.

(iii) If M_1, M_2 are maximal abelian subgroups of G, and $M_1 \neq M_2$, then $M_1 \cap M_2 = \{1\}$.

It is easy to see that conditions (i)–(iii) are indeed equivalent and that a CSA group is commutative transitive.

Proposition A.59 *A tree-free group is a* CSA *group, and hence commutative transitive.*

Proof See Proposition 1.8 and Corollary 1.9 in Bass [4], or Lemma 1.2 in Chapter 5 of Chiswell [10]. □

[2] The definition of CSA-groups appears in Gildenhuys, Kharlampovich, and Myasnikov [21] and in Myasnikov and Remeslennikov [39].

Our next result determines all Λ-free groups with at most two generators; a version for $\Lambda = \mathbb{R}$ in terms of Lyndon length functions can be found in Proposition 5.13 of Harrison [24]. The result for arbitrary Λ was established in Chiswell [9] and in Urbański and Zamboni [51]; see Section 5.2 of [10] for more details and a proof.

Proposition A.60 *Let G be a tree-free group, and let $g, h \in G$ be non-trivial elements. Then $\langle g, h \rangle$ is either free of rank 2 or free abelian of rank at most 2.*

The following result, established by Morgan and Shalen in [35], provides further examples of $\mathbb{R}$-free groups.

Theorem A.61 *The fundamental group of a closed surface is $\mathbb{R}$-free, except for the non-orientable surfaces of genus $1, 2$, or 3 (the connected sum of one, two or three projective planes).*

See Section 4 in Chapter 5 of [10] for an abbreviated account of the proof. For brevity, the fundamental groups of the non-orientable surfaces of genus $1, 2, 3$ will be called *exceptional surface groups*; all other surface groups will be called *non-exceptional*. From what has been said so far it follows in particular that any free product of finitely many groups, each of which is either a finitely generated free-abelian group or a non-exceptional surface group, is $\mathbb{R}$-free. The deepest individual result to date concerning the class $\mathfrak{TF}(\mathbb{R})$ provides a converse to this statement.

Theorem A.62 *A finitely generated $\mathbb{R}$-free group can be decomposed as a free product $G = G_1 * \cdots * G_n$, for some integer $n \geq 1$, where each G_i is either a free-abelian group of finite rank or the fundamental group of a closed surface.*

A proof was outlined by E. Rips during a conference in 1991 at the Isle of Thorns (formerly a conference centre of the University of Sussex, England, but at the time of writing belonging to the organisation Cats Protection). This was never published, but full proofs have appeared in Bestvina and Feighn [6] and in Gaboriau, Levitt, and Paulin [19]; see also Rimlinger [42]. Summarising, we thus have the following characterization of the finitely generated groups in the class $\mathfrak{TF}(\mathbb{R})$.

Theorem A.63 *A finitely generated group G is $\mathbb{R}$-free if and only if it can be written as a free product $G = G_1 * \cdots * G_n$ for some positive integer n, where each G_i is either a finitely generated free-abelian group or a non-exceptonal surface group.*

Given all that has been said so far about the class $\mathfrak{T}\mathfrak{F}(\mathbb{R})$, the following might seem to be a reasonable conjecture concerning the structure of arbitrary $\mathbb{R}$-free groups.

A group is $\mathbb{R}$-free if and only if it is a free product of subgroups of $\mathbb{R}$ and non-exceptional surface groups.

However, counterexamples have been given by Dunwoody [17], and by Zastrow [55]; see pp. 231–233 of [10] for a discussion of Zastrow's example. Further counterexamples have been given more recently in Berestovskii and Plaut [5].

Theorem A.63 provides a non-trivial restriction on subgroups of arbitrary $\mathbb{R}$-free groups. Recall that a group G is called *fully residually free* if, for each positive integer n and every collection of elements $g_1, g_2, \dots, g_n \in G - \{1\}$, there exists a surjective homomorphism $\varphi : G \to H$, where H is some free group, such that $\varphi(g_i) \neq 1$ for all i. Further, G is called *locally fully residually free* if every finitely generated subgroup of G is fully residually free. As an application of Theorem A.63, one can show the following.

Proposition A.64 *An $\mathbb{R}$-free group is locally fully residually free.*

In fact, every locally fully residually free group is Λ-free, where Λ is some non-standard model of $\mathbb{Z}$. The fundamental group of the non-orientable surface group of genus 3, although not (locally) fully residually free, is Λ-free, where $\Lambda = \mathbb{Z} \oplus \mathbb{Z}$ with the lexicographic ordering. For more information, and examples of locally fully residually free groups which are not $\mathbb{R}$-free, see Section 5 in Chapter 5 in [10].

A.7 Exercises

A.1. Show that a $\mathbb{Z}$-metric space (X,d) is geodesic if and only if there exists a graph Y such that $X = V(Y)$ and such that d is the path metric on X. [Hint: Y is obtained by joining two elements $x, y \in X$ by an unoriented edge if and only if $d(x,y) = 1$.]

A.2. Let (X,d) be a Λ-tree. If x, v, $y \in X$ and x_m denotes the point at distance m from v in $[v,x]$ where $0 \leq m \leq d(v,x)$ (with a similar definition for the point

y_n on $[v,y]$ at distance n from v), show that

$$d(x_m, y_n) = m + n - 2\min\{m, n, d(v,w)\},$$

where $[v,x] \cap [v,y] = [v,w]$.

A.3. If $v \in S$, where S is a subset of a Λ-tree, show that the subtree spanned by S is equal to $\bigcup_{x \in S}[v,x]$. [Hint: use the observations after Definition A.5.]

A.4. Show that every closed subtree of a Λ-tree X is closed in the natural topology of X (that is, the topology induced by the balls $B(x,r)$ with $x \in X$ and $r \in \Lambda$, where $r > 0$). [Hint: use Lemma A.15 to show that the complement is open.]

A.5. Verify that the group $\mathbb{Q} \otimes_{\mathbb{Z}} \Lambda$, as described at the beginning of Section A.2, is an ordered abelian group with subgroup $\frac{1}{2}\Lambda$.

A.6. Establish the assertion in Remark A.16.

A.7. Establish properties (iv)–(vi) of a (Λ-valued) length function, as defined at the start of Section A.3.

Appendix B
Some open problems

The careful reader of Chapters 2–11 will have had ample opportunity to formulate a number of questions deserving further investigation. A sample of such open problems is collected in this appendix, together with some comments.

1. Associativity of reduced multiplication The proof that reduced multiplication when restricted to the set of reduced functions is associative, as presented in Chapter 2, is long, involved, and rather technical. Is there perhaps a simpler, more elegant approach to this problem? For instance, one could think of trying to embed $\mathscr{RF}(G)$, equipped with reduced multiplication, homomorphically into some group, without assuming the associativity of reduced multiplication. Such an approach might also shed light on the problem of presenting $\mathscr{RF}(G)$ in terms of generators and defining relations.

2. Hyperbolic elements which are products of elliptics As we saw in Corollary 9.9, $\mathscr{RF}(G)$ is never generated by its elliptic elements. However, by Lemma 3.22, the subgroup $E(G)$ contains an abundance of hyperbolic elements. How can one distinguish the hyperbolic elements in $E(G)$ from those in $\mathscr{RF}(G) - E(G)$? More generally, one might ask for a *set-theoretic characterization* of the elements of $E(G)$. For instance, is it true that a function $f \in \mathscr{RF}(G)$ is contained in $E(G)$ if and only if there exist a natural number k, functions $f_1, f_2, \ldots, f_{k+1} \in \mathscr{RF}(G)$, and elements $g_1, g_2, \ldots, g_k \in G_0 - \{\mathbf{1}_G\}$ such that:

(i) $L(f_j) > 0, \quad 2 \leq j \leq k,$

(ii) $f_1 f_2 \cdots f_{k+1} = \mathbf{1}_G,$

(iii) $f = f_1 g_1 \circ f_2 g_2 \circ \cdots \circ f_k g_k \circ f_{k+1}$?

Such a characterisation might allow one, for example, to deduce that the quotient group $\mathscr{RF}(G)/E(G)$ is torsion-free (as one would strongly suspect to be the case).

3. The substitution problem Let $w(x_1,\ldots,x_n)$ be a reduced word of a free group F with basis $\{x_1,x_2,\ldots,x_n\}$, let G be any group, and let $g \in G$ be an arbitrary element. Then one can ask whether there exists a homomorphism $\varphi : F \to G$ such that $\varphi(w) = g$; in other words, does g have the form

$$g = w(u_1,\ldots,u_n)$$

for some elements $u_1,u_2,\ldots,u_n$ of G? In its most general form, that is, for an arbitrary word w and any $g \in G$, this is called the *substitution problem* for the group G. In the case where G itself is free, results concerning the characterisation of commutators and other 2-variable words were obtained by Wicks; see [52] and [53]. For instance, Wicks shows that *an element g of a free group G is a commutator if and only if*

$$g \equiv uvwu^{-1}v^{-1}w^{-1}$$

for some reduced words u,v,w in G. Can results of this type be obtained for the groups $\mathscr{RF}(G)$? For instance, it follows immediately from Part (i) of Lemma 3.15 that a hyperbolic element $f \in \mathscr{RF}(G)$ is a square if and only if it is of the form

$$f = t \circ f_1 \circ f_1 \circ t^{-1},$$

where $t, f_1 \in \mathscr{RF}(G)$ and f_1 is cyclically reduced and of positive length. What about commutators or other types of words?

4. Classifying normal form relations In the context of Section 3.7, can one in some suitable sense classify the normal form relations of $E(G)$? The larger aim would be to elucidate the structure of this prominent normal subgroup of $\mathscr{RF}(G)$ and to obtain a presentation of $E(G)$.

5. Splitting Does $\mathscr{RF}(G)$ split over $E(G)$ (or, more generally, over any of its proper normal subgroups)?

6. Characterising $\mathscr{RF}$-groups Characterise $\mathscr{RF}$-groups among the class of all groups. Such a characterisation would probably have to involve the fact that $\mathscr{RF}(G)$ acts transitively on an $\mathbb{R}$-tree.

7. Existence of simple groups Does $\mathscr{RF}(G)$ have any infinite simple subgroups not contained (up to isomorphism) in G itself?

8. Induced free substructures Let G be a group, and suppose that

$$G \cong \pi_1(G(-),Y)$$

is the fundamental group, in the sense of Bass and Serre, of a graph of groups $(G(-),Y)$. Is it then true that $\mathscr{RF}(G)$ contains (in a suitable sense) a corresponding free construct, for the same graph Y, which is not contained in the subgroup G_0? See Exercise 2.4 for this.

9. Explicitly embedded surface groups Morgan and Shalen [35] showed that the fundamental group of a closed surface is $\mathbb{R}$-free (that is, has a free action on some $\mathbb{R}$-tree), except for the non-orientable surfaces of genus 1, 2, and 3 (the connected sum of 1, 2, or 3 real projective planes); see also Section 5.4 in Chiswell [10]. In view of the universality property of $\mathscr{RF}$-groups described in Chapter 4, it would be interesting to be able to exhibit explicitly such surface groups embedded as hyperbolic subgroups in $\mathscr{RF}(G)$. Do all non-exceptional surface groups occur for every non-trivial G? It is conceivable that the subgroup $\mathscr{S}(G)$ of step functions over G (as introduced in Exercise 9.15) could be the right setting for such an investigation.

10. Can one describe the global structure of the group $\mathscr{S}(G)$ or, at least, obtain a presentation for it?

11. Image of the map e_G Can Proposition 5.10 be improved by determining the precise image of the map $e_G : \mathscr{RF}(G) \to \prod_{g \in R_G} \mathbb{R}$ introduced in Section 5.5?

12. Around the classification problem As we saw in Proposition 6.3, we have

$$G \cong H \implies \mathscr{RF}(G) \cong \mathscr{RF}(H) \tag{B.1}$$

(this follows immediately from the functoriality of $\widetilde{\mathscr{RF}}(-)$). Does the converse of (B.1) hold? More specifically, one might ask whether all isomorphisms between $\mathscr{RF}$-groups are *induced* via the functor $\widetilde{\mathscr{RF}}(-)$; that is, whether the induced maps

$$\widetilde{\mathscr{RF}}(-)_{(G,H)} : \mathrm{Iso}_{\widetilde{\mathbf{Groups}}}(G,H) \to \mathrm{Iso}_{\widetilde{\mathbf{Groups}}}(\mathscr{RF}(G), \mathscr{RF}(H))$$

are surjective. The core of this problem, the *classification problem for $\mathscr{RF}$-groups*, appears to be to find a characterisation of the subgroup G_0 of $\mathscr{RF}(G)$ in purely group-theoretic terms, that is, without reference to the length function or the $\mathbb{R}$-tree action of $\mathscr{RF}(G)$.

A related (and slightly more general) problem is whether the functor $\widetilde{\mathscr{RF}}(-)$

is *full*, that is, whether the induced maps

$$\widetilde{\mathscr{RF}}(-)_{G,H} : \mathrm{Mor}_{\widetilde{\mathbf{Groups}}}(G,H) \to \mathrm{Mor}_{\widetilde{\mathbf{Groups}}}(\mathscr{RF}(G),\mathscr{RF}(H))$$

are surjective. Identifying the group G with G_0 and H with H_0, we clearly have

$$\widetilde{\mathscr{RF}}(\iota)|_{G_0} = \iota, \quad \iota \in \mathrm{Mor}_{\widetilde{\mathbf{Groups}}}(G,H);$$

that is, the maps $\widetilde{\mathscr{RF}}(-)_{G,H}$, and hence also their restrictions $\widetilde{\mathscr{RF}}(-)_{(G,H)}$, are injective. In other words, the functor $\widetilde{\mathscr{RF}}(-)$ is *faithful*. Consequently, if the maps $\widetilde{\mathscr{RF}}(-)_{G,H}$ are surjective then the functor $\widetilde{\mathscr{RF}}(-)$ is fully faithful, implying the surjectivity of the maps $\widetilde{\mathscr{RF}}(-)_{(G,H)}$ by a standard result in category theory; see Proposition 4.1.5 in Schubert [44].

More generally (and somewhat more vaguely), one might ask which properties of a group G are determined by the isomorphism type of $\mathscr{RF}(G)$. For instance, it follows from Corollary 3.26 that the isomorphism type of $\mathscr{RF}(G)$ determines the list of (isomorphism types of) finite subgroups of G; in particular, if $\mathscr{RF}(G) \cong \mathscr{RF}(H)$ and G is torsion-free, then so is H.

13. Adjoints Does any of the functors introduced and studied in Chapter 6 have an adjoint?

14. Characterising sets of strong periods Characterise those sets $A \subseteq \mathbb{R}$ of real numbers satisfying $A = \Omega_f^0$ for some hyperbolic function $f \in \mathscr{RF}(G)$. (The characterisation should be independent of the non-trivial group G.)

15. Classifying centralisers of hyperbolic elements Is it possible to decide, in terms of their sets of strong periods, when two hyperbolic functions $f_1, f_2 \in \mathscr{RF}(G)$ have isomorphic centralisers in $\mathscr{RF}(G)$? Note that, since each non-trivial subgroup of the additive reals occurs as a centraliser of a hyperbolic function $f \in \mathscr{RF}(G)$, such a criterion, in conjunction with a solution to Problem 13, would lead to a classification, up to isomorphism, of all real groups in terms of a certain type of generating system.

16. Redefining periods Study the effects of redefining periods via equation (8.39) on the results of Chapter 8. Does the main result on the centralisers of hyperbolic elements (Theorem 8.16) remain valid? See Remark 8.27 for some background on this problem.

17. Subgroups of small index Does $\mathscr{RF}(G)$ contain proper subgroups of finite index? It is possible that the centraliser partition property might help with this problem; cf Proposition 8.23 and Corollary 8.24. What about subgroups $\mathscr{H} \leq \mathscr{RF}(G)$ of index satisfying $1 < (\mathscr{RF}(G) : \mathscr{H}) \leq 2^{\aleph_0}$?

18. Determine the subgroup of $\mathscr{RF}(G)$ generated by all test functions.

19. Probability measures on $\mathscr{RF}(G)$ Let G be a non-trivial group. Is it possible to show that, with respect to an appropriately chosen (and, one hopes, rather natural) probability measure, "almost all" functions $f \in \mathscr{F}(G)$ are test functions; so that, in particular, "almost all" $f \in \mathscr{F}(G)$ are cyclically reduced?

20. Presenting $\mathscr{RF}(G)$ We have seen that $\mathscr{RF}(G)$ is generated (as a semigroup) by it hyperbolic elements; see Exercise 9.13. Can one give defining relations for $\mathscr{RF}(G)$ in terms of this or another non-trivial generating system? Perhaps the theory developed in Basarab [3] might help here.

A related problem concerns the lack of known homomorphisms involving $\mathscr{RF}$-groups. So far, our supply consists only of the embeddings induced by the functor $\widetilde{\mathscr{RF}}(-)$ (see Proposition 6.3), the exponent maps introduced in Chapter 5 (see Proposition 5.7), and the maps λ_f defined in Chapter 9 (see Theorem 9.8). Knowledge of a presentation of $\mathscr{RF}(G)$ might help to exhibit further useful homomorphisms.

21. Embedded free products In Chapter 10, the question was asked: which free products of real groups embed as hyperbolic subgroup in $\mathscr{RF}(G)$ for a given group G? This was answered in Theorem 10.5. One can ask a similar question concerning surface groups: which free products of (non-exceptional) surface groups embed as hyperbolic subgroup in $\mathscr{RF}(G)$ for given G? More generally, one may ask the same type of question about free products of (non-trivial) real groups and non-exceptional surface groups. The answer would be an interesting generalisation of Theorem 10.5.

22. Generalisation to other densely ordered abelian groups Does the theory developed in this book generalise to cover $\mathscr{RF}(\Lambda, G)$, as defined in the introduction, for a larger class of densely ordered abelian groups including, say, the groups $\Lambda = \mathbb{R}^n$ for $n \geq 1$?

23. Test functions in a groupoid setting Does the theory of test functions as developed in Chapters 9–10 generalise to the groupoids $\mathscr{ARF}(S,G)$ constructed in Chapter 11? A solution of this problem, apart from being interesting in its own right, is bound to have important consequences; for instance, it should lead to a computation of the cardinality of $\mathscr{ARF}(S,G)$, as well as to insight into the structure of these groupoids.

24. Universality of $\mathscr{ARF}(S,G)$ Does the universality property of the groups $\mathscr{RF}(G)$ and their associated $\mathbb{R}$-trees $\mathbf{X}_G$, discussed in Chapter 4, generalise to the groupoids $\mathscr{ARF}(S,G)$ and their $\mathbb{R}$-tree actions?

References

[1] R. C. Alperin and K. N. Moss, Complete trees for groups with a real-valued length function, *J. London Math. Soc.* **31** (1985), 55–68.

[2] R. Baer, The subgroup of the elements of finite order of an abelian group, *Ann. Math.* **37** (1936), 766–781.

[3] S. A. Basarab, On a problem raised by Alperin and Bass. In: *Arboreal Group Theory* (ed. R. C. Alperin), MSRI Publications vol. 19, pp. 35–68, Springer-Verlag, New York, 1991.

[4] H. Bass, Group actions on non-archimedean trees. In: *Arboreal Group Theory* (ed. R. C. Alperin), MSRI Publications vol. 19, pp. 69–131, Springer-Verlag, New York, 1991.

[5] V. N. Berestovskii and C. P. Plaut, Covering $\mathbb{R}$-trees, $\mathbb{R}$-free groups, and dendrites, *Adv. Math.* **224** (2010), 1765–1783.

[6] M. Bestvina and M. Feighn, Stable actions of groups on real trees, *Invent. Math.* **121** (1995), 287–321.

[7] G. Cantor, Über eine Eigenschaft des Inbegriffes aller reellen algebraischen Zahlen, *J. Reine Angew. Math.* **77** (1874), 258–262. Also in: *Georg Cantor, Gesammelte Abhandlungen* (ed. E. Zermelo), Julius Springer, Berlin, 1932.

[8] G. Cantor, Über unendliche lineare Punktmannigfaltigkeiten V, *Math. Ann.* **21** (1883), 545–591.

[9] I. M. Chiswell, Harrison's theorem for Λ-trees, *Quart. J. Math. Oxford* (2) **45** (1994), 1–12.

[10] I. M. Chiswell, *Introduction to Λ-Trees*, World Scientific, Singapore, 2001.

[11] I. M. Chiswell, A-free groups and tree-free groups. In: *Groups, Languages, Algorithms* (ed. A. V. Borovik), Contemp. Math. vol. 378, pp. 79–86, Providence (RI), Amer. Math. Soc., 2005.

[12] I. M. Chiswell and T. W. Müller, Embedding theorems for tree-free groups, *Math. Proc. Cambridge Philos. Soc.* **149** (2010), 127–146.

[13] I. M. Chiswell, T. W. Müller, and J.-C. Schlage-Puchta, Compactness and local compactness for $\mathbb{R}$-trees, *Arch. Math.* **91** (2008), 372–378.

[14] D. E. Cohen, *Combinatorial Group Theory: A Topological Approach*, London Mathematical Society Student Texts vol. 14, Cambridge University Press, 1989.

[15] P. M. Cohn, *Algebra (Second Edition)*, Volume 3, John Wiley & Sons, Chichester, 1991.

[16] M. Coornaert, T. Delzant, and A. Papadopoulos, *Géométrie et Théorie des Groupes*, Lecture Notes in Mathematics vol. 1441, Springer, Berlin, 1990.

[17] M. J. Dunwoody, Groups acting on protrees, *J. London Math. Soc.* **56** (1997), 125–136.

[18] L. Fuchs, *Abelian Groups*, Pergamon Press, Oxford, 1960.

[19] D. Gaboriau, G. Levitt, and F. Paulin, Pseudogroups of isometries of $\mathbb{R}$ and Rips' Theorem on free actions on $\mathbb{R}$-trees, *Israel J. Math.* **87** (1994), 403–428.

[20] E. Ghys and P. de la Harpe, *Sur les Groupes Hyperboliques d'après Mikhael Gromov*, Birkhäuser, Boston, 1990.

[21] D. Gildenhuys, O. Kharlampovich, and A. G. Myasnikov, CSA groups and separated free constructions, *Bull. Austral. Math. Soc.* **52** (1995), 63–84.

[22] L. Greenberg, Discrete groups of motions, *Can. J. Math.* **12** (1960), 414–425.

[23] M. Gromov, Hyperbolic groups. In: *Essays in Group Theory* (ed. S. M. Gersten), Mathematical Sciences Research Institute Publications vol. 8, pp. 75–263, Springer-Verlag, New York, 1987.

[24] N. Harrison, Real length functions in groups, *Trans. Amer. Math. Soc.* **174** (1972), 77–106.

[25] P. J. Higgins, *Notes on Categories and Groupoids*, Van Nostrand Reinhold, London-New York-Melbourne, 1971. (Reprinted with a new preface by the author: *Repr. Theory Appl. Categ.* **7** (2005), 1–178 (electronic).)

[26] W. Imrich, On metric properties of tree-like spaces. In: *Beiträge zur Graphentheorie und deren Anwendungen* (ed. Sektion MARÖK der Technischen Hochschule Ilmenau), pp. 129–156, Oberhof, 1977.

[27] A. Kertész, *Einführung in die Transfinite Algebra*, Birkhäuser Verlag, Basel–Stuttgart, 1975.

[28] F. Levi, Arithmetische Gesetze im Gebiete diskreter Gruppen, *Rend. Palermo* **35** (1913), 225–236.

[29] G. Levitt, Constructing free actions on $\mathbb{R}$-trees, *Duke Math. J.* **69** (1993), 615–633.

[30] R. C. Lyndon and P. E. Schupp, *Combinatorial Group Theory*, Springer-Verlag, Berlin–Heidelberg, 1977.

[31] W. Magnus, A. Karrass, and D. Solitar, *Combinatorial Group Theory. Presentations of Groups in Terms of Generators and Relations*, reprint of the 1976 second edition, Dover, Mineola, NY, 2004.

[32] G. A. Margulis, *Discrete Subgroups of Semisimple Lie Groups*, Springer-Verlag, Berlin–Heidelberg, 1991.

[33] J. C. Mayer, J. Nikiel, and L. G. Oversteegen, Universal spaces for $\mathbb{R}$-trees, *Trans. Amer. Math. Soc.* **334** (1992), 411–432.

[34] J. W. Morgan and P. B. Shalen, Valuations, trees and degenerations of hyperbolic structures: I, *Ann. of Math.* (2) **122** (1985), 398–476.

[35] J. W. Morgan and P. B. Shalen, Free actions of surface groups on $\mathbb{R}$-trees, *Topology* **30** (1991), 143–154.

[36] T. W. Müller, A hyperbolicity criterion for subgroups of $\mathscr{RF}(G)$, *Abh. Math. Semin. Univ. Hambg.* **80** (2010), 193–205.

[37] T. W. Müller, Some contributions to the theory of $\mathscr{RF}$-groups. In preparation.

[38] T. W. Müller and J.-C. Schlage-Puchta, On a new construction in group theory, *Abh. Math. Semin. Univ. Hambg.* **79** (2009), 193–227.

[39] A. G. Myasnikov and V. N. Remeslennikov, Exponential groups, II: extensions of centralizers and tensor completion of CSA-groups, *Internat. J. Algebra Comput.* **6** (1996), 687–711.

[40] A. G. Myasnikov, V. N. Remeslennikov, and D. Serbin, Regular free length functions on Lyndon's free $\mathbb{Z}[t]$-group $F^{\mathbb{Z}[t]}$. In: *Groups, Languages, Algorithms* (ed. A. V. Borovik), Contemp. Math. vol. 378, pp. 33–77, Providence (RI), Amer. Math. Soc., 2005.

[41] M. H. A. Newman, On theories with a combinatorial definition of "equivalence", *Ann. of Math.* (2) **43** (1942) 223–243.

[42] F. S. Rimlinger, $\mathbb{R}$-trees and normalisation of pseudogroups, *Exper. Math.* **1** (1992), 95–114.

[43] H. L. Royden, *Real Analysis*, Macmillan, New York, 1963.

[44] H. Schubert, *Categories*, Springer-Verlag, Berlin–Heidelberg, 1972.

[45] J.-P. Serre, *Trees*, Springer-Verlag, Berlin–Heidelberg, 1980.

[46] H. Short, Notes on word hyperbolic groups. In: *Group Theory from a Geometrical Viewpoint* (eds. E. Ghys, A. Haefliger, and A. Verjovsky), World Scientific, Singapore, 1991.

[47] W. Sierpiński, *Cardinal And Ordinal Numbers*, Monographs of the Polish Academy of Science vol. 34, Warsaw, 1958.

[48] H. J. S. Smith, On the integration of discontinuous functions, *Proc. London Math. Soc.* **6** (1875), 140–153.

[49] T. Szele, Ein Analogon der Körpertheorie für abelsche Gruppen, *J. Reine Angew. Math.* **188** (1950), 167–192.

[50] J. Tits, A 'theorem of Lie–Kolchin' for trees. In: *Contributions to Algebra: A Collection of Papers Dedicated to Ellis Kolchin*, Academic Press, New York, 1977.

[51] M. Urbański and L. Q. Zamboni, On free actions on Λ-trees, *Math. Proc. Camb. Phil. Soc.* **113** (1993), 535–542.

[52] M. J. Wicks, Commutators in free products, *J. London Math. Soc.* **37** (1962), 433–444.

[53] M. J. Wicks, A general solution of binary homogeneous equations over free groups, *Pacific J. Math.* **41** (1972), 543–561.

[54] D. L. Wilkens, Group actions on trees and length functions, *Michigan Math. J.* **35** (1988), 141–150.

[55] A. Zastrow, Construction of an infinitely generated group that is not a free product of surface groups and abelian groups, but which acts freely on an $\mathbb{R}$-tree, *Proc. Royal Soc. Edinburgh* (A) **128** (1998), 433–445.

Index